Piezocone and Cone Penetration Test (CPTu and CPT) Applications in Foundation Engineering

Piezocone and Cone Penetration Test (CPTu and CPT) Applications in Foundation Engineering

Abolfazl Eslami
Professor
Dept. of Civil and Environmental Engineering
Amirkabir University of Technology (AUT)

Sara Moshfeghi
M.Sc. Graduate
Dept. of Civil and Environmental Engineering
Amirkabir University of Technology (AUT)

Hossein MolaAbasi
Assistant Professor
Dept. of Civil Engineering
Gonbad Kavous University

Mohammad M. Eslami
Ph.D. GeoPentech Inc.

ELSEVIER
Butterworth-Heinemann
An imprint of Elsevier

Butterworth-Heinemann is an imprint of Elsevier
The Boulevard, Langford Lane, Kidlington, Oxford OX5 1GB, United Kingdom
50 Hampshire Street, 5th Floor, Cambridge, MA 02139, United States

Notices
Knowledge and best practice in this field are constantly changing. As new research and experience broaden our understanding, changes in research methods, professional practices, or medical treatment may become necessary.

Practitioners and researchers must always rely on their own experience and knowledge in evaluating and using any information, methods, compounds, or experiments described herein. In using such information or methods they should be mindful of their own safety and the safety of others, including parties for whom they have a professional responsibility.

To the fullest extent of the law, neither the Publisher nor the authors, contributors, or editors, assume any liability for any injury and/or damage to persons or property as a matter of products liability, negligence or otherwise, or from any use or operation of any methods, products, instructions, or ideas contained in the material herein.

Library of Congress Cataloging-in-Publication Data
A catalog record for this book is available from the Library of Congress

British Library Cataloguing-in-Publication Data
A catalogue record for this book is available from the British Library

ISBN: 978-0-08-102766-0

For information on all Butterworth-Heinemann publications visit our website at https://www.elsevier.com/books-and-journals

Publisher: Matthew Deans
Acquisition Editor: Matthew Deans
Editorial Project Manager: Emma Hayes
Production Project Manager: Nirmala Arumugam
Cover Designer: Mark Rogers

Typeset by TNQ Technologies

Contents

Preface ix
Acknowledgment xiii

1 Geotechnical engineering 1
1.1 Introduction 1
1.2 Topics in geotechnical engineering 1
1.3 Geotechnical investigations: assessments, aims, and applications 3
1.4 Steps of performing geotechnical investigations 4
1.5 Extent of geotechnical investigation in foundation engineering 6
1.6 Sources of geotechnical data 10
1.7 In situ penetration tests 16
1.8 Geotechnical reports 20
1.9 Remarks 20
1.10 Book organization and scope 21
References 22

2 Background to foundation engineering 25
2.1 Introduction 25
2.2 Foundation analysis and design considerations 26
2.3 Foundation classification 29
2.4 Intermediate trends 31
2.5 Overall step-by-step procedure for foundation design 39
2.6 Basic soil mechanics for foundation engineering 40
2.7 Uncertainty in foundation engineering 48
2.8 The role of CPT in reduction of uncertainty or increasing reliability 49
References 51

3 CPT equipment, performance, and records 55
3.1 Introduction and background 55
3.2 Electrical cones 58
3.3 Piezocone penetration test 58
3.4 Equipment 60
3.5 Data presentation 61

3.6 Factors influencing CPT measurements and interpretation **63**
3.7 Special cones **65**
3.8 Mini-cone and CPT in physical modeling **71**
3.9 Codes, standards and software **75**
References **77**

4 Geotechnical parameters from CPT records **81**
4.1 Introduction **81**
4.2 Unit weight **82**
4.3 Relative density **84**
4.4 Undrained shear strength **90**
4.5 Over consolidation ratio **91**
4.6 Stiffness **91**
4.7 Constrained modulus **94**
4.8 Shear wave velocity and shear modulus at small strain **95**
4.9 Sensitivity **95**
4.10 *CPT* correlations with *SPT* **97**
4.11 Permeability **98**
4.12 Liquefaction **99**
References **108**

5 Soil behavior classification (SBC) using CPT and CPTu records **111**
5.1 Introduction **111**
5.2 Brief survey of soil profiling methods **111**
5.3 The Eslami–Fellenius CPTu profiling and soil type classification method **119**
5.4 Comparison between the Eslami–Fellenius and Robertson (1990) methods **121**
5.5 Comments on current methods **126**
5.6 UniCone **128**
5.7 Problematic soils evaluation via SBC **129**
5.8 Case records for problematic soils **129**
5.9 Analysis and comparison **131**
5.10 Triangular chart **136**
References **142**

6 CPT in foundation engineering; scale effect and bearing capacity **145**
6.1 Scale effect in correlation between pile, foundation, and CPT **145**
6.2 CPT and pile scale effect for toe capacity **146**
6.3 Scale effect in estimation of shaft capacity **157**
6.4 Bearing capacity of shallow foundations **160**
6.5 Analytical approach for shallow foundation bearing capacity: direct CPT method **166**
6.6 Comparison of direct CPT methods via databases **174**
References **178**

7 CPT and CPTu applications for deep foundation–bearing capacity 183
7.1 Introduction **183**
7.2 Pile design **185**
7.3 CPT records for pile installation **186**
7.4 Bearing capacity of deep foundations **186**
7.5 Pile axial–bearing capacity from indirect CPT–based approaches **189**
7.6 Commonly used direct CPT–based methods for pile axial–bearing capacity **194**
7.7 Piles under uplift loading **215**
7.8 Comments on the methods **215**
7.9 Case study: Urmia Lake causeway **224**
References **228**

8 CPT and CPTu for foundation settlement and load–displacement (P-Δ) estimation 233
8.1 Introduction **233**
8.2 Basic equations for settlement estimation **234**
8.3 Background to Janbu (1967) approach **236**
8.4 Modified stress exponent and modulus number **239**
8.5 CPT-based nonlinear stress–strain approach to evaluate foundation settlement **241**
8.6 Pile group settlement **249**
8.7 CPT-based load–displacement behavior of foundations **249**
References **259**

9 CPT & CPTu for ground modification 263
9.1 Introduction **263**
9.2 General description, function, and application of the methods **263**
9.3 CPT and CPTu application in ground improvement practice **268**
9.4 Quantitative analysis of soil geotechnical parameters **270**
9.5 Comparison of pre- and postmodification **278**
9.6 Soil behavior classification approach **285**
References **294**

10 CPT-based soft computing implementations in foundation engineering 299
10.1 Introduction **299**
10.2 Review of GMDH-type neural network method **299**
10.3 Review of support vector machine method **301**
10.4 Case Study: pile-bearing capacity **302**
10.5 Case Study No. 4: shallow foundation settlement **308**
10.6 Case Study No. 5: friction angle, cohesion, and shear stress **310**
10.7 Case Study: liquefaction potential assessment **310**
References **316**

11 Uncertainty and reliability in foundation engineering: CPT-Based approach **319**
11.1 Introduction **319**
11.2 Uncertainty sources in geotechnical engineering **320**
11.3 Principles: random analysis techniques in geotechnical engineering **324**
11.4 Case studies **328**
11.5 LRFD application for pile design (new approach) **345**
References **349**

12 CPT and CPTu databases in foundation engineering **353**
12.1 Introduction **353**
12.2 Databases of soil parameters and CPTu **354**
12.3 Databases for shallow foundations and CPT **355**
12.4 Databases for pile and CPT **355**
12.5 AUT: GEO-CPT&Pile database **359**
12.6 Implementation of the developed database **364**
References **366**

Index **371**

Preface

Geotechnical investigation is the procedure of acquiring information and data on subsurface soil conditions and synthesizing them in order to determine geomaterial parameters required for design. Despite notable progress, many solutions are still approximate, which is mainly due to the natural inherent inhomogeneity of soils and dominant environmental and geologic conditions. Among different sources of data productions available to overcome the uncertainties and guided engineering judgment for interpretations, in-situ penetration tests are employed as a complementary and versatile tool to laboratory testing and are considered to be fast, accurate, and economical sources for collecting data.

In light of cost saving for geotechnical investigations and considering the time-consuming process of conventional procedures such as drilling, sampling, transferring samples, and laboratory testing, it is more favored to take the measurement equipment to the attributed project site. In addition, development of more equipped in-situ tests has facilitated more comprehensive data compilations and proper understanding of soil behavior. In-situ tests can be employed for directly defining soil profiling and geotechnical soil parameters and characteristics.

Accordingly, in the realm of foundation engineering, due to the variety of conditions and parameters affecting soil behavior, the analysis and design of foundation systems for determining bearing capacity, settlement estimation, and stability control are knowledge-based and sophisticated. In this regard, the versatile cone penetration test, CPT, or piezocone, CPTu, provides valuable data combining tip resistance (q_c), sleeve friction (f_s) and excess pore pressure (u), and a few other supplementary records. Representation of most soil in-situ characteristics can play an important role to select and interpret various required parameters for Performance-Based Design.

Moreover, the reliability of any foundation system design depends heavily on the trustworthiness of input values of soil properties in the analyses, which in turn depends on the level of accuracy and carefulness of the site investigation and characterization carried out. In this regard, an initial step for the engineering process is the development of a subsurface profile of soil types and their geotechnical properties, in which the cone penetration test (CPT) is an ideal tool. CPT supplies continuous records with depth and allows a variety of sensors to be accompanied and is performed under field stresses and boundary conditions, actual conditions that the foundation system will encounter during its construction and lifetime functioning. The choice of pile installation method as well as pile material type is influenced by subsurface conditions, structural and geometrical characteristics of the project, in which the practical and economical aspects are essential.

Among commonly used approaches for estimating the bearing capacity of deep foundations, such as theoretical solutions, static analysis, and full-scale loading tests, employment of in-situ records is realized as a supplementary of static analysis. Evaluating the bearing capacity of foundations from CPT data is one of the earliest applications of this test and includes two main approaches; direct and indirect methods. Direct CPT methods apply the measured values of cone bearing for bearing resistance with some modifications regarding scale effects, i.e., the influence of foundation width to the cone diameter ratio. Indirect CPT methods employ friction angle and undrained shear strength values estimated from CPT data based on bearing capacity or cavity expansion theories.

Due to similarities between CPT and pile, cone penetrometer data can be employed for the determination of pile capacity by applying corrections related to geometry differences and limited displacement of pile under load bearing. The attempts to predict pile toe and shaft capacities from CPT data were generally successful. However, there is still some skepticism regarding the application of a small-scale cone penetrometer to large-scale foundations, which is referenced to as scale effects. Regarding these issues and in view of the scale effect to relate cone penetrometer to foundations, either shallow or deep, critical factors have been investigated through experimental and theoretical approaches. For both toe and shaft capacity determination, the focus is mainly on diameter, embedment depth, penetration rate, strain level, homogeneity condition, influence zone, data processing, and ultimate pile capacity interpretation.

For achieving suitable application of CPT and CPTu records for safe and optimum foundation design and focusing on reliability approaches, the materials of this book are organized in the following four major parts:

I Fundamentals of Geotechnical and Foundation Engineering via In-situ Testing
II CPT and CPTu; Correlations, SBC, and Scale Effect to Foundations
III CPT and CPTu Implementation in Geotechnical Design of Foundations
IV CPT-based soft computing, reliability and data-based design approaches

In the first part, an initial introduction to geotechnical engineering has been presented in Chapter 1 including procedures and common sources of data acquisition. Chapter 2 presents principles of Foundation Engineering, types and classifications, in addition to major aspects for analysis and design. Additionally, uncertainties in foundation engineering are recognized, and the role of CPT and CPTu data for a more reliable design is presented. Chapter 3 deals with introducing the Cone Penetration Test, detailed discussions about the apparatus, testing procedures, performance, and data presentations.

In the second part starting with Chapter 4, a variety of correlations are covered for estimating soil properties such as strength and stiffness parameters, particularly employed in foundation engineering by means of CPT measurements. In Chapter 5, the classification of soils by in-situ tests is demonstrated vastly. Then, various soil behavior classification (SBC) charts and approaches in terms of CPT measurements, especially for challenging deposits, are reviewed and compared. In addition, to relate penetrometer to foundation, Chapter 6 starts with introducing factors pertinent to scale effects. Then, it describes the application of CPT in geotechnical design of shallow foundations in terms of bearing capacity.

For the third part, Chapter 7 discusses deep foundation data-based design, and realizing cone penetrometer as a model pile deals with a wide range of direct CPT-based methods in estimating the pile axial bearing capacity. Additionally, it provides examples of direct as well as in-direct implementation of CPT records in calculation of pile bearing capacity. Chapter 8 encompasses various approaches for estimating foundation settlement by CPT-based correlations and load-displacement behavior using CPT records. Chapter 9 introduces challenging soils commonly encountered in geotechnical practice and provides criteria for recognizing and classifying these soil types by means of CPT and CPTu charts. Furthermore, this chapter introduces applications of CPT for evaluating and comparing the soil properties before and after soil improvement as an indicator of improvement effectiveness. The comparisons are presented in the form of various case studies worldwide.

In the last part, Chapter 10 introduces the applicability of soft computing techniques, such as the Group Method of Data Handling (GMDH)-type Neural Network (NN), based on CPT records for predicting pile bearing capacity and shallow foundation settlement. Chapter 11 deals with the topic of sources of uncertainty in geotechnical engineering, and reliability-based approaches are introduced. Some examples are presented on implementing reliability-based approaches in evaluating the performance of CPT-based methods of estimating piles bearing capacity. Finally, Chapter 12 demonstrates a wide range of CPT databases in the literature and their applications. At the end, an extensive recently developed database is presented, including records of CPT and CPTu piling records along with geotechnical information.

Typical examples, explanations, applications, and implementation for enhancing foundation geotechnical design via predictive methods, particularly for deep foundations, are illustrated and interpreted through a few sophisticated data banks including foundation loading test records and nearby performed CPT or CPTu data.

Acknowledgment

I would like to express my sincerest gratitude toward Professors Bengt Fellenius and Dick Campanella for various technical discussions and insights on CPT and Deep Foundation Design, as well as nontechnical talks, mentorship, and their guidance over the past 25 years beginning in Ottawa, ON, and later in Vancouver, BC, Canada during and after my PhD studies. Additionally, I gratefully acknowledge Dr. J.A. Infante, my fellow at the University of Ottawa, Civil Engineering Department, for cooperation and development of the UniCone software (Eslami-Fellenius method), used and employed for pile geotechnical design, presented in the early stages of this millennium.

Parts of this work were initially developed in the environment of the civil engineering group at Amirkabir University of Technology (AUT) and Guilan University. I am grateful to many colleagues and counterparts in the geotechnical engineering groups for their collaboration, discussions, and comments on various technical sections of the chapters, and generally over the years. Additionally, I would like to thank Amirkabir University of Technology IT group for assisting in creating the AUT: Geo-CPT&Pile database, including more than 600 case records for piles and CPT, accessible for worldwide usage.

Writing a textbook involves significant time commitment, which could only be accomplished with the collective effort of my colleagues and students. With the help of numerous graduate and undergraduate students at Guilan University and AUT, extensive research has been carried out over the years on the contents of this book; *CPT, CPTu, and Foundations*. This text, to a high degree, is a compilation of the efforts of my former and current students, and their work is greatly acknowledged. I wish to present my special thanks to Dr. Sara Heidarie Golafzani for invaluable contribution to contents of Chapter 11, Dr. Fatemeh Valikhah for precious aid in Chapters 8 and 9, Mr. Mohammad Esmaeilzadeh for sharing data in Chapter 4, and Mr. Ramin Ezzatdoust for priceless time and effort on preparing the graphical contents of the chapters.

I would like to express my gratitude toward my coauthors in this book, starting as being dedicated students in my classrooms, and now experts in the field whom I've had the opportunity to work alongside in preparation of the book chapters and many technical projects over the recent years. Ms. Sara Moshfeghi, for her continuous specialized effort and organization of all chapter materials; Dr. Hossein MolaAbasi, for initiating this project and developing the soft computing sections; and Dr. Mandro

M. Eslami, for providing critical comments, constructive points, and expert peer review.

I extend my most profound gratitude toward other authors and publishers who contributed and provided many tables, charts, plots, and their underlying data and permission for reuse in the book chapters. Without their cooperation, this work would not have been possible. Finally, I would like to thank the Elsevier publishing team for continuous assistance and collaboration in each step of preparing this book.

It is hoped that this book on CPT and CPTu applications in Foundation Engineering will serve its intended purpose and, hence, prove useful to readers as a resource in understanding and addressing problems in teaching, research, and engineering practice.

Abolfazl Eslami
October 2019

Geotechnical engineering

1

1.1 Introduction

Geotechnical engineering is the systematic application of techniques which allows construction on, in, or with geomaterials, i.e., soil and rock. Every civil engineering structure and construction is related to soil in some way, and subsequently, its design will depend on properties of the soil or rock. Geotechnical operations are of importance with respect to soil sampling, investigating geomaterials properties, controlling groundwater level and flow as well as environmental and hydrological interactions. Foundation engineering, excavations and supporting ground structures, underground structures, dams, natural or artificial fills, roads and airports, subgrades and ground structures, and slope stability assessments are examples of geotechnical engineering applications in practice.

Despite notable progress in geotechnical engineering, many solutions are still approximate, which is mainly due to the natural inherent inhomogeneity of soils and dominant environmental conditions. Additionally, soils are more sensitive to local environmental conditions compared to other prefabricated building materials such as steel or concrete. Consequently, it would be necessary to have comprehensive understanding of natural soil deposits, environment interactions, and response to local conditions to allow more accurate prediction of geomaterials engineering performance and behavior in projects.

Fig. 1.1 demonstrates several branches related to geotechnical engineering and their overlap. Geotechnical engineering activities are a part of a team effort involving other disciplines including geology, structural engineering, construction management, hydraulics, earthquake and transportation engineers, and other pertinent branches. The final design of any project reflects a collaboration of these professions.

1.2 Topics in geotechnical engineering

Importance, application, and serviceability of structures, along with broad application of natural geomaterials in construction projects, necessitate investigation, reconnaissance, collection as well as measurements of geomaterial properties reported in literature, laboratory, and field. This has led to the development of the branch of "geotechnical exploration" or "geotechnical site investigation" in geotechnical engineering.

The following themes are addressed in geotechnical engineering:

1. Characterizing and defining the location and thickness of subsurface soil and rock strata

Piezocone and Cone Penetration Test (CPTu and CPT) Applications in Foundation Engineering
https://doi.org/10.1016/B978-0-08-102766-0.00001-8

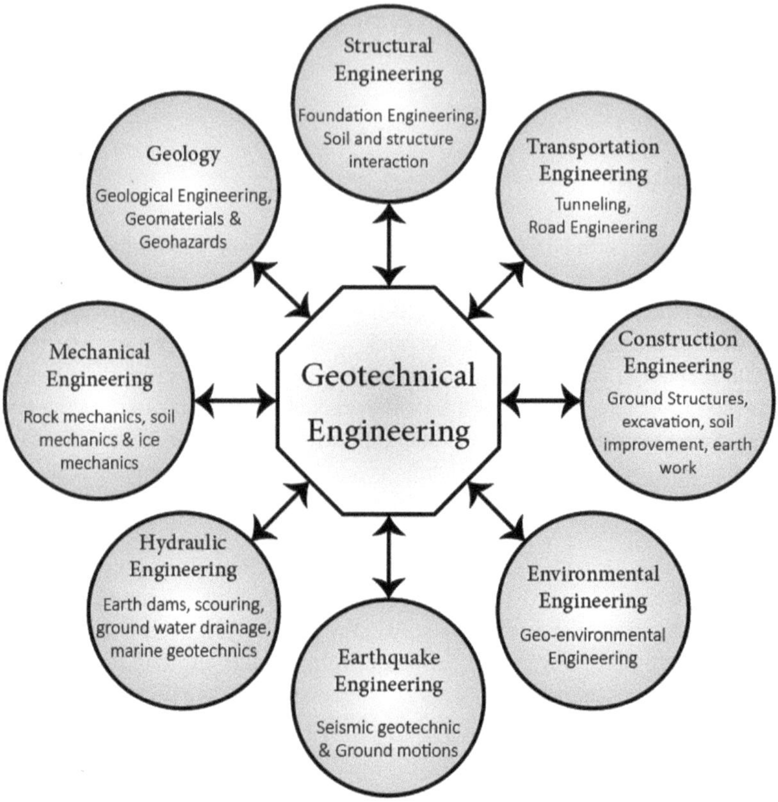

Figure 1.1 Overlap of geotechnical engineering with other sciences.

2. Sample recovery for observation, testing, and evaluation
3. Determining groundwater conditions and consequences of possible variations for projects in future
4. Sufficiency of subsurface geomaterials for safe bearing of superstructure loads
5. Consequences and complications caused by construction, earth works, or excavation
6. Selecting and design of appropriate foundation systems for a given project
7. Considerations and requirements for stabilization of artificial and natural deposits and slopes
8. Necessity of employing ground or underground structures in projects and their design
9. Site response to seismic loads or any other natural or artificial geohazards
10. Geoenvironmental issues, health and safety hazards, and management strategies

For some rudimentary projects with common environmental conditions, these items are relatively straightforward and there would be little need for a geotechnical engineer to intervene. For more sophisticated projects, however, geotechnical investigations are more comprehensive and require expensive exploration, testing, and insight analyses. In sites with problematic soils, geotechnical concerns might dictate and control the feasibility of projects from technical or economical aspects.

In addition to design aspects, geotechnical engineering plays an important role in construction phase as well. Geotechnical services during construction phase include:

- Assessing the condition of geomaterials (soil and rock) observed in explorations and comparing them with what has been anticipated in design phase. This is especially applicable for projects which include vast drilling and excavations. At times, the condition encountered during construction might be so different from design phase that dictates serious modifications or changes to initial design.
- Comparing actual performance to what was anticipated in design: it facilitates installation of special equipment for measuring and recording displacements, groundwater elevation, and other characteristics. This procedure, known as observational method, allows for applying amendments proportional to actual conditions to design phase.
- Quality assurance, specifically in compacted fills, and geotechnical services for most foundations continue until the final stages of construction. For instance, sites prone to long-term settlement require several months, even years, of monitoring after construction. Postconstruction activities may either pertain to installations with unsatisfactory performance or help improve the measurements.

1.3 Geotechnical investigations: assessments, aims, and applications

1.3.1 Assessments

Prior to running any field reconnaissance and subsurface exploration, the designer team including geotechnical, structural, and construction engineering must compile the following information for the proposed development:

- Form, type, location of the project
- Geometrical dimensions in plan and height
- Load combination of structures and buildings
- Method of construction
- Existing topography and ground leveling
- Accessibility to roads and facilities
- Presence of previous or ongoing developments

One of the primary and main stages in investigations and design in projects, namely geotechnical investigation, is assessment and evaluation of soil and bedrock characteristics which can be acquired by the following approaches:

- Remote sensing via assessing conventional aerial photos and modern net-based photographs
- Reviewing published sources including geological and geographical maps, soil and previous geotechnical reports, geological hazard maps/earthquake seismicity, and liquefaction potential maps
- Site survey and visual investigation
- Local experience, observations, and ongoing construction activities
- Geophysical and seismic assessments
- Drilling, sampling, and groundwater monitoring
- In situ testing
- Laboratory testing

- Analysis and assessment of environmental indices
- Instrumentation and monitoring

1.3.2 Aims

Adequate serviceability, engineering functionality, stability, and being economical are the main objectives in civil projects. Optimum design and construction of a foundation system requires accurate application of soil or rock mechanics, experience, and engineering judgment.

Explorations for common projects include general in situ assessments, drilling and sampling, and laboratory testing. While, for more important projects, special subsurface conditions, site environmental conditions, as well as project serviceability define the extent of geotechnical investigation, and necessitate knowledge-based analysis and design along with optimum construction of projects.

1.3.3 Applications

- Overall assessment of the site in terms of use and construction of the proposed project
- Determining strength properties of soil and bedrock
- Quantifying stiffness properties of soil and bedrock
- Assessing compressibility
- Choosing an optimum type of foundation system (shallow, semideep, deep)
- Deciding whether or not site requires soil improvement
- Selecting method of constructing foundation
- Recognizing borrow sources
- Defining how artificial materials be employed along with natural materials for more appropriate performance
- Predicting the performance of structures at different stages and conditions including construction, against seismic and dynamic loads, short-term unanticipated conditions, serviceability, and lifetime performance
- Measurement and monitoring substructure performance, and its interaction with superstructure, back analysis, and optimizing design
- Assessment of probable ground problems such as instability and any type of geohazards confronting measures
- Identifying geoenvironmental problems and proposing appropriate mitigation measures
- Technical, constructional, and economical (optimum) design aspects of projects

1.4 Steps of performing geotechnical investigations

Generally, site investigations are performed according to the following major steps:

I. Data collection and literature review
II. Site visit and nondestructive testing
III. Field operations including drilling, sampling, and in situ testing
IV. Laboratory testing
V. Data synthesis and geotechnical report writing
VI. Instrumentation and monitoring

Step I: The first step is collecting initial data regarding site plan and defining the location of structural elements (such as columns, walls, and foundation), the pressure applied by superstructure, type, size and importance of structure, number of structure stories and underground floors, structure height, loading condition, restrictions due to serviceability (including settlement, displacement, and tilts anticipated in structural design), previous geotechnical reports for the area or adjacent sites, condition of existing structures, topographic maps, aerial photos, geological maps, reports by local agencies, hydrology and climatology, condition of existing mines near the site, seismic hazard maps, historic groundwater maps, and shear wave velocity maps.

Step II: Preliminary investigations mainly comprise nondestructive testing (geophysical tests) and site visit so as to prepare topographical and geological site maps and acquire information on previous constructions, earthquake or landslide hazards, performance of existing structures, site geomaterial condition and subsequently applicability of drilling equipment, natural and artificial obstacles and complications during site investigations and construction. At this stage, it is necessary to perfect the collected information and data from Step I by comparing them with current site observations.

Step III: In spite of providing some insight of subsurface conditions from geophysical approaches, geotechnical explorations must proceed via operations in site including drilling and digging trenches or excavating galleries. Sampling, coring, and in situ tests are performed to provide more accurate and detailed soil profiling including layer thickness, extent of soil and rock layers in plan within the structure influence zone, and groundwater fluctuations.

In situ tests are accompanied by carrying special equipment to the field, inserting into the ground, and testing on the existing soil or rock in the field.

Step IV: Transferring the prepared samples to a soil laboratory and performing required tests related to the project. Important soil and rock properties are measured through laboratory testing. Major laboratory studies focus on index, compressibility, stiffness, strength, and durability characteristics.

Step V: Preparing geotechnical reports includes providing a clear description of subsurface soil types and profiles, methods of site exploration, methods of performed laboratory tests and the obtained results, groundwater conditions, and information on unusual or problematic conditions that may occur during construction or in the long term. Finally, with synthesis of the obtained data and in situ or laboratory measurements, and review of possible geological hazard scenarios, geotechnical design parameters and recommendations are provided in the geotechnical investigation report.

Step VI: Subsequent to finishing geotechnical investigations and start of construction, it would be necessary to employ instrumentation to monitor the geotechnical–structural interactions, which helps in planning and preparation for possible anticipated scenarios during construction.

The data and information pertaining to the adjacent structures can provide a general overview of site subsurface condition. The extent of geotechnical measures at any of the abovementioned stages depends on importance of project, soil condition as well as previous information and experience the geotechnical engineer.

1.5 Extent of geotechnical investigation in foundation engineering

The location and order of drilling and sampling depend primarily on importance of structure, soil and bedrock stratification, notable changes in layers, applied loads, as well as investigation phase. In general, it would be more favored to continue drilling more than the minimum required depth to provide better description of subsurface geological condition, depth, and properties of soft or loose layers and reassures sufficient subsurface information when structural demands and design are not available at the investigation phase.

1.5.1 Number and spacing of borings

First, it should be mentioned that there is no specific recommendation for the number and depth of borings required; however, this can be a function of site conditions, geological complexities, project type, available equipment, previous similar experience, and engineering judgment. In general, the number and depth of borings must be opted in a way that provides sufficiently accurate soil profile within the influence zone of the superstructure. During initial investigations, less number of borings with more spacing can be considered. However, for the complementary and final investigations, the number and depth of borings will need to increase. The higher the transferred stress from the superstructure, the lesser the borings spacing and the deeper borings are required. For projects such as roads with uniform ground condition, the boring spacing of approximately 500−1000 m would suffice in normal and planar ground conditions. For structures such as powerhouses and dams, a boring spacing of about 20−100 m is more common. More borings would be required if the ground condition variation is more considerable in the horizontal direction than depth. One boring is never sufficient, even for small projects or special projects with vertical structures such as antenna towers. A minimum of three borings are usually considered in these cases. However, if the soil condition is assessed as uniform, the number of borings can be reduced to two.

1.5.2 Depth of investigations

The geotechnical assessments must be performed within the area whose geomaterials are affected by the superstructure or construction process. The following guideline can be employed in this regard:

As a common rule proposed by Canadian Engineering Manual (CFEM, 2006) and American Society of Civil Engineers (ASCE, 1993), the minimum depth of investigation borings is a depth at which the net induced stress by the superstructure equals to either 10% of the transferred stress from foundation to subsoil ($q_{applied}$), or 5% of the soil effective stress (σ_z). The deeper depth obtained from these criteria is opted as the minimum investigation depth. In case of encountering rock or considerably dense soil layers at shallower depth than the one derived from this criterion, the investigation depth can be revised and shallower investigation depth can be considered. If the subsurface soil is normally consolidated clay and prone to settlement, deeper depths may be considered in investigations for more comprehensive assessments. It is also important to allow for the effect of adjacent structures on the net stress increase of soil when determining the investigation depth. The adjacent structures may induce additional stresses, and subsequently, increase the required investigation depth.

Assuming the 2:1 stress distribution with depth and equating the total area of single footings with a virtual mat foundation with dimension size of B*L as a result of stress overlaps for single footings, the minimum boring depth according to the 5% criterion would be as shown in Eq. (1.1).

$$0.05\gamma H = \frac{q_{avg} \cdot B \cdot L}{(B+H)(L+H)} \tag{1.1}$$

where B and L are the width and length of equivalent mat foundation (or footprint dimensions of structures), respectively. q_{avg} is average pressure from superstructure, H is minimum investigation depth based on 5% effective stress criterion. By substituting "x"for 0.05γ, Eqs. (1.2) and (1.3) are derived:

$$xH = \frac{q_{avg} \cdot B \cdot L}{(B+H)(L+H)} \tag{1.2}$$

$$xH(B+H)(L+H) = q_{avg} \cdot B \cdot L \tag{1.3}$$

Then by dividing the equation by B^3x:

$$\frac{H\left(BL + BH + HL + H^2\right)}{B^3} = \frac{q_{avg}}{B} \times \frac{L}{B} \times \frac{1}{X} \tag{1.4}$$

$$\left(\frac{H}{B}\right)^3 + \left(\frac{H}{B} + 1\right)\left(\frac{H}{B}\right)^2 + \frac{L}{B} \times \frac{H}{B} - \frac{q_{avg}}{B} \times \frac{L}{B} \times \frac{1}{X} = 0 \tag{1.5}$$

Assuming $L/B = n$, $H/B = m$, and $x = 0.05\gamma$:

$$m^3 + (n+1)m^2 + mn - n\left(\frac{q_{avg}}{B}\right) \times \frac{1}{0.05\gamma} = 0 \tag{1.6}$$

By considering the soil unit weight value and the applied stress (q_{avg}) on foundation, the minimum investigation depth can be calculated. Subsequently, the extent of investigations on plan can be determined by Eqs. (1.7) and (1.8).

$$\frac{L'}{L} = 1 + \left(\frac{H}{L}\right) \tag{1.7}$$

$$\frac{B'}{B} = 1 + \left(\frac{H}{B}\right) \tag{1.8}$$

where H is investigation depth, L' and B' are length and width of investigation area, respectively. In case the groundwater elevation is at shallow depths, buoyant/submerged unit weight (γ') must be used in calculations, and as a result, the investigation depth increases. The results of Eqs. (1.4)–(1.6) are presented in Tables 1.1 and 1.2. Also, the graphical presentation of the results is depicted in Figs. 1.2 and 1.3. Thus, the values of investigation depth (H), width (B'), and length (L') of investigation area can be determined in terms of average applied pressure (q_{avg}), width (B), and length (L) of foundation.

Example 1.1: Considering a shallow foundation system for a 10-story building with the dimensions of 15*25m, what would be an appropriate extent, in depth and in plan, for exploratory borings if the soil unit weight is 1.8 ton/m^3 and the groundwater level is 25 m (the average dead and live load for each story is assumed 1 ton/m^2).

Step 1: Calculating q_{avg}:

$$q_{avg} = 10 * 1 = 10 \ \ \text{ton}/\text{m}^2$$

Step 2: Determining the appropriate exploratory depth by Fig. 1.3.

$$\frac{L}{B} = \frac{25}{15} = 1.7, \quad \frac{q_{avg}}{B} = \frac{10}{15} = 0.67\frac{ton}{m^2} \Rightarrow \frac{H}{B} = 1.56$$

$$H = 1.56 \times 15 = 23.4 \text{ m}$$

Table 1.1 H/B values for determining investigation depth (H) based on 5% effective stress criterion for $\gamma_{avg} = 16$ kN/m^3 (Eslami, 2007).

	L/B						
P_{avg}/B	**1**	**2**	**3**	**4**	**6**	**8**	**10**
0.05	0.345	0.381	0.396	0.405	0.414	0.419	0.422
0.1	0.533	0.601	0.633	0.652	0.673	0.684	0.692
0.15	0.672	0.767	0.814	0.842	0.874	0.892	0.904
0.2	0.785	0.905	0.964	1.000	1.043	1.068	1.083
0.30	0.969	1.128	1.210	1.261	1.323	1.359	1.383
0.4	1.12	1.309	1.411	1.476	1.555	1.602	1.633
0.50	1.243	1.465	1.584	1.66	1.756	1.813	1.851
0.60	1.354	1.602	1.737	1.825	1.934	2.000	2.05
0.70	1.454	1.725	1.874	1.972	2.095	2.171	2.222
0.80	1.545	1.838	2.00	2.108	2.244	2.328	2.386
0.90	1.629	1.942	2.117	2.233	2.382	2.474	2.537
1.00	1.707	2.040	2.226	2.351	2.511	2.611	2.680

B and L are the width and length of equivalent mat foundation, respectively.

Table 1.2 H/B values for determining investigation depth (H) based on 5% effective stress criterion for $\gamma_{avg} = 18$ kN/m^3 (Eslami, 2007).

	L/B						
P_{avg}/B	**1**	**2**	**3**	**4**	**6**	**8**	**10**
0.05	0.320	0.350	0.364	0.371	0.379	0.383	0.386
0.1	0.497	0.558	0.586	0.603	0.622	0.632	0.638
0.15	0.629	0.716	0.757	0.783	0.811	0.827	0.838
0.2	0.737	0.846	0.90	0.933	0.971	0.993	1.01
0.3	0.912	1.059	1.134	1.181	1.237	1.269	1.290
0.4	1.06	1.233	1.33	1.385	1.457	1.499	1.527
0.5	1.175	1.381	1.490	1.561	1.647	1.699	1.733
0.60	1.281	1.512	1.637	1.717	1.817	1.877	1.918
0.70	1.380	1.630	1.768	1.858	1.971	2.04	2.086
0.80	1.464	1.738	1.889	1.988	2.113	2.189	2.241
0.90	1.550	1.838	2.000	2.108	2.244	2.328	2.386
1.00	1.620	1.930	2.105	2.220	2.367	2.458	2.521

B and L are the width and length of equivalent mat foundation, respectively.

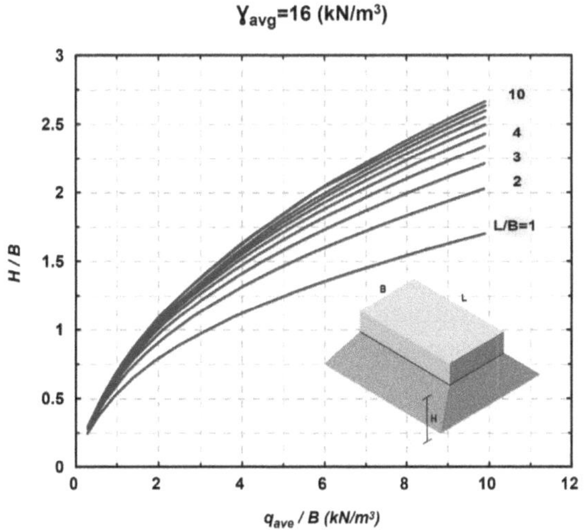

Figure 1.2 Values of investigation depth (H) based on 5% effective stress criterion for $\gamma_{avg} = 16$ kN/m^3 (Eslami, 2007).

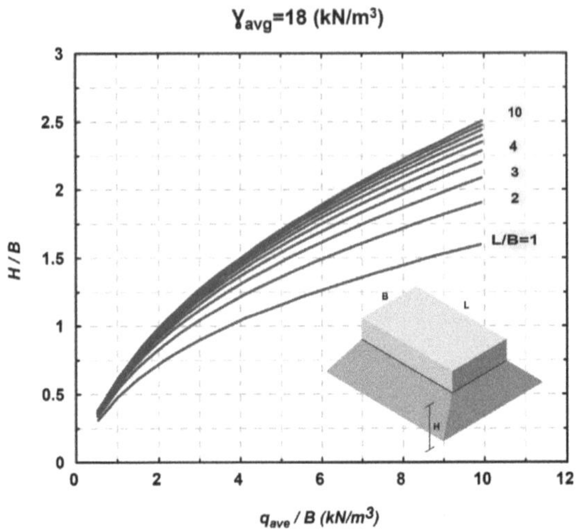

Figure 1.3 Values of investigation depth (H) based on 5% effective stress criterion for $\gamma_{avg} = 18\ kN/m^3$ (Eslami, 2007).

Step 3: Determining the exploration extent in plan by assuming 2:1 stress distribution in depth.

$$\frac{B'}{B} = 1 + \frac{H}{B} = 1 + \frac{23.4}{15} = 2.56 \Rightarrow B' = 38.4\ \text{m}$$

$$\frac{L'}{L} = 1 + \frac{H}{L} = 1 + \frac{23.4}{25} = 1.94 \Rightarrow L' = 48.5\ \text{m}$$

1.6 Sources of geotechnical data

Unlike structural engineers, geotechnical engineers mostly encounter natural geomaterials which involve variable and unknown properties. Therefore, due to natural spatial variability, safe and reliable design of foundations and to minimize uncertainties, it is necessary to rely on different compiled and measured records and information. This major purpose can be accomplished via employment of data sources which are available through the following approaches:

1. Maps and technical literature review
2. Site visit
3. On-surface in situ testing
4. Geophysical surveying
5. Drilling operations
6. In situ penetration tests
7. Laboratory element tests
8. Physical modeling
9. Full-size field tests
10. Instrumentation and monitoring

Each of these categories is briefly reviewed in the following sections.

1.6.1 Maps and technical literature review

The first step for site exploration can be attributed to the remote sensing. In geotechnical engineering, this can be done via using aerial photos. For closer assessment of project location, different types of relevant maps can be used such as geological, geographical, hydrological, topographical maps, as well as geological hazard maps such as landslides, liquefaction, and earthquake/fault maps.

Other than site-specific literature review that may or may not be available, any sort of literature, maps, or site investigation reports available for adjacent sites or areas are also beneficial as a primary tool for better understanding regional geology and subsurface conditions, and can provide a better idea to the engineer how to plan the new set of site explorations.

1.6.2 Site visit

Site visit or field reconnaissance, visual assessment, and manual testing at the project site are also quite valuable preliminary studies for a given project. A few typical pieces of information can be gathered by the engineer such as:

- Surface soil, rock, and drainage conditions
- Occurrence of any potential geohazards
- Performance of previous construction activities
- Interaction with nearby structures
- Access roads or paths for investigation and future construction

1.6.3 On-surface in situ testing

Usually subsequent to site visit, a few visual and manual tests are performed to verify the above-referenced procedures and planning for detailed and optimized site investigation program. They can be relevant to the following topics:

- Color and odor of near-surface soils or rocks
- Texture and fabric
- Preliminary assessments of soil gradation and plasticity (coarse grained or fine grained)
- Moisture and water flow or traffic compaction
- Shallow strength which can be measured by any movable vehicles, sondage, pocket penetrometer, etc.
- Any signs of chemical activities of soil or water, noticeable erosion, or corrosion phenomena

1.6.4 Geophysical test

Geophysical surveying or nondestructive tests (NDT) provide qualified and significant information for important projects with no disturbance. The main advantage of geophysical tests is rapid three-dimensional (3D) covering of a vast area. However, these tests are economically justifiable if the investigated area is large enough and the other quick methods such as boring or penetration tests are neither practical nor economical. In addition, excavation measures including well, boring, test pit (TP), gallery, and trenches only cover a limited zone of the investigated site and are best not to generalize to the entire site. In general, geophysical tests do not provide any direct measurements of soil shear strength, compressibility, grain size characteristics. Indeed, their best application is determining depth and location of bedrock, in marine structures or dams for instance, where by geophysical assessments and by means of physical principles, specific soil layers properties and stratification can be evaluated. Geophysical tests can be divided into three categories: electrical tests, seismic tests, magnetic tests.

Electrical tests are based on the difference of electrical resistance or conductivity in various soils. In this approach, apparent electric resistance is determined by means of electrical flow, and subsequently, soil type or bedrock can be distinguished. The seismic approach depends on the differential seismic wave velocity in various soils. Accordingly, the velocity of waves caused by stimulation is measured and consequently, the soil type is distinguished. In addition, the bedrock depth can be estimated based on waves' reflection. The magnetic approach which is limited than the other methods, relies on variation of the earth magnetic field. Generally, a few geophysical test records such as shear wave velocity and ground resistivity are measured by some penetration tests, which enrich the interpretation of underground conditions measured for strength and stiffness purposes.

Overall, geophysical tests are considered as a part of initial investigation phase and provide supplementary 3D information based on sporadic observations (large distances between borings, TPs, etc.).

1.6.5 Drilling operations

The office works, site visit, and performing NDT is followed up by operations on site or exploratory borings. This is achieved by operations such as drilling vertical boreholes (BHs), TPs, horizontal excavations (galleries). The major tools used for drilling depend on depth, width, and type of geomaterial formation. They include hand-operated augers, portable power-operated vehicles, and heavier drilling rigs.

1.6.6 Element laboratory testing

One of the main aims of drilling is acquiring soil and rock samples for performing laboratory tests and determining the subsurface soil profile. The following two approaches are employed for sampling:

a) Undisturbed samples, where characteristics such as soil texture, structure, natural stress condition, and physical parameters such as water content and unit weight are similar to the in situ conditions.
b) Disturbed samples, in which the soil initial properties and conditions have changed during drilling and sampling procedures.

During sampling, specific measures are used to conserve the in situ soil structure using special tools and equipment. Of course, it is best to obtain and test undisturbed samples; however, the following reasons lead to samples being disturbed during drilling and sampling:

- Compressibility and shearing in soil during sampling
- Volume change due to stress relaxation
- Friction between soil sample and sampling apparatus
- Evaporation of soil moisture and consequently change of the water content of soil sample
- Elimination of air bubble pressure in soil
- Climatology conditions (cold or warm weather, humidity, wind) can affect soil sample
- Elimination and changes in vertical and horizontal stresses within the soil skeleton
- Shaking soil sample in some drilling methods and during transfer to laboratory

In practice, by means of some methods, such as preparing large or block samples and using thin-wall coring pipes, disturbance can be reduced in sampling of cohesive soils. However, these approaches are not pragmatic for cohesionless soils, unless some other supplementary techniques such as ground freezing are applied.

Some common categorized laboratory tests include:

- Identification and index tests such as water content and density, gradation (sieve analysis, hydrometer test, and wash test), Atterberg limits
- Volume change or compressibility tests including consolidation, ramming tests, CBR,[1] and inflation
- Water flow tests such as permeability, consolidation, inflation, sand boil, and capillary
- Chemical tests to determine corrosivity and understanding the sulfate and chloride contents, water and soil pH, and electrical resistivity
- Shear strength tests including uniaxial, triaxial, direct shear, and simple shear tests
- Dynamic tests such as cyclic triaxial, resonant column test, and cyclic simple shear

Despite the fact that laboratory tests provide varied platforms regarding stress path, sample size, drained or undrained conditions, static or cyclic loading, consolidation condition, etc. for analysis and design, they are accompanied by some limitations such as sampling, modeling actual in situ stress conditions, sample size, and neglecting some actual soil structural conditions. In this regard, physical modeling and in situ testing can be employed to compensate for some of these restrictions.

1.6.7 Physical modeling or medium-scale testing

Due to the complexities and difficulties of performing large-scale physical modeling tests for geostructures, researchers usually attempt constructing small- or medium-scale prototype laboratory models that allow for simulating actual conditions that represent physical and mechanical behavior of the structure and its surrounding environment. In general, physical models for assessing behavior of footings and deep foundations (piles) in laboratory are meant to, along with employing small-scale

[1] California Bearing Ratio.

piles or footings, provide the condition in the laboratory as similar as possible to in situ conditions in terms of confining stresses, state parameter, other geotechnical parameters, and aspect ratio. Simple chambers (1g), calibration chambers (CCs), centrifuge apparatus (ng), frustum confining vessel (FCV) are examples of physical modeling apparatuses used for studying the behavior of foundations and piles in the laboratory.

Among commonly used physical modeling tools, the FCV is developed more recently for simulation of in situ stress states in depth. An FCV combines the advantages of centrifuges and CCs in one apparatus. In fact, it can create linear vertical and horizontal stress gradients at depth, increasingly from zero at the top of the vessel to the stress proportional to the applied pressure at the base. FCVs produce stress conditions highly similar to those distributed in the field, while being less expensive to operate than a centrifuge test.

The first FCV was constructed in 1996 at McMaster University (Horvath and Stolle, 1996; Sedran, 1999). The second one was constructed in 1999 at the University of South Florida (Mullins et al., 2001). In 2012, the third and largest FCV was constructed at Amirkabir University of Technology. The FCV-AUT has dimensions larger than those of the existing FCVs, allowing for modeling of larger piles. Thus, the errors and limitations related to the effects of scales and boundaries can be minimized (Zare and Eslami, 2014; Zarrabi and Eslami, 2016; Khazae and Eslami, 2017).

As illustrated in Fig. 1.4, the FCV-AUT is a 1000-mm-high frustum with two vertical wings above and beneath with a height of 100 and 200 mm, respectively. For ease of modeling, the frustum is made of two 500-mm detachable pieces. The diameter of the top of the FCV-AUT is 300 mm, and the bottom diameter is 1300 mm.

The vertical and horizontal stress changes are recorded by the soil pressure meters at the central axis of the vessel. These stress variations along the depth are for base pressures of up to 200 kPa in loose relative density (D_r) = (20%−25%), medium (D_r = 45%−50%), and dense (D_r = 65%−70%) states of sands. The vertical load is applied in a stepwise manner by a hydraulic jack. A load cell and LVDT with the high precision of 0.001 mm are located on the loading piston to evaluate the total vertically imposed load and the corresponding displacement of the pile cap.

1.6.8 Full-scale testing in the field

For most of geotechnical engineering projects, such as dams, excavations, combined foundations, pile groups, highways, and airports, the full-scale testing is somehow difficult, complex, time-consuming, and quite expensive to carry out. However, for some separate elements such as single footing, single piles, and anchors, full-scale testing upon a few regulations is mandatory and performed frequently. This is more pronounced in a special development site, where the project involves a large number of geostructural units. Via observations during testing in vertical, horizontal, compression, or tension loading, the actual performance of the considered element, i.e., footing, pile, anchor, etc., a lot of data and information can be acquired. Accordingly, not only via real testing for structural units surrounded by actual soil condition, the desired function can be studied and quantified. Therefore, a source of data can be provided by direct measurements instead of indirect or even irrelevant investigations.

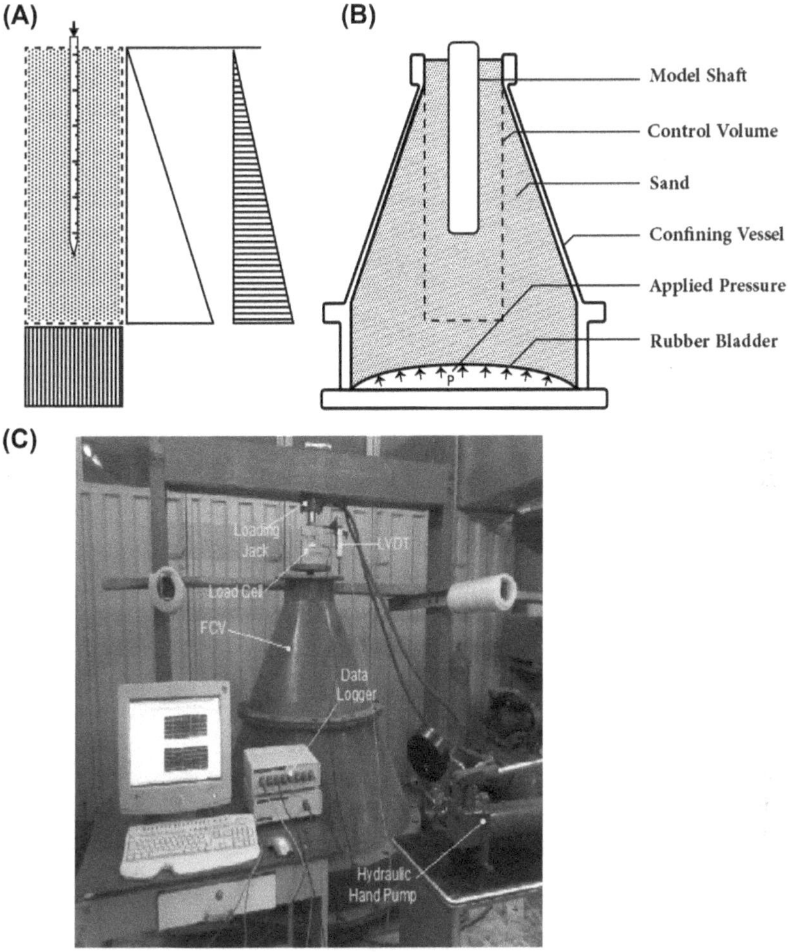

Figure 1.4 Frustum confining vessel (FCV) apparatus: (A) linear distribution of confined stress on the shaft of a pile or a penetrometer installed in sand prior to loading, (B) schematic view of device components and instruments in simulating stress, (C) operational system (Zare and Eslami, 2014).

1.6.9 Instrumentation and monitoring

Designers of geotechnical structures encounter more complicated problems compared to designers of steel or concrete structures. It is because materials used in concrete and steel structures are man-made, and consequently, they have more specified properties and behavior, while there are difficulties and ambiguities regarding determining actual properties and performance of geomaterials. Thus, monitoring and surveillance are of significant importance in geotechnical design, and it is crucial to employ different tools for recording displacements and stresses in structures and monitoring behavior of geotechnical structures at different stages of construction and serviceability.

Monitoring of such structures are done by installing specific tools, namely fine instruments. At the next step, transducers are used to digitize the changes recorded by fine instruments.

Monitoring is the process of acquiring necessary and accurate data for design and practical aspects. Indeed, fine instruments are a tool for measuring and monitoring soil parameters during construction, postconstruction, and lifetime. The reasons for the necessity of instrumentation and monitoring are as follows:

- Cost reduction: An appropriately instrumented project can significantly reduce the design risk and hazards in sites with sensitive conditions such as compressible or problematic soils. Lower risk leads to cost saving.
- Safety factor reduction: Without instrumentation and suitable monitoring, selection of geotechnical parameters would be based on some assumptions, some of which are conservative and subsequently, lead to higher safety factor and expense.
- Design modification: In case of insufficient design for existing project condition, data from instrumentation at initial phases of project can help distinguish the need for modifying the design.
- Control of operation procedure: Fine instruments provide information regarding progress of operations such as soil improvement and preloading. The sufficiency of the number of drains used is one example for controlling the procedure.
- Safety: Instrumentation can warn of the possible hazards in advance, especially in projects regarding slope stability.

1.7 In situ penetration tests

In light of cost saving in geotechnical investigations and considering time-consuming process of traditional methods such as drilling, sampling, transferring samples to laboratory, and performing tests, it is more favored to take the testing equipment to the site. In addition, development of more equipped in situ tests has facilitated more comprehensive and proper understanding of soil behavior.

In the last two decades, there has been significant progress in the realm of in situ testing, specifically in determining soil strength and stiffness properties. In situ tests have shown more applicability in cohesionless soils where there are limitations in soil sampling and disturbance. However, it should be mentioned that in situ tests suffer from some limitations as well, including the lack of control on drainage conditions, stress path. Thus, it would be more appropriate to employ both laboratory tests and in situ tests simultaneously to reach a more accurate understanding and evaluation of soil properties. This way, each approach compensates for the disadvantages of the other, providing more reliable information and properties for the subsurface soil. In other words, laboratory and in situ testing are complementary in geotechnical investigations.

In situ test can be employed for directly defining soil profiling, geotechnical soil parameters and characteristics. More commonly used in situ tests in geotechnical practice include the following:

- Standard penetration test (SPT)
- Cone and piezocone penetration test (CPT and CPTu)
- Pressure meter test (PMT)

- Vane shear test (VST)
- Dilatometer test (DMT)
- Plate load test (PLT), penetrometer plate load test (PPLT)
- Becker penetration test (BSPT)

Each of the abovementioned tests applies a different form of loading and measures the corresponding soil response by which soil strength and stiffness properties can be assessed and determined by empirical approaches. Fig. 1.5 demonstrates the various in situ testing equipment.

Performing SPT, PMT, and VST requires drilling a hole and pertinent rotary drilling rig and equipment. The CPT, CPTu, and DMT use direct penetration technology and thus need no BH. Special types of PMT (i.e., fully displacement) and VST can be carried out without BHs. In these tests, the drilling rod is directly transmitted to the desired depth by means of mobile hydraulic systems or standard drilling towers. Table 1.3 depicts the applicable common range of in situ tests in terms of various soil characteristics.

In situ test can be divided into two general categories: logging tests and specific tests. Logging tests are primarily penetration tests which are mainly fast and economical. These tests are repeatable and provide continuous records of soil measurements in depth (e.g., CPT). Specific tests, however, are generally used to determine a specific soil parameter such as shear strength or deformation modulus (e.g., PMT), which are more time consuming and expensive compared to logging tests. Albeit, logging tests and specific tests are best used in combination of one another.

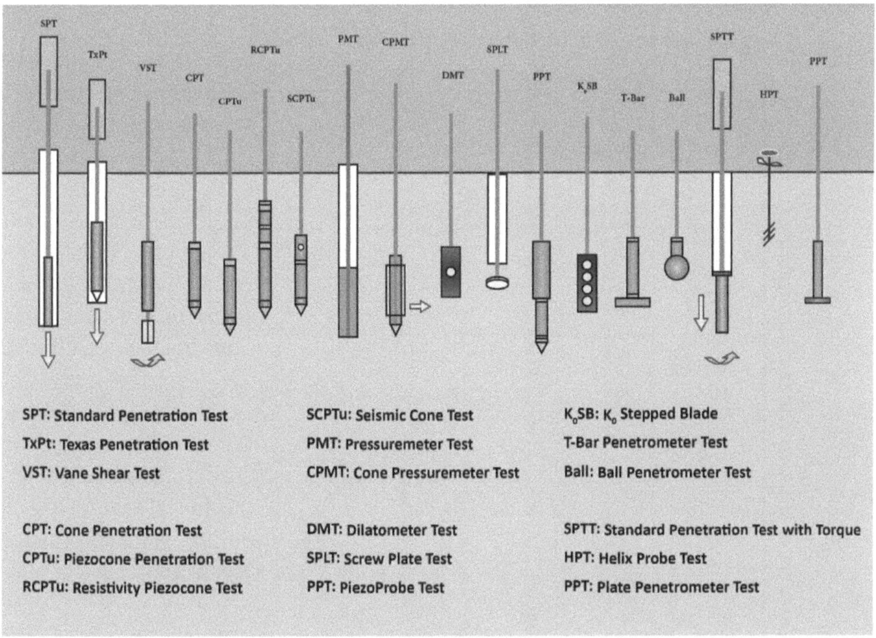

Figure 1.5 Typical in-situ test schematic performance.

Table 1.3 Different in situ tests and their applicability (Lunne et al., 1997).

		Soil parameters													Ground type						
Group	Device	Soil type	Profile	u	*ϕ'	S_u	ID	m_v	c_v	k	G_0	σ_h	OCR	σ-ε	Hard rock	Soft rock	Gravel	Sand	Silt	Clay	Peat
Penetro meters	Dynamic	C	B	–	C	C	C	–	–	–	C	–	C	–	–	C	B	A	B	B	B
	Mechanical	B	A/B	––	C	C	B	C	–	–	C	C	C	–	–	C	C	A	A	A	A
	Electric (CPT)	B	A	–	C	B	A/B	C	–	–	B	B/C	B	–	–	C	C	A	A	A	A
	Peizocone (CPTU)	A	A	A	B	B	A/B	B	A/B	B	B	B/C	B	C	–	C	–	A	A	A	A
	Seismic (SCPT/ SCPTU)	A	A	A	B	A/B	A/B	B	A/B	B	A	B	B	B	–	C	–	A	A	A	A
	Flat dilatometer (DMT)	B	A	C	B	B	C	B	–	–	B	B	B	C	C	C	–	A	A	A	A
	Standard penetration test (SPT)	A	B	–	C	C	B	–	–	–	C	–	C	–	–	C	B	A	A	A	A
	Resistivity probe	B	B	–	B	C	A	C	–	–	–	–	–	–	–	C	–	A	A	A	A
Pressure meters	Prebored (PBP)	B	B	–	C	B	C	B	C	–	B	C	C	C	A	A	B	B	B	A	B
	Self-boring (SBP)	B	B	A(1)	B	B	B	B	A(1)	B	A(2)	A/B	B	A/B(2)	–	B	–	B	B	A	B
	Full displace ment (FDP)	B	B	–	C	B	C	C	C	–	A(2)	C	C	C	–	C	–	B	B	A	A

Others	Vane	B	C	—	—	A	—	—	—	—	—	—	B/C	B	—	—	—	—	—	A	B
	Plate load	C	—	—	C	B	B	B	C	C	A	C	B	B	B	A	B	B	B	A	A
	Screw plate	C	C	—	C	B	B	B	C	C	A	C	B	—	—	—	—	A	A	A	A
	Borehole permeability	C	—	A	—	—	—	—	B	A	—	—	—	—	A	A	A	A	A	A	B
	Hydraulic fracture	—	—	B	—	—	—	—	C	C	—	B	—	—	B	—	—	—	—	A	C
	Crosshole/ downhole/ surface seismic	C	C	—	—	—	—	—	—	—	A	—	B	—	A	A	A	A	A	A	A

Applicability: *A*, high; *B*, moderate; C = low; −, none.
*ϕ, Will depend on soil type; (1), Only when pore pressure sensor fitted; (2), Only when displacement sensor fitted.
Soil parameter definitions: *u*, in situ static pore pressure; *ϕ'*, effective internal friction angle; S_u, undrained shear strength; *ID*, density index; *mv*, constrained modulus; c_v, coefficient of consolidation; *k*, coefficient of permeability; *G0*, shear modulus at small strains; σ_h, horizontal stress; *OCR*, overconsolidation ratio; σ-ε, stress−strain relationship.

1.8 Geotechnical reports

Upon completion of site investigation, the geotechnical engineer is responsible for synthesis and interpretation of the data, construction of a model of the site geology, selection of appropriate geomaterials design parameters, and engineering analyses for design.

In general, two types of geotechnical report are prepared (Brown et al., 2010): geotechnical site investigation report and geotechnical design report.

A geotechnical site investigation report documents the field and laboratory investigations performed and presents the obtained data. Boring logs, rock coring logs, CPT soundings, and exploration logging should be prepared in accordance with standard procedures and formats according to any related codes or agencies. This type of report typically does not include interpretations of the subsurface conditions or design recommendations. More information on what data are presented in a geotechnical investigation report is described by Brown et al. (2010).

A geotechnical design report, also called a foundation report, typically provides an assessment of existing subsurface and potential geological hazard conditions at the project site and presents geotechnical analyses. It also provides appropriate recommendations for design of the foundation system and relevant earth structures, retaining walls, or other necessary facilities.

The report must identify each soil and rock unit of engineering significance and recommend design parameters for each of these units. This requires a summary and analysis of all factual data to justify the recommended design properties. Groundwater conditions are particularly important for both design and construction and, accordingly, they need to be assessed carefully.

For any design issue, especially foundations axial resistance, response to lateral loading, settlement analysis, group behavior, etc., must be addressed in accordance with the recommended methodologies.

A site location plan should be provided for reference on a regional or local-scale map. Locations of all field tests, sampling, and exploratory studies should be shown clearly on the scaled map of the site. Preferably, the plan should be a topographic map with well-delineated elevation contours and a properly established benchmark. Additionally, site location maps can be plotted directly on aerial photos, indicating true north (N) direction.

Geotechnical reports are normally accompanied by the presentation of subsurface profiles developed from field and laboratory test data. Longitudinal profiles are typically developed along the project alignment, and a limited number of transverse profiles may be included for key locations. Subsurface profiles, coupled with judgment and an understanding of the geologic setting, aid the geotechnical engineer in the interpretation of subsurface conditions between the investigation sites.

1.9 Remarks

As reviewed in this chapter, sources of collecting and synthesizing data for assessment of soil parameters in order to achieve a comprehensive design in geotechnical engineering consist mainly of maps, site visits, technical reports, geophysical testing, drilling, sampling, laboratory testing, in situ testing, and physical or full-scale testing. Additionally, analytical and numerical models can be used as supplementary tools along with the abovementioned tests to ensure an optimum design.

Although laboratory and in situ testing have played a significant role in design procedures over the last century, as depicted by Lacasse (1985), in recent years, with increasing the reliability of geophysical tests, and more significantly analytical and numerical models, these approaches have been more extensively used in design. Moreover, it can be noticed that engineering judgment, especially in geotechnical engineering, which is engaged with parts of empiricism and materials complexities, has invariably governed the design procedure regardless of the design approaches.

1.10 Book organization and scope

To assist the reader, the structure of the book chapters is reviewed in the following:

- An initial introduction of geotechnical engineering has been presented in Chapter 1. Site investigation procedures, common sources of subsurface data acquisition, and report writing in geotechnical engineering have been briefly discussed.
- Chapter 2 presents principles of foundation engineering, types and classifications, in addition to principles for analysis and design. Moreover, the fundamental aspects of soil mechanics for foundation engineering are reviewed. Additionally, uncertainties in foundation engineering are recognized, and the role of CPT and CPTu data for a more reliable design is discussed.
- Chapter 3 deals with introducing the cone penetration test; detailed discussions about the apparatus and testing procedures are presented. In addition, various equipment to accompany the CPT for characterizing additional soil properties such as pore pressure, shear wave velocity are demonstrated.
- In Chapter 4, a variety of correlations for estimating soil properties such as strength and stiffness parameters particularly employed in foundation engineering by means of CPT measurements are presented.
- In Chapter 5, classifications of soils by in situ tests are briefly reviewed. Then, various soil classification charts and approaches in terms of CPT measurements are presented and compared. Additionally, the characterization of a few complex deposits via different soil behavior classification charts is compared.
- Chapter 6 starts with introducing factors pertinent to scale effect between penetrometer and foundations and their effects on bearing capacity. Then, it describes the application of CPT in geotechnical design of shallow foundations in terms of bearing capacity, mainly focusing on scale effects. Some direct and indirect methods for determining the bearing capacity of shallow foundations are reviewed. And an analytical method is presented. The chapter is closed by an example of determining the bearing capacity of a shallow foundation.
- Chapter 7, realizing cone penetrometer as a model pile, deals with a wide range of direct CPT-based methods in estimating the pile axial bearing capacity and provides examples of direct as well as indirect implementation of CPT records in calculation of pile bearing capacity.
- Chapter 8 encompasses various approaches of estimating shallow foundation settlement by CPT-based correlations. In addition, several CPT-based correlations in estimating stiffness properties of soil are presented.

- Chapter 9 provides criteria for recognizing and classifying problematic soil types by means of CPT and CPTu charts as presented in Chapter 5. Furthermore, this chapter introduces applications of CPT for evaluating and comparing the soil properties before and after soil improvement as an indicator of improvement effectiveness. The comparisons are presented in the form of various case studies worldwide.
- Chapter 10 introduces applicability of soft computing techniques such as the group method of data handling-type neural network, and support vector machine based on CPT records for predicting pile bearing capacity, shallow foundation settlement. Additionally, several examples of employing these techniques are presented to shed light on their effectiveness in realm of foundation engineering.
- Chapter 11 deals with the topic of sources of uncertainty in geotechnical engineering, and reliability-based approaches are introduced. Some examples are presented on implementing reliability-based approaches in evaluating performance of CPT-based methods of estimating piles bearing capacity.
- Chapter 12 demonstrates a wide range of CPT databases in the literature and their applications. The databases in the realm of liquefaction assessments, correlating SPT and CPT data, soil improvement assessments, and piling and in situ test evaluations are examples of databases presented in this chapter. At the end, an extensive recently developed database is presented including records of CPT and CPTu piling records and geotechnical information.

References

ASCE, 1993. Bearing capacity of soils. In: Technical Engineering and Design Guides. US Corps of Engineers. No. 7.

Brown, D.A., Turner, J.P., Castelli, R.J., 2010. Drilled Shafts: Construction Procedures and LRFD Design Methods. US Department of Transportation, Federal Highway Administration (FHWA).

Canadian Foundation Engineering Manual, CFEM, fourth ed., 2006. Canadian Geotechnical Society, BiTech Publishers, Vancouver. 488 pp.

Coduto, D.P., 2002. Foundation Design: Principles and Practices, second ed. Prentice Hall, Inc, Upper Saddle River, NJ.

Eslami, A., 2007. Foundation Engineering, Design and Construction. Building and Construction Research Press. No. 437.

Horvath, R.G., Stolle, D., 1996. Frustum for testing model piles. Canadian Geotechnical Journal 33, 499–504.

Khazaei, J., Eslami, A., 2017. Postgrouted helical piles behavior through physical modeling by FCV. Marine Georesources & Geotechnology 35 (4), 528–537.

Lacasse, S., 1985. Design parameters of clays from in-situ and lab tests. In: Proc. Symposium on New Concepts in Geotechnical Engineering, Rio de Janeiro; also Norwegian Geotechnical Inst. Report No. 52155-50, Oslo.

Lunne, T., Robertson, P.K., Powell, J.J., 1997. Cone Penetration Testing in Geotechnical Practice. Blackie Academic and Professional, London.

Mullins, G., Dapp, S., Fredrerick, E., Wagner, R., 2001. Pressure Grouting Drilled Shaft Tips—phase I Final Report. Final Rep. Submitted Florida Department of Transportation, FL.

Sedran, G., 1999. Experimental and Analytical Study of a Frustum Confining Vessel. Doctoral Thesis. McMaster University.

Zare, M., Eslami, A., 2014. Study of deep foundation performances by frustum confining vessel (FCV). International Journal of Civil Engineering (IUST University, Tehran, Iran).
Zarrabi, M., Eslami, A., 2016. Behavior of piles under different installation effects by physical modeling. International Journal of Geomechanics 16 (5), 04016014.

Background to foundation engineering

2

2.1 Introduction

Foundation engineering is a branch of geotechnical engineering which applies soil mechanics, structural engineering, and project serviceability requirements for design and construction of foundations for onshore, offshore, and in-land structures. Foundation engineering can be realized as an "artistic" approach rather than a routine procedure because well-designed and constructed foundations continue to perform efficiently during the lifetime of a project. The major task and goal of a foundation engineer is to create a technically sound, construction-feasible, and economical (avoiding costly and overdesign) design of the foundation system to support the superstructure.

Foundation elements or systems are structural units that transfer various load combinations from the superstructure to the underlying soils or rocks (i.e., geomaterials). Foundation units may tolerate the loads individually or by contribution of other elements such as basement walls, floors, or slabs.

The major role of the foundation is to spread and moderate the highly concentrated stresses in the structural units (i.e., wall, column, or piers) with the normal magnitude of 10−200 MPa and transfer them to the subsoil with the usual tolerable compression stresses of 0.05−0.5 MPa. In this regard, the road pavement, baseplate of steel columns, and roots of trees and plants can be considered as footings or foundations.

Typical examples of structures are illustrated in Fig. 2.1, including bridges, airport structures and subbase, skyscrapers, marine platforms, etc. In addition, various transitional elements (between structures and underground bearing strata) as foundation systems are presented in Fig. 2.2.

As depicted by Coduto (2002), foundation engineering must be realized as a multidisciplinary process of knowledge-based and interactive practice among structural, geotechnical, and constructional engineers as illustrated in Fig. 2.3.

For structural issues, since the foundation supports a structure, the following items are very significant: source and nature of loads and the tolerance of any civil engineering project to the foundation displacements. Moreover, any footing or foundation system can be realized as a structural member for safe load transfer from superstructure to the subsoil. Therefore, the proper understanding of internal stress distribution and employment of suitable materials are important for design process.

In view of geotechnical engineering approach, any foundation interact with the ground, hence, the design process must reflect the engineering properties and behavior of adjacent geomaterials.

At last but not least, for foundation construction the building materials and equipment are determinant. However, deliberate dealing with geomaterials and performance

Piezocone and Cone Penetration Test (CPTu and CPT) Applications in Foundation Engineering
https://doi.org/10.1016/B978-0-08-102766-0.00002-X

Figure 2.1 Various types of structures in civil engineering.

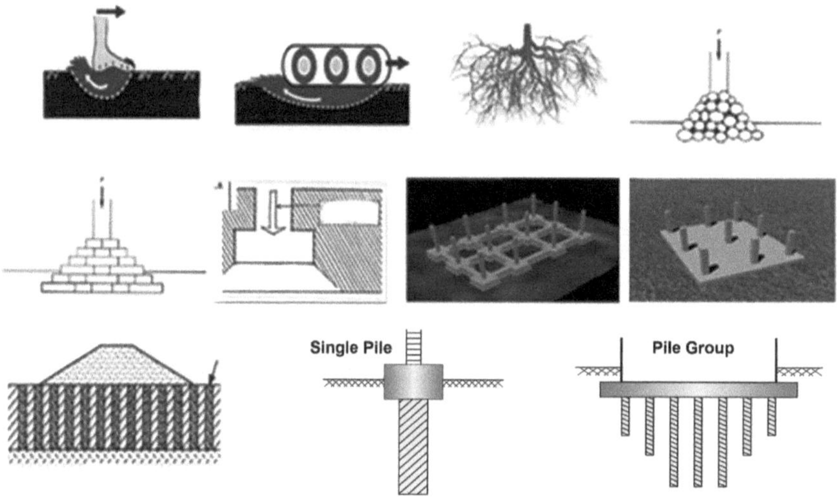

Figure 2.2 Various elements acting as foundation systems.

of surrounding soil, especially, in difficult conditions can lead to efficient and economical practice.

2.2 Foundation analysis and design considerations

Overall, the following issues are realized in efficient foundation engineering practice including technical, practical, and economical aspects, mainly focused on performance-based design:

1. Bearing capacity
2. Serviceability (settlement and torsion)
3. Structural design
4. Stability control

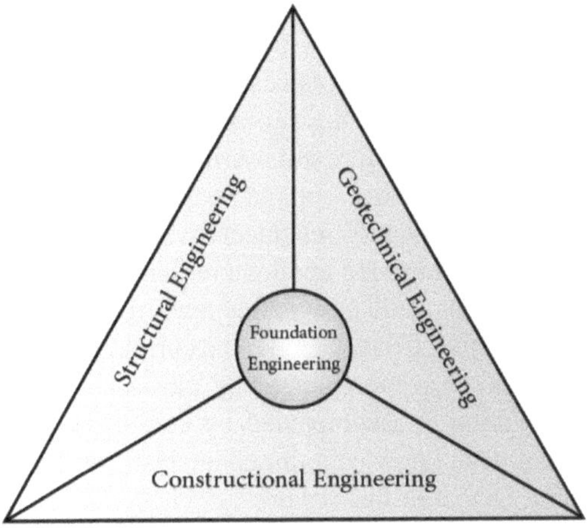

Figure 2.3 Interaction of geotechnical, constructional, and structural engineering in foundation engineering.

5. Full or model scale testing
6. Constructional aspects
7. Durability
8. Economical requirements

Actually, the safety, serviceability, and economic issues must be covered. Accordingly, the bearing capacity, structural and stability design are related to safety, while settlement and durability are relevant to serviceability concerns.

Basically, the bearing capacity addresses the capability of below-foundation soil or rock for tolerating the loads without any type of failure. In reality, it is governed by the geomaterial's shear strength parameters, foundation geometry, and embedment, as well as loading conditions. Various equations are recommended in textbooks or foundation design handbooks. In this book, subsequent to a review of principle equations for shallow and deep foundations, CPT-based relations are presented and reviewed in details.

The foundation structural design or internal design is related to stress mobilization in the foundation element caused by external applied forces from the structure, and the underneath reaction imposed by supporting soil. In current practice, where the majority of foundations are made of reinforced concrete, the magnitude and direction of internal stresses owing to flexural and shear forces need to be determined. Afterward, appropriate concrete and longitudinal or transversal steel bars have to be designed. Structural design and analysis depending on foundation type, construction materials, and regional codes are conducted through either of ASD (Allowable Stress Design), USD (Ultimate Stress Design), or LRFD (Load and Resistance Factor Design) approaches.

The serviceability criterion is attributed to deformation and durability controls in foundation design. Deformation related to the foundation serviceability limits involve a few occurrences such as settlement, heave, angular distortion, tilt, lateral displacement, and vibration. Failure to satisfy these limits and requirements does not generally lead to major failure or loss. However, it can cause higher maintenance costs, aesthetic issues, downgraded functionality of the project, and other environmental aspects (Coduto, 2002). Among various deformation criteria for foundations, settlement or downward movement due to vertical applied loads is the more significant aspect than others. Therefore, the foundation engineer must be able to estimate the total and differential settlements considering applied pressure, foundation geometry and rigidity, subsoil behavior against static or dynamic loading, time history as well as geomaterial stiffness characteristics. Consequently, the correlated effects must be tolerable for the superstructure and to be accommodated by substructure. Principally, the settlement analysis and calculation is accompanied by complexity; therefore, employing comprehensive source of data, knowledge-based analysis, and elaborated engineering judgment can be crucial factors in this regard.

As pronounced by Fellenius (2015), the analysis and design of foundations are an iterative process inasmuch as the amount of imposed loads, corresponding settlement, and foundation geometry are interactive, being affected by geotechnical capacity, structural capacity, and settlement requirements.

Stability control for foundations is incorporated in situations where other than usual vertical loads such as horizontal, bending moments or uplift forces are induced to the foundations. They can be initiated from various circumstances, for example, soil, water lateral pressure, earthquake and wind loads pullout, and vessel or vehicle collision loads. The stability check and control include aspects of sliding, overturning, overall slope stability, and interface tension between foundation and supporting soil or rock. One or more unsatisfactory stability conditions can lead to major damage and failure simultaneously to the foundations and on built structures.

Full scale or even model scale in situ foundation testing through safety design stage is highly desirable particularly in case of massive production like numerous piles or footings mainly for unusual sites. This is to verify and demonstrate that the actual foundation behavior is consistent with the design assumptions. This usually can be done by some form of testing via full, medium, or even small scale or footings. If the behavior differs from that of assumed, then it is necessary that the foundation design be revised. Based on the number of tests, quality, and quantity of instrumentation, type and method of interpretation in most cases, this procedure decreases conventional factor of safety and consequently leads to efficient design. For making appropriate decisions, the foundation engineer must be aware that foundation testing involves only individual elements of the substructure system, and that the foundation elements in the system, i.e., the piles, single footings, and the raft within the system will interact as a hybrid. The overall foundation behavior may thus not be able to be assessed directly from the foundation test results without consideration of the foundation–soil interaction effects.

In the durability related to serviceability limits for foundations, it is intended to guarantee and ensure that foundations can resist against various physical, chemical, and biological deterioration. Over the design lifetime of a structure, it is expected that foundations provide suitable performance. The durability limit states often related to corrosion or other time-dependent weakening or damage of foundations. Therefore, the susceptibility of foundation material made of concrete, steel, or wood against any environmental attack including soil, water, or atmosphere must be prevented, diminished, or protected. Coduto (2002) investigated this topic in detail, which is mostly attributed to the points considered in site investigation, structural design, and construction stages.

The constructability performance requirements is interconnected to the contractor usual work and practices which means the foundation system can be built and installed without using any extraordinary method or equipment. Decision-making depends on factors such as the project site, accessory, type of foundation, near or offshore environments, ground conditions, groundwater level, installation equipment, specified cost, timing, and adjacent structures or utilities. Therefore, it is necessary to keep in mind that foundation engineering and design is a knowledge-based, multidisciplinary, and team working task.

In view of economical aspects, it is obvious in foundation design to incorporate a higher level of conservatism compared to the design of the superstructure. This is attributed to the uncertainty of soil and rock characterization, relatively weaker underground construction in the presence of excavated soil mixture with water and formworks, probable damage during installation, and relatively complex load transfer among SFSI (soil, foundation, and structure interaction). However, the degree of conservatism must be reasonably justified. Particularly, overdesign can lead to overconsumption of material and equipment as well as requiring more time-consuming procedures, which can impose unnecessary costs to the project. Therefore, efficient foundation design with appropriate conservatism must be accompanied by technical and practical engineering features to be in proper alignment with value engineering.

Above all, for appropriate foundation performance as stated by Burland et al. (2012) and illustrated in Fig. 2.4, a cyclic perspective pattern of data, design, and performance involvement is governed, not a conventional simple and routine linear rule.

2.3 Foundation classification

Based on geometry in plan and embedment depth, load transfer mechanism, supporting the structure as individually or group, foundations are classified into two categories: shallow and deep foundations.

Shallow foundations are located on or near ground surface and transmit the loads to the soils immediately below them. The normal reaction is mobilized through contact area between the soil and the bottom of the foundation. The most commonly used type is single foundation or spread footing, which distributes the applied load over enough soil area to maintain reaction soil stresses within a tolerable limit. Shallow foundations are mostly used for light to medium loads on sites where the soil conditions are relatively good. Single footings are typically used for each column and in case of multiple closely spaced columns that support a wall, they can be combined in the form of continuous or strip footings. On the occasion of two directional strip foundations, grid foundations are formed

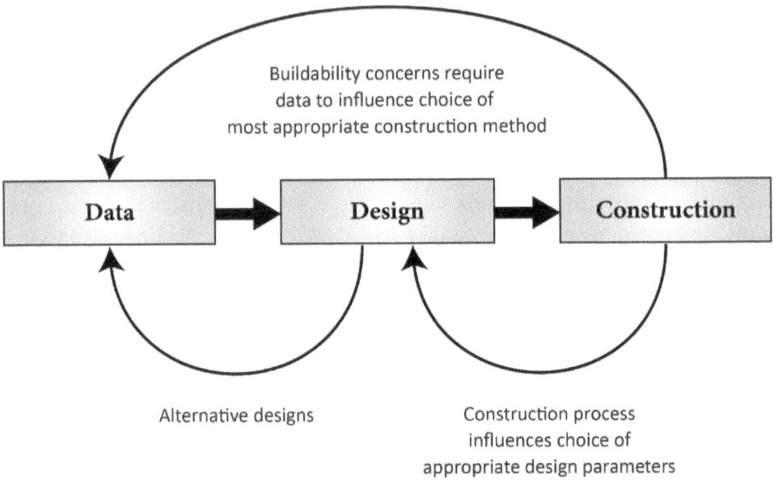

Figure 2.4 Cycle of data, design, and performance (Burland et al., 2012).

which can act more efficiently than a total of isolated spread foundations. Upon the increase of applied loads and when weaker surface soils conditions are present, single and combined footings are merged together over most or the entire footprint of the superstructure, called mat or raft foundations. In comparison to other types of shallow foundations, the mats have the advantage of routine construction, providing structural continuity and rigidity, basement waterproofing, more resistant against instability conditions as well as distributing the loads over an extended area. Fig. 2.5 presents various commonly used types of shallow foundations.

In contrast, deep foundations or piles transfer the applied load to the deeper soil strata through toe and shaft reactions. They are long structural members which can

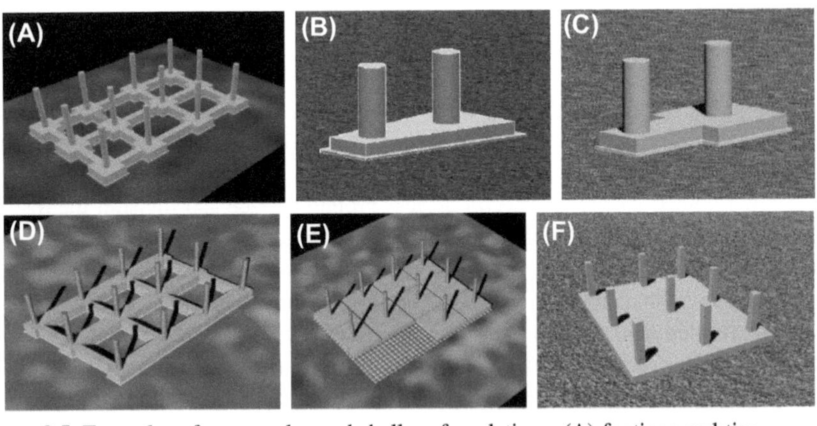

Figure 2.5 Examples of commonly used shallow foundations: (A) footings and ties, (B) combined, (C) strip, (D) grid, (E) attached single footing, (F) mat.

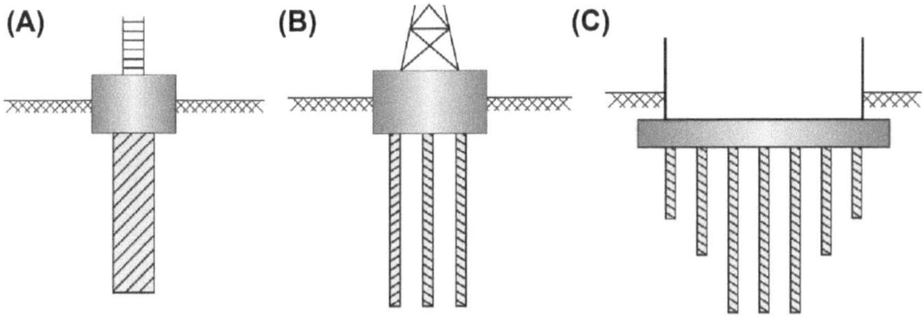

Figure 2.6 Applications of deep foundations: (A) single shaft or caisson, (B) pile group, (C) piled raft foundation.

be installed precast elements by driving, push, and vibrations such as cast in place. It is inevitable the employment of deep foundations for situations such as heavy structures, tall buildings, on or offshore structures, existing weak and compressible deposits at surface area, and stability controls. Since, the carrying capacity and stiffness of soils generally increase with depth, the hybrid system of piles and surrounded soils is capable to mobilize huge geotechnical capacity, low settlement, and high stability. Fig. 2.6 illustrates typical application of deep foundations.

2.4 Intermediate trends

2.4.1 Semideep foundations

Shallow foundations transmit structural loads to the bearing soil strata at a relatively small depth (D_f), compared to foundation width (B). With rapid growth of urban areas, bridges, marine structures, high-rise buildings, and towers, there is an excessive dependence on deep foundations ($D/B \geq 10$) for construction of infrastructure. This is mainly because of lack of confidence in other foundation types. Moreover, it is well experienced that shallow foundations suffer from several limitations and shortcomings such as low bearing capacity, excessive settlement, and low stability. Meanwhile, piles are the most widely used option in heavy, tall, and important structures. However, the construction equipment, cost, installation time, and environmental problems related to drilled shafts and pile driving restrict the wide application of deep foundations. In this regard, two alternatives including semideep foundations and ground modifications are intermediate solutions.

The shape and dimension of foundations, embedment depth, surrounding soil characteristics, and load combination, all affect the geotechnical and structural performance of foundations. Recent studies have shown that for foundations with large breadth and D/B equal to 2, kinematic mechanisms and failure modes are different from two main conventional types of foundations (i.e., shallow and deep). Therefore, D/B factor must not be considered as the only factor for defining foundation type (Rezazadeh and Eslami, 2018).

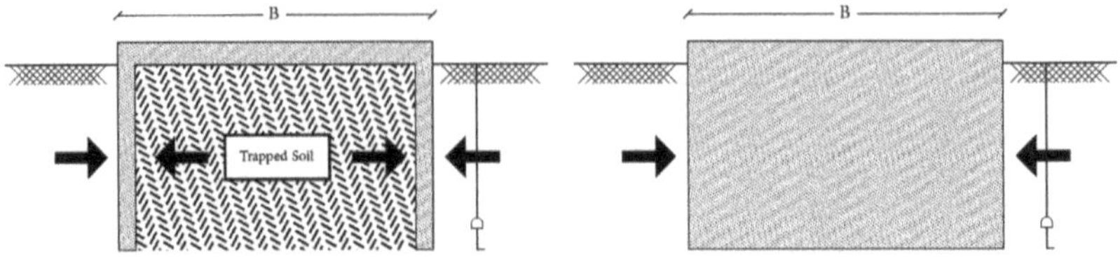

Figure 2.7 Hybrid confinement in soil and foundation.

These foundations lead to an increase in bearing capacity and improve load-displacement behavior by hybrid confinement of soil and foundation as depicted in Fig. 2.7.

"Semideep foundations" are another foundation category distinguished based on their performance. This relatively new concept is becoming increasingly an optimum alternative for shallow and deep foundations to support structures especially in accessible depth. As illustrated in Fig. 2.8, different types of semideep foundations consist of (Rezazadeh and Eslami, 2018):

- Bucket foundation or suction caisson,
- Spudcan,
- Hybrid foundation,
- Floating and box foundation,
- Well foundation,
- Ring foundation,

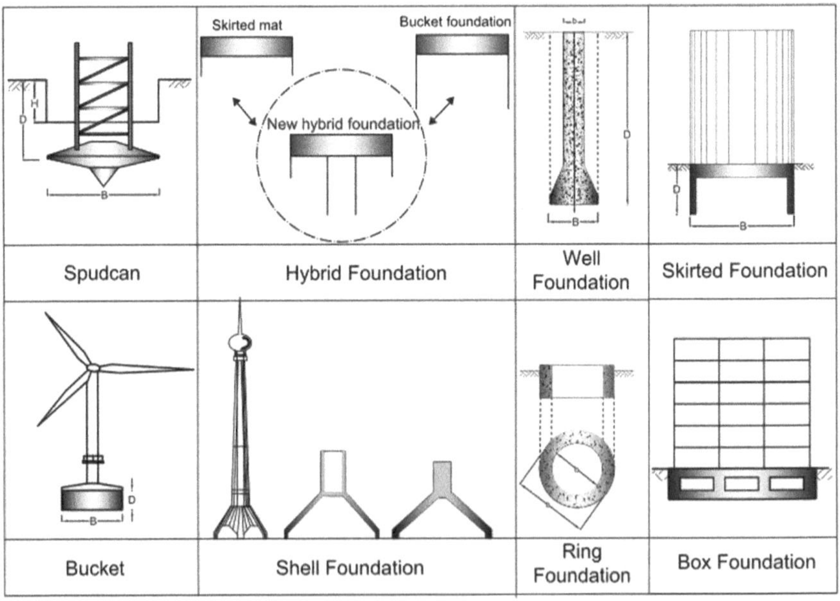

Figure 2.8 Different types of semideep foundations.

- Shell foundation,
- Skirted foundation.

Bucket foundations are used as embedded foundations and cost-reducing substructures for wind turbines and offshore oil and gas facilities. Spudcans are the common foundations for jack-up platforms. Roughly circular in plan, spudcans typically have a shallow conical underside (in the order of 15–30 degrees to the horizontal). The hybrid foundation aims to provide additional horizontal and moment capacity by optimizing the amount of steel required and minimizing structural design complexity associated with long and large diameter skirted foundation. It consists of a skirted mat with (an) internal caisson compartment(s) (Bienen et al., 2012).

Floating and box foundations are the other category of semideep foundations used to support heavily loaded structures resting on soft and weak deposits where the net pressure on the soil beneath the foundation is further reduced by higher embedment depth of the foundation and simultaneous construction of a basement wall which leads to the settlement reduction.

Well foundation is a kind of underreamed drilled shaft. When a homogeneous stiff clay, hardpan, or soft cohesive rock exists at a relativity shallow depth, the underream can be easily constructed and is the least expensive type of foundation.

A skirted foundation is a continuous semideep circular foundation with thin skirts around the circumference. The basic functions of the skirt are mainly confining soft surface soils and transfering the loads down to harder underlying layers to resist lateral loads and moments effectively against sliding and overturning. Also, they are used to improve protection against piping and scour. Furthermore, by use of peripheral skirts the soil beneath foundation can be prevented from squeezing out and any probable damage due to excavations for adjacent construction works. The relative ease of installation provides a significant economic incentive particularly in areas with difficulties related to pile construction. Moreover, they can transfer load through end resistance at the tip level and shaft friction alongside the skirt (Rezazadeh and Eslami, 2018). Various types of semideep foundations are illustrated schematically in Fig. 2.8.

2.4.2 Ground modification approaches

With urbanization, there have been increased demands for the use of land for better living and transportation. More and more houses, commercial buildings, high-rise office buildings, highways, railways, tunnels, levees, and earth dams have been constructed and will be continuously built in the future. As suitable construction sites with favorable geotechnical conditions become less available, the need to utilize unsuitable or less suitable sites for construction increases. Engineers have faced increased geotechnical problems and challenges, such as bearing failure, large total and differential settlements, instability, liquefaction, erosion, and water seepage. The options to deal with problematic geomaterials and geotechnical conditions include the following:

1. Avoiding the site,
2. Designing superstructures accordingly,

3. Removing and replacing problematic geomaterials with better and nonproblematic geomaterials,
4. Improving geomaterial properties and geotechnical conditions.

When superstructures are to be built on ground, there are five foundation options for bearing on natural ground, replaced surface deposits, compacted/consolidated strata, composite ground and piles to deeper stratum.

Ground improvement modification operation is done to reach many goals such as:

- Increase strength,
- Reduce erodibility,
- Reduce distortion under stress,
- Reduce compressibility,
- Control shrinking, swelling and permeability,
- Reduce water pressures,
- Redirect seepage,
- Prevent the detrimental physical or chemical changes due to environmental conditions,
- Mitigate susceptibility to liquefaction and natural variability of borrow materials or foundation soils.

Various methods and techniques are commonly used for ground modifications or soil improvements, which can be classified into:

- Earthworks
- Densification
- Physical and chemical modification
- Hydraulic modification
- Reinforcement

Also, the soil improvement methods can be classified to different categories according to the improvement in depth (shallow and deep methods); pre, during, and post induced construction loading (statics and dynamical methods); time-dependent procedures; manner of implementation (extrusion and intrusion); and the velocity of the operation (fast and slow methods).

In the following section, different categories of ground improvement methods are presented briefly.

2.4.3 Earthworks

Overexcavation and replacement is one of the traditionally but still commonly used ground improvement methods in practice. The basic concept of this method is to remove a problematic geomaterial and replace it with nonproblematic fill. Replacing fills are often rock, gravel, and sand.

This method is often cost-effective to improve problematic geomaterials when their area and depth are limited and fill materials are readily available. Also it is simple, reliable, and well established. It does not require specialty contractors and special equipment except excavators and rollers if no temporary shoring and dewatering is not required. Depending on site conditions, this method may be limited by deep

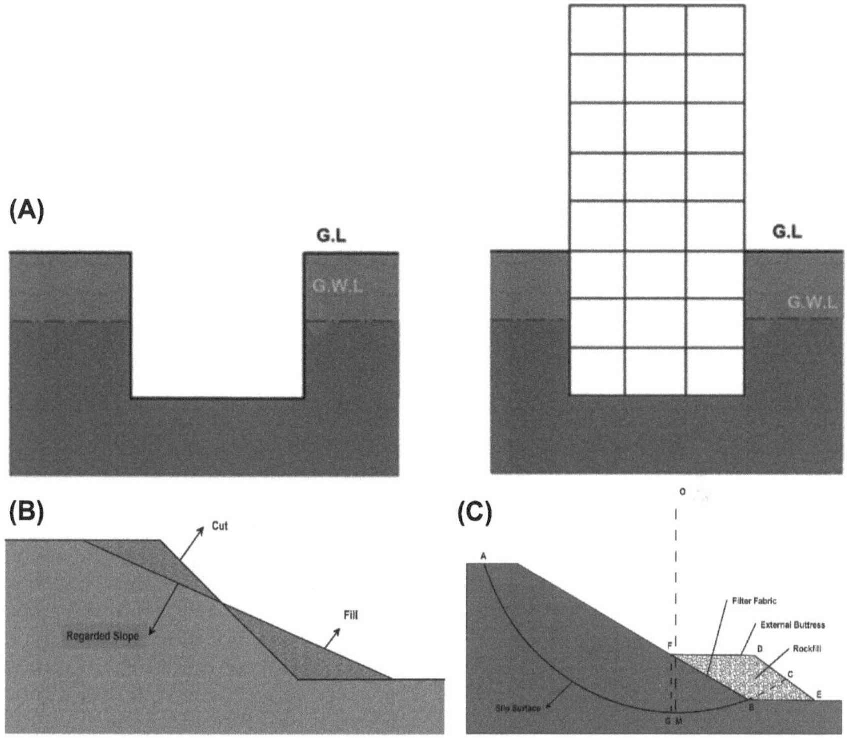

Figure 2.9 (A) Unloading and Flotation, (B) Slope regrading using available onsite soils (cut and fill), and (C) External Buttress supporting the base of a circular arc.

excavation required, high groundwater table, on-site or nearby existing structures and utility lines, limited truck access to the site, long distance for hauling fill material, and disposing of problematic geomaterial and time. In Fig. 2.9, some typical approaches of earthwork to modify the ground are shown.

2.4.4 Densification

Shallow and deep compaction methods have been commonly used to improve geomaterial properties near surface and at depth through a densification process by vibration, pressure, kneading, and/or impact on ground surface. This technology is effective to improve cohesionless geomaterial or cohesive geomaterial with low plasticity. Conventional plate or roller compaction has been used for many years, and it densifies geomaterial to a shallow depth by repeated passing of a vibratory plate or a roller on a relatively thin lift, as shown in Fig. 2.10A. This is one of the most commonly used ground improvement methods in practice for earthwork. To achieve better densification, proper compaction equipment should be chosen, geomaterials should be prepared at appropriate lift thickness and moisture content, which is close to an optimum moisture content, and sufficient compactive energy should be applied.

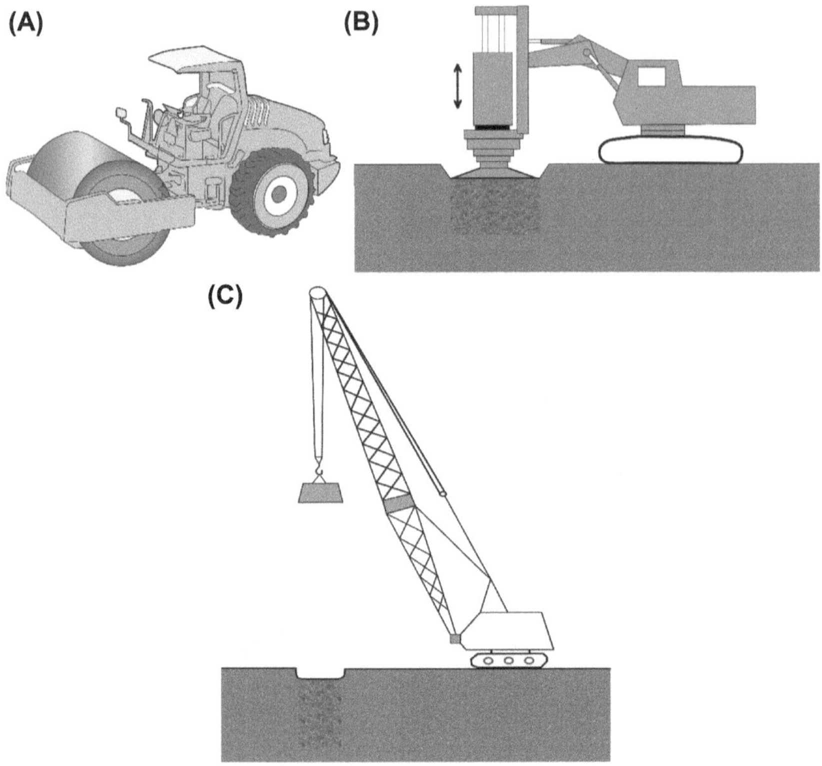

Figure 2.10 Different compaction equipment: (A) Roller, (B) rapid impact compaction (RIC), and (C) deep dynamic compaction.

Rapid impact compaction (RIC) is an intermediate compaction technology from shallow to deep compaction, and it rapidly applies impact on ground surface using a hydraulic hammer. The RIC equipment is shown in Fig. 2.10B. Vibro compaction densifies cohesionless soil by driving a vibrating probe into the ground to apply lateral vibratory forces which rearrange particles into a dense state. For saturated cohesionless soil or when water is injected into the ground, vibration can also cause liquefaction to the soil, and the soil is densified after the dissipation of excess pore water pressure.

Deep dynamic compaction extends the depth of geomaterial densification to a greater depth by applying high-energy impact through repeated dropping of a large and heavy weight on ground surface, as shown in Fig. 2.10C. The process of dynamic replacement involves tamping, backfilling, and continued tamping until stone columns are formed.

2.4.5 Physical and chemical modification

The most common artificial additives are (in order of usage) portland cement (and cement-fly ash), lime (and lime-fly ash), and bitumen and tar. The reason of their

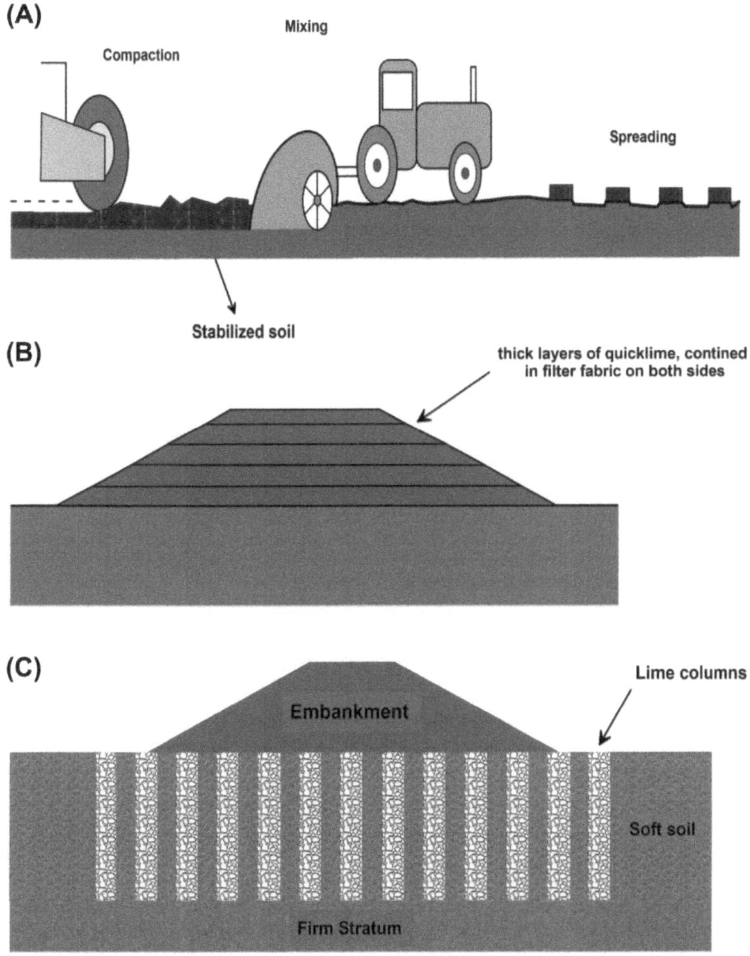

Figure 2.11 Examples of ground modification with admixtures: (A) Surface stabilization for roads, (B) Embankment construction using quicklime sandwich, and (C) Lime columns below embankment.

popularity is that they are applicable to a considerable range of soil types, they are widely available, their costs are relatively low, and they are environmentally acceptable. The purpose of mixing additives is to increase strength and durability; reduce permeability, deformability, and erodibility; provide volume stability and control variability. In Fig. 2.11, examples of physical ground stabilization are shown.

2.4.6 Hydraulic modification

Dewatering of soil in civil engineering or mining projects is carried out for one or more of the following reasons:

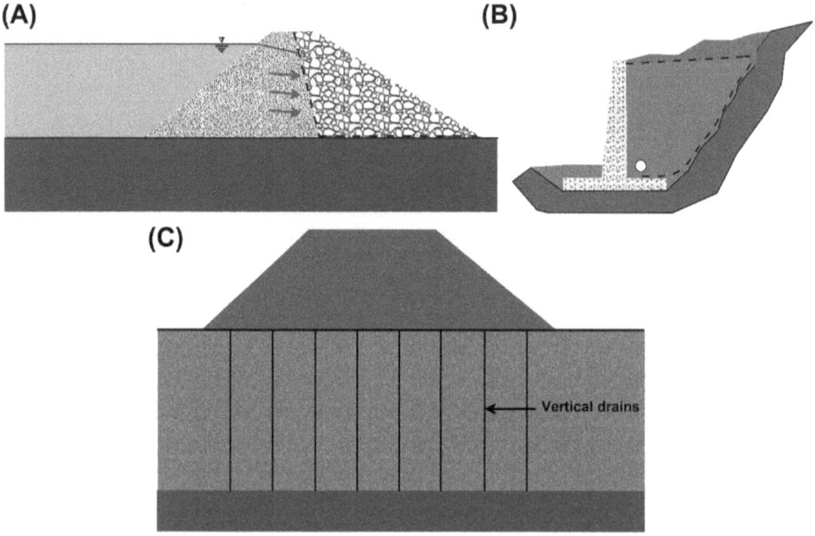

Figure 2.12 Examples of drainage applications: (A) Chimney drain for dam or levee, (B) Retaining wall drain, and (C) Vertical drains in soft soil.

- Provide a dry working area
- Stabilize constructed or natural slopes
- Reduce lateral pressures on foundations or retaining structures
- Reduce the compressibility of granular soils
- Increase the bearing capacity of foundations
- Improve the workability or hauling characteristics of borrow materials
- Prevent liquefaction potential during earthquakes
- Prevent soil particle movement by groundwater
- Prevent surface erosion
- Prevent or reduce damage due to frost heave

Drains have been used for many different applications. Fig. 2.12 shows a few examples of these applications. Retaining wall drains (Fig. 2.12A) is to reduce lateral earth pressures on wall facing. Chimney drains (Fig. 2.12B) are often installed in dams or levees to lower water heads between upper stream and lower stream. Vertical drains (Fig. 2.12C) are to accelerate the dissipation of excess pore water pressure in soft soil.

2.4.7 Reinforcement

In situ ground reinforcement is a technique to stabilize existing unstable ground due to the change of geotechnical conditions by nature and/or human activities. Ground anchors, soil nails, micropiles, and slope stabilizing piles have been used as in situ ground reinforcement techniques to mitigate the preceding problems as shown in Fig. 2.13. Also, steel or plastic geogrids are used to reinforce earth. The effectiveness of the reinforcement is governed by its tensile strength and the bond it develops with

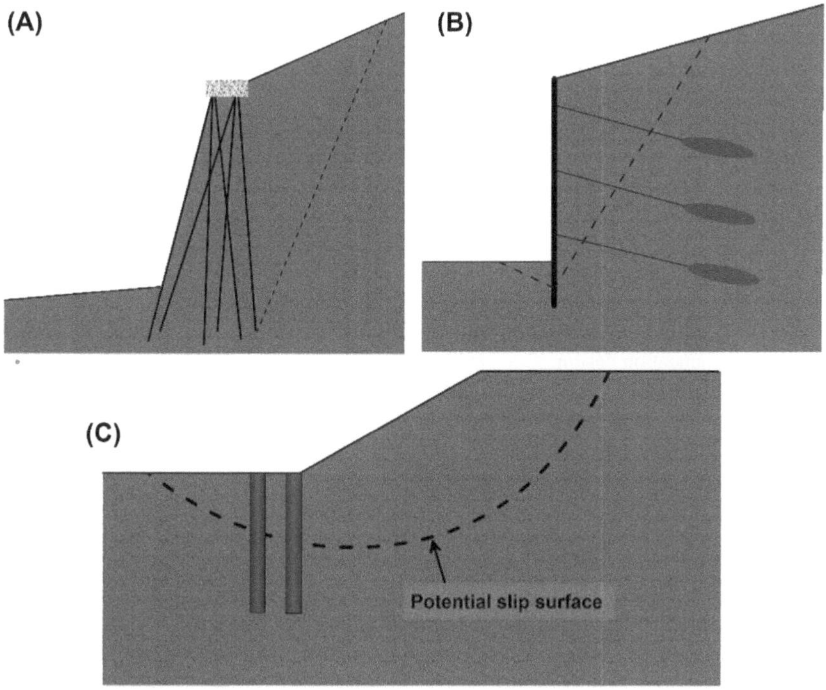

Figure 2.13 Types of in situ ground reinforcement: (A) micropiles, (B) ground anchor and (C) slope stabilizing piles.

the surrounding soil. The presence of reinforcement noticeably improves the mechanical properties of granular soil and, depending on confining pressure, various modes of failure can be observed.

2.5 Overall step-by-step procedure for foundation design

The process of foundation design is well established and generally involves the following steps (Poulos et al., 2017; Viggiani et al., 2014):

1. A desk study for collection of the geological and hydrogeological evidence and information of the area in which the site is located.
2. Planning and execution of the site investigation to assess site stratigraphy and variability.
3. In situ testing to assess appropriate engineering properties of the key strata. Laboratory testing is to supplement the in situ testing and to obtain more detailed information on the behavior of the key strata than may be possible with in situ testing.
4. The formulation of a geotechnical model for the site, incorporating the key strata and their engineering properties. In some cases where ground conditions are variable, a number of models for different parts of the site may be necessary to allow proper consideration of the variability over the site.

5. Collection and synthesis of geological and geotechnical evidences; available information on the subsoil; planning, execution, and interpretation of in situ testing site and laboratory investigations on the subsoil, or underlain layers; development of a geotechnical characterization of the subsoil.
6. Preliminary assessment of foundation requirements and choice of foundation, based upon a combination of experience and relatively simple methods of analysis and design. In this assessment, considerable simplification of both the geotechnical profile(s) and the structural loading is necessary (bearing capacity criteria).
7. Refinement of the design, based on more accurate representations of the structural layout, the applied loadings, and the ground conditions, prediction of settlement, differential rotation, and other displacement. Check these items with admissibility and function of structures from this stage and beyond, close interaction with the structural designer is an important component of successful foundation design.
8. Detailed design, in conjunction with the structural engineer. According to the applied load, foundation base, and amount of rigidity or flexibility, the internal stresses are determined and consequently proper material is allocated. *Note: foundation design is an iterative and interactive procedure regarding bearing capacity, settlement and structural capacity aspects.*
9. Definition of the installation methods and preparation of the technical specifications.
10. Evaluation of the cost, also to assist in the choice between possible alternative solutions.
11. In situ foundation testing at or before this stage is highly desirable, if not essential, in order to demonstrate that the actual foundation behavior is consistent with the design assumptions. This usually takes the form of testing of prototype or near-prototype piles or footings. If the behavior deviates from that expected, then the foundation design may need to be revised. Such a revision may be either positive (a reduction, in foundation requirements) or negative (an increase in foundation requirements). In making this decision, the foundation engineer must be aware that foundation testing involves only individual elements of the foundation system, and that the foundation system, the piles and the raft within the system will interact. The overall foundation behavior may thus not be able to be assessed directly from the foundation test results without consideration of the foundation−soil interaction effects.
12. Monitoring of the performance of the building during and after construction. At the very least, settlements at a number of locations around the foundation should be monitored, and ideally, some of the footings, piles, and sections of the raft should also be monitored to measure the sharing of load among the foundation elements or contact pressure. Such monitoring is becoming more accepted as standard practice for high-rise buildings, but not always for more conventional structures. As with any application of the observational method, if the measured performance violates significantly from the design expectations, then a contingency plan should be implemented to address such departures. It should he pointed out that departures may involve not only settlements and differential settlements that are greater than expected but also those that are smaller than expected.

2.6 Basic soil mechanics for foundation engineering

Since foundation design is directly dependent on the underlain soil, the characteristics, behavior, and performance of supporting soils play a major role in foundation engineering and must be well understood. Consequently, a few relevant topics of soil mechanics for foundation analysis and design will briefly be reviewed.

2.6.1 Origin of soils

Civil engineering projects are made with, in or on top of the soil. Geomaterials including soils or rocks have been used via two categories. First, most projects are built on them, where they must tolerate and bear the loads transferred from the superstructure. Second, they are applied as borrow material in various conditions such as production of natural materials, engineered fills, embankment dams, subbase and base of roads, retaining wall backfills, drainage, barriers, and masonry elements.

Soils are originated from physical and chemical decomposition of rocks. The physical processing initiated by erosions by wind and water, glacier, gravity, fall, or disintegration by cycles of freeze and thaw or wet and dry of fluids existing in cracks and voids of rocks. In this type of soils, the original formation of soil and native rock are the same, such as sands which are obtained from sandstone or quartz. The coarse-grained soils, i.e., gravel and sands are in this category.

In chemical decomposition process, the mineral of native rock has been changed by a few agents such as dioxide carbon, oxygen, water in alkane or acidity environments. Accordingly, the chemical structure of native rock alters, for example, Kaolin clay is made by feldspar decomposition by presence of water and CO_2.

Often, fine-grained soils, such as silt and clays, are categorized in this category and include the sheet or planar microstructure with electric bonds.

2.6.2 Soil identification and classification

In foundation engineering applications, soils are divided into two broad categories: cohesive and cohesionless, which can be nominated generally to fine-grained and coarse-grained soils, respectively. Recently, in new trends in geotechnical engineering, soils are categorized into three major categories: clays, granular, and intermediate geomaterials.

Cohesive soils are subjected to interparticle forces that make the particles stick together. In foundation engineering, this type of soil designs are mainly focused on low to medium strength, difficult compaction, relatively medium to high plasticity, prone to consolidation (depends on time, water content, and load), and low permeability, while coarse-grained soils involve medium to high strength and stiffness, immediate settlement under loading, relatively low plasticity, high permeability. Moreover, due to workability and densification the coarse-grained soils are realized more as foundation base layers and constructional materials.

This is important to note that the distinction behaviors cohesive and cohesionless are not exactly the same as fine-grained and coarse-grained per definition of most soil classification systems. Moreover, many real soils do not fall routinely into either category and most have their own behavior (Mitchell and Tseng, 1990).

Soils can be identified in the field or laboratory via various criteria. For reconnaissance and preliminary identification indices such as color, odor, visual inspection, shaking behavior, dry strength, and stiffness, test by hand. Conversely, by some laboratory testing and analysis, soils can be better characterized.

Two major measurements are used for soil classifications, i.e., particle size distribution (gradation curve) and plasticity chart of Casagrande (1948) (Atterberg limits), which have been covered in detail in soil mechanics textbooks.

Among various soil classification approaches, the USCS (Unified Soil Classification System) is commonly used in foundation engineering. In USCS soils are classified into major groups of gravel, sand, silt (mita), and clay.

There are three major divisions: coarse-grained soil (gravel and sand (G, S)), fine-grained soil (silt and clay), and highly organic soil (O and pt). These are further divided in 15 basic soil groups and subgroups. The group or subgroup depends on grain-size distribution and Atterberg limits.

2.6.3 Water in soil

As pronounced in classification and identification of soils via manual, virtual, or laboratory assessments, the presence of water in geomaterials is very important. Existing water in soil significantly affects the engineering behavior of most deposits, especially fine-grained ones. In most geotechnical engineering design and construction projects, the role of water in soil particle system can be considered as existing in molecular structured, absorbed, or moisturized static (underground) or dynamic (water flow or seepage) conditions.

Regardless of the type of presence, the soil−water interaction can include following phenomena that may involve difficulties in geotechnical engineering to some extent:

- Reduction of effective stress
- Capillary action
- Swelling
- Frost
- Physical dispersity
- Chemical dispersity
- Shrinkage
- Collapsibility
- Sensitivity
- Piping
- Consolidation
- Liquefaction
- Sand boil
- Uplift and buoyant forces
- scouring

Due to the variety of effects mentioned above, Terzaghi (1943) stated that problems in geotechnical engineering are solely associated with the fluids in soils, pores, and soil mass.

In foundation engineering, investigation of water content, groundwater fluctuations, and water flow, either in static or dynamic conditions, is very important. This is because all these interactions influence foundation subsoil, bearing capacity, settlement, and internal stability.

The water pressure in soil for calculating in geosources involves three following items:

$$u = u_h + u_e \pm u_i \tag{2.1}$$

where,

u_h = hydrostatic water pressure = $Z_w.\gamma_w$,
Z_w = the water depth from ground level,
γ_w = water unit weight,
u_e = excess pore water pressure,
u_i = depending on upward or downward water gradient = $i\gamma_w$,
i = hydraulic gradient.

2.6.4 Stresses in soil

For foundation analysis, the knowledge of stresses in soil is required. The sources of stress can be attributed in three kinds:

- Geostatic
- Horizontal
- Induced

The geostatic stresses are as a result of the force of gravity acting directly on the soil mass. The most important component is the vertical compressive stress and can be calculated as follows:

$$\sigma_z = \sum \gamma h \tag{2.2}$$

where

σ_z = geostatic vertical total stress,
γ = total unit weight of soil stratum,
h = thickness of soil stratum.

According to effective stress principle in the soil mass, part of the vertical total stress is carried by the solid particles, and the rest is carried by the pore water, i.e.,

$$\sigma_z = \sigma'_z + u \tag{2.3}$$

where

σ_z = vertical total stress,
σ'_z = effective stress,
u = pore water pressure.

2.6.4.1 Horizontal stress

The geostatic horizontal stress is also important for many engineering analyses. Regardless of total or effective, upon the following equation, it can be determined from in situ vertical stress.

$$\sigma'_h = K\sigma'_z \tag{2.4}$$

where K is the coefficient of lateral earth pressure.

The K_0 value upon soil lateral displacement condition can be expressed at three states: at rest, active, or passive. For classic and routine conditions, the K value can be calculated as follows (Holtz et al., 2011):

$$K_o = (1 - \sin\varphi)OCR^{\sin\varphi} \tag{2.5}$$

$$K_a = \tan^2(45 - \varphi/2) \tag{2.6}$$

$$K_p = \tan^2(45 + \varphi/2) \tag{2.7}$$

In condition of soil–structure interaction such as pile, the attributed K is named K_c, which represents construction effects on the interface.

2.6.4.2 Induced stress

The induced or applied stresses are caused by external loads such as structural loads, fills, or any other type of surface loading. These types of stresses are changed and decreased with depth and are important in foundation engineering. In view of soil–structure interaction, analyses focus on the subsurface soil response to these types of stress magnitude and distribution.

For rigid foundation, the contact pressure, q, due to concentrated load, *P*, is:

$$q = \frac{P}{A_f} \tag{2.8}$$

where

q = contact or bearing pressure,
P = applied vertical load, and
A_f = the foundation area.

An approximate solution for q_z (varying of induced stress with depth is the 2:1 method of distribution with depth), accordingly, can be found, for foundation $B \times L$ and contact pressure of q:

$$q_z = \frac{q \times B \times L}{(B + z)(L + z)} \tag{2.9}$$

In foundation engineering, different relations and charts are presented to find the I_z value, based on the following equation (Holtz et al., 2011; Coduto, 2002):

$$\Delta\sigma_z = q_z = I_z \cdot q \tag{2.10}$$

Both q (contacted pressure) and q_z are considered in foundation design for bearing capacity and settlement aspects.

2.6.5 Compressibility and settlement

The vertical deformation of soils beneath foundation loading is called settlement. Compressibility of soils involves stress–strain relationships. In this regard, the trend of stress–strain curves plus soil stiffness plays an important role in settlement calculations. Generally, due to static or dynamic loading by foundation and consequence of induced stresses distributed in depth, these types of settlement are diagnosed as following:

- Immediate (elastic) or distortion settlement:

It is the result of lateral movement of soil due to change of vertical stress. This type of settlement almost occurs in all types of soil.

- Consolidation settlement:

It is related to time-dependent change of volume of voids within the soil due to increase of σ'_z and excess pore pressure dissipation. Accordingly, in saturated soils, the soil particles and fluids in voids are incompressible, and the void volume change occurs if some pore water is squeezed out from the soil mass skeleton in response to external loading. This settlement is dependent on time, and fine-grained saturated soil is the most important source of settlement.

- Secondary settlement:

This is a form of creep and happens in constant σ'_z. It is also time dependent and significantly occurs in highly plastic and organic soils.

Therefore the total settlement below a foundation is the sum of three components:

$$S = S_e + S_c + S_s \tag{2.11}$$

where

S_e = elastic settlement,
S_c = consolidation settlement,
S_s = secondary settlement.

Upon basic stress–strain relationships, the settlement can be formulated as:

$$S = kqB \tag{2.12}$$

where

k = factor depending on soil stiffness and foundation rigidity,
q = contact pressure,
B = foundation width.

However, two major approaches are recommended for settlement estimation:

- Modulus-based method

In this approach, the vertical compression is:

$$\varepsilon_z = \frac{\Delta\sigma_z}{M} \quad (2.13)$$

where M is the constrained modulus (secant or tangent) equal to:

$$M = \frac{E(1-\nu)}{(1+\nu)(1-2\nu)} \quad (2.14)$$

And the common equation is:

$$S_d = \frac{qB(1-\nu^2)}{E_s} I \quad (2.15)$$

- e-logσ approach:

$$\varepsilon = \frac{\Delta e}{1+e_0} \quad (2.16)$$

$$\varepsilon_z = m_v \Delta\sigma' \cdot H \quad \text{or} \quad (2.17)$$

$$\varepsilon_z = \frac{C_c}{1+\varepsilon_0} \log\frac{\sigma'_1}{\sigma'_0} \quad (2.18)$$

where

C_c = compression index,
m_v = coefficient of volume change,
e_0 = initial void index,
σ'_0 = initial vertical stress,
σ'_1 = final vertical stress,
H = compressive soil thickness or influence zone of the applied pressure.

$$\Delta\sigma' = \sigma'_1 - \sigma'_0$$

Overall, the following factors dominate the foundation settlement:

1. Foundation shape, depth, geometry, and rigidity
2. Soil stress–strain behavior
3. Stiffness characteristics
4. Poisson effect
5. Stress distriction in depth
6. Drainage condition
7. Static or dynamic loading
8. Stress history
9. One-dimensional or three-dimensional deformation
10. Plasticity and organic content

2.6.6 Shear strength of soils

Most failure patterns in geotechnical or foundation engineering are based on shear stress development in soil mass due to induced loading or unloading caused by construction practices.

If the induced shear stresses exceed the underneath soil shear strength, failure occurs. Accordingly, in analysis and design evaluation of stress initiation and magnitude with soil strength is an important task for foundation engineering.

In geotechnical practice, the Mohr–Coulomb failure criterion is commonly used for determining soil shear strength. Generally, the failure criterion is based on following equation:

$$\tau = c + \sigma' \tan \varphi \tag{2.19}$$

where

τ = shear strength,
c = effective or total cohesion,
φ = effective or total friction angel,
σ' = effective or mean confining stress acting on the shear surface.

The following conditions influence the Mohr–Coulomb failure criteria, especially on c, φ, and σ':

- Drained, undrained, and intermediation conditions
- Volume change during loading via mean normal stress or deviator stress
- Saturated or unsaturated soils
- Time-dependent behavior
- Dynamic or static loading
- Effective or total stress analysis (ESA or TSA)
- Normally consolidated or over consolidated situation
- Slow maintained or rapid loading
- Confinement condition
- Sensitivity and thixotropy
- Contractive or dilative behavior

- Softening or hardening
- Fine-grained or coarse-grained soils
- Stress path pattern
- Physicochemical alterations

Therefore, due to a variety of conditions and parameters affecting soil shear strength behavior, the analysis and design of foundation for bearing capacity and stability are knowledge based and sophisticated. In this regard, the versatile CPT or CPT-u provides valuable data from tip resistance (q_c), sleeve friction (f_s), and excess pore pressure (u). Representation of most soil in situ characteristics can play an important role to select and interpret various required parameters for performance-based design.

2.7 Uncertainty in foundation engineering

In spite of many advancements in foundation engineering, there are still many gaps in our understanding and the importance of consideration of possible multihazard events for the reliable performance evaluation of structures (Coduto, 2002).

The primary objectives of engineering design are safety, serviceability, and economy. Safety and serviceability can be improved by increasing the design margins or levels of safety to reduce the probability of failure. However, this increases the cost of the structure. Considerations of overall economy in design involve balancing the increased cost associated with increased safety against the potential losses that could result from unsatisfactory performance and failure.

Regardless of the design philosophy and approach used, the basic design criterion is that the capacity or resistance of the system must be greater than the demand or loads applied to the system for an acceptable or required level of safety. Failure or unsatisfactory performance occurs when the demand on a system exceeds the capacity of the system. The basis of design is to achieve a state that lies in the safe region during the lifetime of the structure.

Design criteria for safety lies above the failure boundary surface; levels of safety are defined as a measure of the distance from the failure surface. The level of safety used depends on the class and importance of the structure and consequences of failure. The more important the structure and the more serious the consequence of failure, the higher the level of safety necessary in the design process.

Geotechnical design process starts off with the project description (e.g., a building with specific capacity and serviceability requirements based on the client's needs). A basic design issue, from the perspective of geotechnical engineers, is related to determining the most appropriate type and size of foundation units (e.g., what width of footing is required to safely and economically support the building and satisfy the design criteria).

Significant and varying degrees of uncertainty are inherently involved in the design process. Allowances must be made for these uncertainties. The source of uncertainty in foundation design can be grouped into four main categories:

1. Uncertainties in estimating the loads: Despite the loads transferred from superstructure to foundation, one of the main reasons for unsatisfactory performance of almost geotechnical

elements/structures is disregarding/ignoring the extreme events such as flood, scour, earthquake, landslide, and hurricane (Deng et al., 2015).

2. Uncertainties associated with variability of the ground conditions at the site: Soil is a complex engineering material that has been formed by a combination of various geologic, environmental, and physicochemical processes. Because of these natural processes, all soil properties in situ will vary vertically and horizontally. On the other hand, the process of measuring soil properties by some physical means introduces additional variability into the soil data, and this kind of error arises from equipment, procedural operator, and random testing effects (Phoon and Kulhawy, 1999a).
3. Uncertainties in evaluation of geotechnical material properties: The direct measurement from a geotechnical test typically is not directly applicable to design. Instead, a transformation model is needed to relate the test measurement to an appropriate design property. Some degree of uncertainty will be introduced because most transformation models are obtained by empirical data fitting (Phoon and Kulhawy, 1999b).
4. Uncertainties associated with the degree to which the analytical model represents the actual behavior of the foundation, structure, and the ground that supports the structure due to the simplifications and assumptions made (Eslami et al., 2016).

It is well recognized that uncertainties are unavoidable in geotechnical engineering and that the quantification of these uncertainties is necessary. The gross errors including human errors or omissions that occur in practice (e.g., imperfect design and construction method, and lack of inspection and maintenance) are seldom quantified in design (Whitman, 2000; Wang, 2009).

In contrary, the uncertainties associated with the natural variability of ground conditions and with the evaluation of the geotechnical properties are usually the greatest, as a consequence of the complex geological processes involved with the deposition and formation of soil and rock (Tan et al., 1991; Kulhawy and Phoon, 1993; Bolton, 1981) and can be reduced by increasing the quality and quantity of site investigation tests.

2.8 The role of CPT in reduction of uncertainty or increasing reliability

The reliability of a geotechnical design depends heavily on the reliability of input values of soil properties in the analyses, which in turn depends on the level of carried out site investigation and characterization. In this regard, an initial step of geotechnical engineering process is the development of a subsurface profile of soil types and their geotechnical properties, and the cone penetration test (CPT) is an ideal tool. CPT supplies continuous records with depth and allows a variety of sensors to be accompanied and is performed under field stresses and boundary conditions (Schneider et al., 2008; Eslami and Fellenius, 1997).

Among prevailing in situ−based tests, CPT is distinctive due to its capability in recording several different parameters such as cone resistance (q_c), friction resistance (f_s), shear wave velocity (V_s), excess pore water pressure (u_2), etc., under field stress conditions. These parameters help the geotechnical engineer to overcome prudent designs because of less error gained by a cone penetrometer apparatus. As Table 2.1 and

Table 2.1 Summary of coefficient of variation (V_R) for geotechnical properties and resistances (Becker, 1997).

Geotechnical characteristics	V_R^a	References
Index properties		
Natural variability	0.05–0.15	Kay (1993)
Natural water content	(18)	Kulhawy (1992), Phoon et al. (1993)
Liquid and plastic limits	(0.11)	Kulhawy (1992), Phoon et al. (1993)
Unit weight	0.04–0.16 (0.07)	Cherubini et al. (1993), Kulhawy (1992)
Initial void ratio	(0.2)	Kulhawy (1992)
SPT N penetration resistance	0.15–0.5	Barker et al. (1991), Meyerhof (1993, 1995)
CPT q_c tip resistance	0.15–0.37	Barker et al. (1991), Meyerhof (1993, 1995)
Strength		
Angle of internal friction		
From laboratory tests	0.05–0.25 (0.13)	Cherubini et al. (1993), Meyerhof (1993, 1995), Kulhawy (1992), Manoliu and Marcu (1993)
From CPT correlation for sand	0.15–0.25	Barker et al. (1991)
Undrained shear strength	0.12–0.85 (0.34)	Meyerhof (1993, 1995), Cherubini et al. (1993) Kulhawy (1992)
Deformation		
Elastic modulus	0.2–0.5	Meyerhof (1993, 1995)
Modulus of deformation	0.2–0.4	Meyerhof (1993, 1995)
Compression index, Cc	0.17–0.55 (0.37)	Cherubini et al. (1993), Meyerhof (1993, 1995) Kulhawy (1992)
Resistance models		
Bearing capacity Shallow foundations		
From SPT/CPT correlations	0.5	Barker et al. (1991)
From theory	0.25–0.3	Barker et al. (1991), Meyerhof (1993, 1995)
Pile capacity: From theory		
α, λ method	0.17–0.41	Barker et al. (1991)
β method	0.21	Barker et al. (1991)

Table 2.1 Summary of coefficient of variation (V_R) for geotechnical properties and resistances (Becker, 1997).—cont'd

Geotechnical characteristics	V_R^a	References
SPT correlations	0.5	Barker et al. (1991)
CPT correlations	0.36	Barker et al. (1991)
Drilled shafts	0.15−0.46	Barker et al. (1991)
Pile load tests	0.08−0.3 (0.25)	Kay (1993), Hettler (1993), Matsumoto et al. (1993), Okahara et al. (1993)
Earth pressure	0.15−0.2	Barker et al. (1991), Meyerhof (1993, 1995)
K_A, K_0	0.2	Barker et al. (1991)
Embankment stability	0.14−0.32	Kay (1993)
Design model uncertainty	0.05−0.25	Kay (1993)
Design decision uncertainty	0.15−0.45	Kay (1993)
Construction variability	0.05−0.15	Kay (1993)

*Mean value for V_R provided in parentheses when sufficient information provided by reference sources.
CPT, Cone penetration test.

Heidarie Golafzani (2018) elucidated, high-tech cone penetrometer apparatus is affected less by uncertainties, originating from measurement errors in comparison to other conventional in situ−based methods. On the other hand, the models developed directly based on CPT records have a better estimate of axial pile bearing capacity than other existing SPT (Standard Penetration Test)-based methods and static analyses for a compiled database (Heidarie Golafzani, 2018).

The CPT defines the soil profile with great resolution since it retrieves data continuously with depth and allows a variety of sensors to be incorporated with the penetrometer. It also, enhances an engineer with an illustrative data stratigraphy along with the ability of fine changes. Moreover, the CPT is much less prone to error due to differences in equipment and technique, and thus is more repeatable and reliable than the SPT, and this resulted in a vast amount of CPT-based correlations in foundation engineering practice (Clayton, 1995; Fellenius, 2015; Coduto, 2002).

References

Barker, R.M., Duncan, J.M., Rojiani, K.B., Ooi, P.S., Tan, C.K., Kim, S.G., 1991. Manuals for the design of bridge foundations: Shallow foundations, driven piles, retaining walls and abutments, drilled shafts, estimating tolerable movements, and load factor design specifications and commentary.

Becker, D.E., 1997. Eighteenth Canadian geotechnical colloquium: limit states design for foundations. Part II. Development for the national building code of Canada. Canadian Geotechnical Journal 33 (6), 984−1007.

Bienen, B., Gaudin, C., Cassidy, M.J., Rausch, L., Purwana, O.A., Krisdani, H., 2012. Numerical modelling of a hybrid skirted foundation under combined loading. Computers and Geotechnics 45, 127−139.

Bolton, G.C., 1981. Spoils and Spoilers: Australians Make Their Environment 1788−1980. Allen & Unwin, .

Burland, J., Chapman, T., Skinner, H.D., Brown, M., 2012. ICE Manual of Geotechnical Engineering Volume 2: Geotechnical Design, Construction and Verification.

Cherubini, C., Giasi, C.I., Rethati, L., 1993. The coefficients of variation of some geotechnical parameters. In: Lo, S.-C.R (Eds.), Probabilistic methods in geotechnical engineering, A.A. Balkema, Rotterdam, pp. 179−183.

Casagrande, A., 1948. Classification and identification of soils. Transactions of the American Society of Civil Engineers 113 (1), 901−930.

Clayton, C.R., 1995. The Standard Penetration Test (SPT): Methods and Use. Construction Industry Research and Information Association.

Coduto, D.P., 2002. Foundation Design Principles and Practices, second ed. Prentice Hall, Inc, Upper Saddle River, NJ.

Deng, L., Wang, W., Yu, Y., 2015. State-of-the-art review on the causes and mechanisms of bridge collapse. Journal of Performance of Constructed Facilities 30 (2), 04015005.

Eslami, A., Fellenius, B.H., 1997. Pile capacity by direct CPT and CPTu methods applied to 102 case histories. Canadian Geotechnical Journal 34 (6), 880−898.

Eslami, A., Golafzani, S.H., Chenari, R.J., 2016. Assessment of Babolsar concrete pedestrian bridge failure for 1964 flood event and retrofitting practice. Engineering Failure Analysis 68, 101−112.

Fellenius, B.H., 2015. Basics of Foundation Design, Electronic Edition. 432 pp. www.Fellenius.net.

Hettler, A., 1993. Probabilistic approach and partial safety factors for driven piles. In: Proceedings of the International Symposium on Limit State Design in Geotechnical Engineering, Copenhagen, pp. 26−28.

Heidarie Golafzani, S., 2018. Application of Reliability Method in Processing of CPT Data for Determination of Pile Axial Capacity. Ph.D. Thesis. University of Guilan, Rasht, Iran.

Holtz, R.D., Kovacs, W.D., Sheahan, T.C., 2011. An Introduction to Geotechnical Engineering, second ed. Prentice Hall, Inc.

Kay, J.N., 1993. Probabilistic design of foundations and earth structures. In: Proceedings of the Conference on Probabilistic Methods in Geotechnical Engineering, Canberra, Australia, pp. 49−62.

Kulhawy, F.H., 1992. On the evaluation of static soil properties. In Stability and performance of slopes and embankments II. ASCE, pp. 95−115.

Kulhawy, F.H., Phoon, K.K., 1993. Drilled shaft side resistance in clay soil to rock. In: Design and Performance of Deep Foundations: Piles and Piers in Soil and Soft Rock. ASCE, pp. 172−183.

Manoliu, I., Marcus, A., 1993. 25 years of utilization of the limit state concept in the Romanian Code for geotechnical design. In: Proceedings of the International Symposium on Limit State Design in Geotechnical Engineering. Copenhagen, Vol. 2. Sponsored by the Danish Geotechnical Society, pp. 533−542.

Matsumoto, T., 1993. Soil parameter selection for serviceability limit design of a pile foundation in a soft rock. In: Proceedings of the International Symposium on Limit State Design in Geotechnical Engineering, Vol. 1, pp. 141−152.

Meyerhof, GG., 1995. Development of geotechnical limit state design. Canadian Geotechnical Journal. 32 (1), 128−136.

Meyerhof, GG, 1993. Development of geotechnical limit state design. In: proceedings of the international symposium on limit state design in geotechnical engineering. Sponsored by the Danish Geotechnical Society, Copenhagen, Vol. 1, May 26-28, pp. 1–12.

Mitchell, J.K., Tseng, D.J., 1990. Assessment of liquefaction potential by cone penetration resistance. In: Proceedings from the H. Bolton Seed Memorial Symposium Duncan. J. M. BiTech, Vancouver, B. C., pp. 335–350

Okahara, M., Kimura, Y., Ochiai, H., Matsui, K., 1993. Statistical characteristics of bearing capacity of single piles. In: proceedings of the ninth us-japan bridge engineering workshop, Vol. 3230.

Phoon, K.K., Kulhawy, F.H., 1999a. Characterization of geotechnical variability. Canadian Geotechnical Journal 36 (4), 612–624.

Phoon, K.K., Kulhawy, F.H., 1999b. Evaluation of geotechnical property variability. Canadian Geotechnical Journal 36 (4), 625–639.

Phoon, K.K., Kulhawy, F.H., Grigoriu, M.D., 1993. Observations on reliability-based design of foundations for electrical transmission line structures. Limit State Design in Geotechnical Engineering, 2, 351–362.

Poulos, A., Brown, C., McCulloch, D., Cole, J., 2017. U.S. Patent No. 9,791,921. U.S. Patent and Trademark Office, Washington, DC.

Rezazadeh, S., Eslami, E., 2018. Bearing capacity of semi-deep skirted foundations on clay using stress characteristics and finite element analyses. Marine Georesources & Geotechnology 36 (6), 625–639.

Schneider, J.A., Xu, X., Lehane, B.M., 2008. Database assessment of CPT-based design methods for axial capacity of driven piles in siliceous sands. Journal of Geotechnical and Geoenvironmental Engineering 134 (9), 1227–1244.

Tan, C.D., Duncan, J.M., Rojiani, K.B., Barker, R.M., 1991. Engineering manual for shallow foundations. Part 1. In: Manual for the Design of Bridge Foundations, vol. 343. Transportation Research Board, National Co-operative Highway Research Program Report, pp. 1–51.

Terzaghi, K., 1943. Theoretical Soil Mechanics. John Wiley and sons, New York, p. 510p.

Viggiani, C., Mandolini, A., Russo, G., 2014. Piles and Pile Foundations. CRC Press.

Wang, Y., 2009. Reliability-based economic design optimization of spread foundations. Journal of geotechnical and geoenvironmental engineering 135 (7), 954–959.

Whitman, R.V., 2000. Organizing and evaluating uncertainty in geotechnical engineering. Journal of Geotechnical and Geoenvironmental Engineering 126 (7), 583–593.

CPT equipment, performance, and records

3

3.1 Introduction and background

Penetration testing involves pushing or driving a system of a steel cone and rods into the ground, and recording the mobilized resistance to penetration in the soil. Probing with rods through weak soils to locate a firmer stratum was originally developed by Collin in France in the mid-19th century. Early versions of soundings were developed in the 1920s by the Swedish and Danish State Railways (Barentsen, 1936). Significant improvements of penetration techniques were made around 1930 with the development of the dynamic penetration test methods in the United States and the static penetration test methods in Europe (Eslami, 1996).

The first cone penetrometers were made in the Netherlands in 1934 in a form recognizable today (Barentsen, 1936). The method has been referred to as the static penetration test, quasi-static penetration test, Dutch cone penetration, Dutch static cone penetration test, and Dutch deep sounding test. The term quasi-static is used because this type of penetrometer is forced into the ground at a constant speed rather than being subjected to truly static loading conditions (Sanglerat, 1972; Broms, 1988).

Initially, the apparatus consisted of a cone and was followed by a design of a mantle to prevent entrance of soil particles between the cone and the push rods (Vermeiden, 1948). A friction sleeve to measure local side friction over a short length near the cone point was introduced later by Begemann (1953, 1965).

The use of cone penetration test (CPT) is increasing widely as a result of the influence of electronics which has greatly enhanced the accuracy of measurements. The static cone penetration test is becoming the preferred type of penetration test for site investigation. Reasons for the dominance of CPT among other in situ tests are as follows (after Mitchell (1988)):

- The CPT is simple and relatively economical.
- Continuous records with depth are obtained.
- The results are interpretable on both empirical and analytical bases.
- A variety of different sensors can be incorporated with penetrometer point and friction sleeve.
- A large experience-based knowledge is now available.

A cone of 10 cm^2 base area with an apex angle of 60 degree is generally accepted as standard and has been specified in the European and American practice (Fig. 3.1). The friction sleeve located above conical point has a standard area of 150 cm^2. The cone penetrometer is pushed at a standard rate of 20 mm/s. Standard 1 m long rods are used to push the cone penetrometer into the soil as depicted in Fig. 3.2. The standard rods have the same outside diameter as the cone, i.e., 35.7 mm. The measured soil

Piezocone and Cone Penetration Test (CPTu and CPT) Applications in Foundation Engineering
https://doi.org/10.1016/B978-0-08-102766-0.00003-1

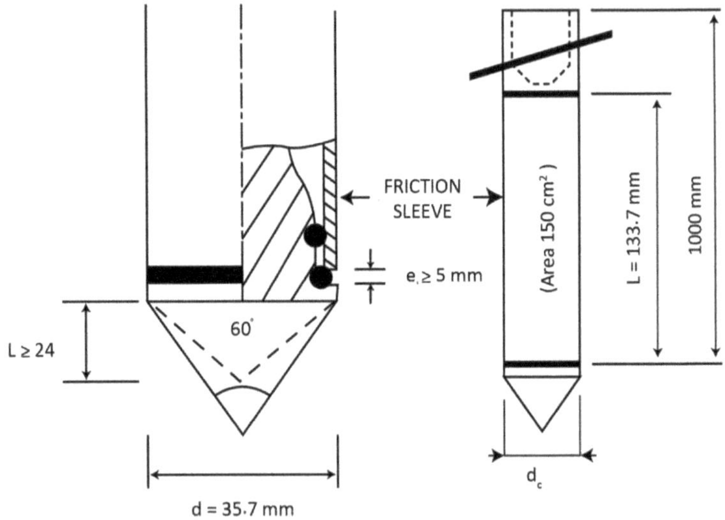

Figure 3.1 Standard dimensions of the cone in cone penetration test.

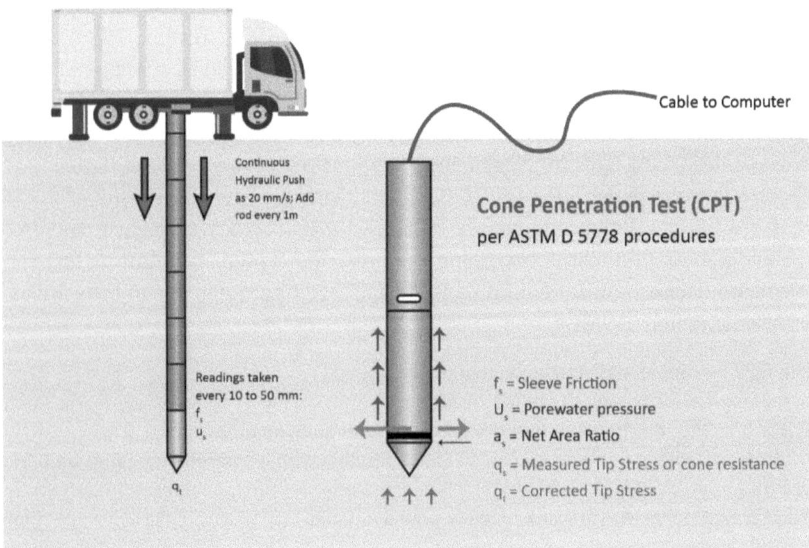

Figure 3.2 Procedure of performing cone penetrating testing.

resistance by data acquisition systems is then transferred through cables to the computers located inside the truck, where data are converted to digital formats.

The common two CPT measurements are cone tip resistance, q_c, and sleeve friction, f_s. The cone tip resistance is the ratio of the vertical force on the projected area of conical point. The sleeve friction is the shear stress acting on a lateral friction sleeve.

Penetrometers are divided into two main categories, mechanical and electrical. A mechanical cone is operated using inner and outer rods. When operating with the mechanical cone penetrometer, the cone advanced ahead of the outer rods to activate the cone bearing. Then, the cone and the outer rods are advanced together to measure the total load (combination of bearing and friction resistance). The sleeve friction value is determined by subtracting the previously recorded cone resistance from the total load. The Begemann type of mechanical cone is illustrated in Fig. 3.3.

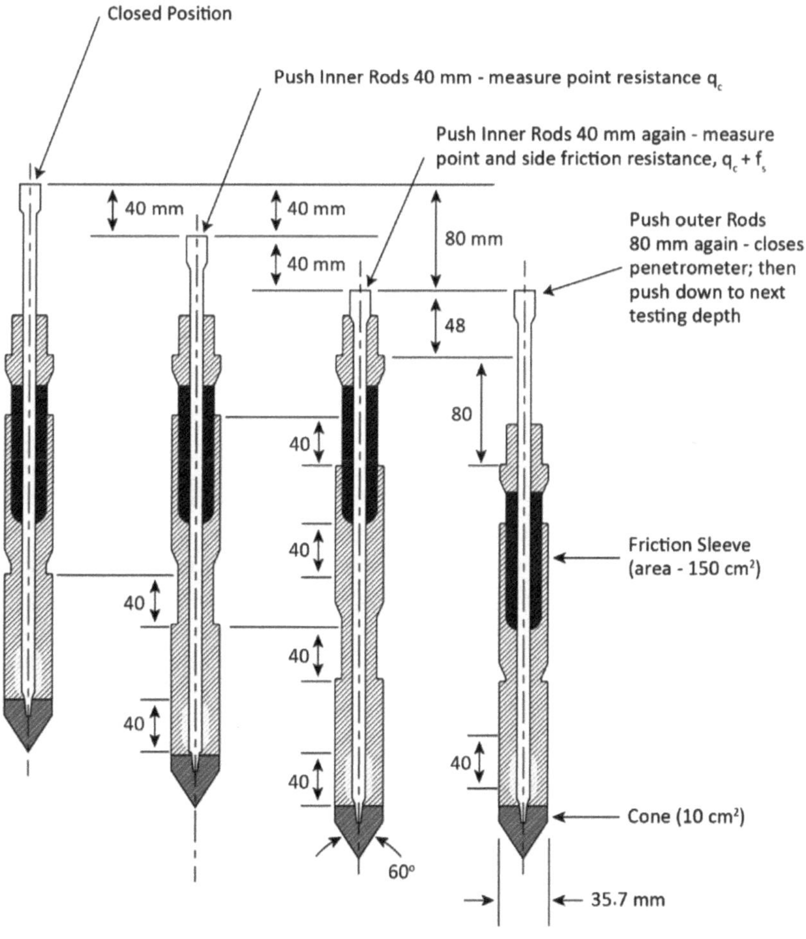

Figure 3.3 Sequence of operation for typical mechanical cone.

3.2 Electrical cones

Electrical versions were developed c.1948 by the Delft Soil Mechanics Laboratory, which offered continuous measurements of tip resistance with depth and direct strip chart plotting of the sounding record (Vlasblom, 1985). Electrical penetrometers with both tip and friction readings were designed as research tools as early as 1949 and became commercially available in the 1960s (De Ruiter, 1971). These solved noted problems associated with poor load readings acquired by mechanical cone systems because of frictional force buildups between the inner and outer sets of rods, primarily as a result of rusting and bending.

The electrical CPTs are also faster to do than mechanical CPTs because they are conducted at a constant rate of push rather than stepped increments. In the electrical systems, the penetrometer is linked by means of a wired cable through the hollow cone rods to a field computer at the surface for automated data acquisition. An inclinometer was incorporated to detect deviations from verticality and thus offer a warning to the user against excessive slope and/or buckling problems.

In the electrical friction-cone penetrometer, the cone and the friction sleeve do not move relative to each other. The cone resistance and sleeve friction are monitored continuously during penetration via separate strain gage load cells mounted on the cone and friction sleeve. Appropriate load cell selection is important in order to maximize sensitivity and prevent over stressing damage. The signals from the strain gages are transmitted to the surface to the data acquisition system via cables which pass through the center of the hollow push rods, see Fig. 3.4.

The mechanical cone offers the advantage of an initial low cost for equipment and simplicity of operation. However, it involves a rather slow incremental procedure (usually every 200 mm), is less effective in soft soils, and requires moving parts. Moreover, the quality of data is poor and operator dependent, and the depth is limited.

In contrast, the electric cone offers more rapid procedure; continuous recording; higher accuracy and repeatability; the possibility of incorporating of other additional sensors in the cone; and automatic data logging, reduction, and plotting (Campanella et al., 1982). However, the electrical cones have an initial high cost for equipment and require skilled operators and adequate backup for technical facilities, calibration, and maintenance.

3.3 Piezocone penetration test

The measurement of pore water pressure during the penetration of a probe into soil (dynamic pore pressure) was introduced in the mid-1970s (Torstensson, 1975, 1982; Wissa et al., 1975). By incorporating piezometer elements into standard electric cone penetrometers, the piezocone (CPTu) was introduced in the early 1980s. The piezocone measures the cone resistance, sleeve friction, and the pore pressure mobilized during penetrating. Also, if required, the rate of dissipation of pore pressure can be measured during any pause in the cone penetration.

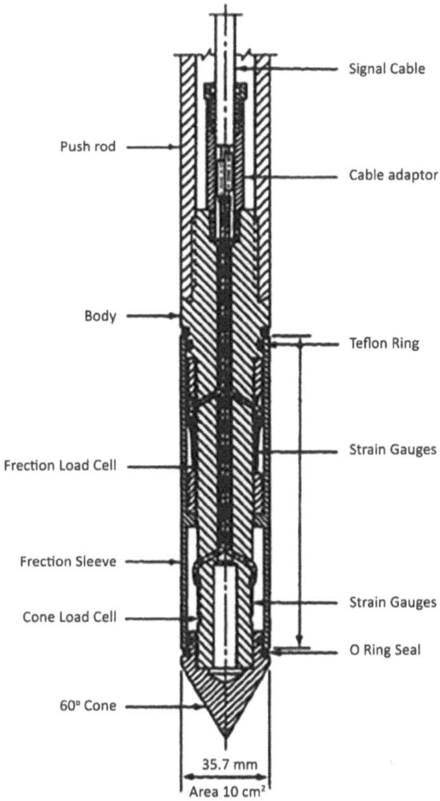

Figure 3.4 Electrical friction-cone penetrometer (Robertson and Campanella, 1983).

The main advantages of the CPTu over the conventional CPT are:

- Ability to distinguish between drained, partially drained, and undrained parameters
- Ability to correct measured cone data to account for unbalanced water forces due to unequal end areas in cone design
- Improved soil profiling and interpretation
- Ability to evaluate consolidation characteristics
- Ability to assess pore pressure gradients
- Improved evaluation of geotechnical parameters
- Assistance in the assessment of stress history and OCR (Over Consolidation Ratio) for cohesive soils

Soil properties, cone geometry, and location of the pore pressure filter influence the pore pressures measured during penetration. In normally consolidated soft clays and silts, pore pressures measured on the face of the cone, u_1, are generally 10% to 20% larger than those measured immediately behind the cone, u_2. In overconsolidated clays, silts, and fine sands, the pore pressure on the face of the cone tends to be large and positive, whereas that of the measured behind the cone may be considerably smaller and possibly negative. This can be attributed to the stress conditions around

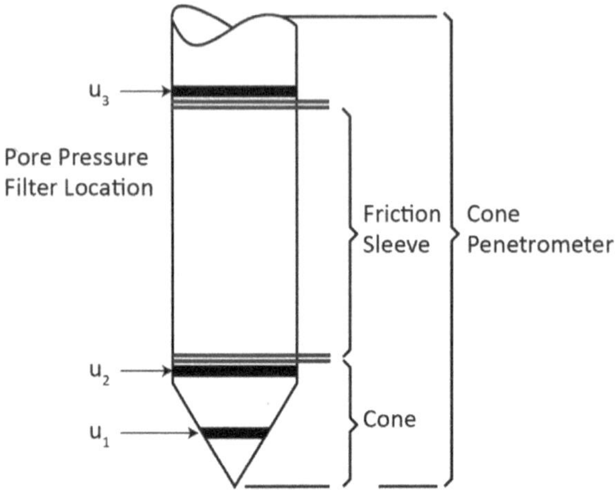

Figure 3.5 Location of porous filters in piezocone.

the cone. The measured pore pressures are affected by normal stress at the face and shear stresses behind the cone.

Measuring pore pressure behind the cone has main advantages such as that a porous filter is less susceptible to damage, measurements are less influenced by filter compressibility, and the position is appropriate for correction due to unequal end areas (Campanella and Robertson, 1988). No single location can provide information for all applications of pore pressure interpretation. Therefore, more advanced cones must have been developed which are able to measure the pore pressure at different locations simultaneously, in addition to measuring the cone resistance and sleeve friction. The three different locations of porous filter in the piezocone are depicted in Fig. 3.5.

3.4 Equipment

The CPT tanks used to advance the cone into the ground mainly consist of a hydraulic jack and a reaction system. In some cases, other equipment such as an anchored auger is incorporated as well. The axial bearing capacity required for the test varies typically between 100 and 200 kN. However, there are also some systems with capacity ranging from 20 to 50 kN, which can be employed in soft soils.

The CPT tanks are usually placed in a truck weighing 15 ton or more, and torsional anchors can be implemented to provide the required reaction force. The power for the hydraulic jack is supplied by the truck's motor. The electrical equipment and computers needed for recording the test's measurements are located inside of the truck to provide an appropriate condition for monitoring the temperature and suitable protection of the electrical equipment. Different examples of CPT rigs and trucks are shown in Fig. 3.6.

Figure 3.6 Examples of cone penetration test trucks (Robertson and Cabal, 2010).

A 20-ton penetrating system usually provides an insertion down to a depth of 30 m in medium dense sands and stiff clays. In weaker soils, penetration to the depths of more than 100 m requires the rods to be advanced adequately vertically. Layers with gravel or boulders, or zones with hard cemented soils considerably restrict the penetration, can damage the cone and rods, and deviate the rod from vertical alignment.

3.5 Data presentation

Generally, the data recorded by CPT can be presented within three categories:

1. Measured parameters
2. Corrected parameters and parameters derived indirectly
3. Additional information

3.5.1 Measured parameters

Typically, in CPT sounding, the cone resistance, q_c, sleeve friction, f_s, and pore pressure response, u at one of the three location, at the face of the cone (u_1), behind the cone (u_2), or behind the friction sleeve (u_3), are measured. Temperature, T, and inclination, I, are also measured simultaneously as the cone penetrometer is advanced in the ground. All channels are continuously monitored and reported digitally at 25 mm intervals.

3.5.2 Calculated parameters

In order to employ the CPT records for different purposes such as soil profiling, estimating soil stiffness and strength parameters, determining pile capacity, etc., some secondary parameters are calculated based on the primary CPT measured data. Some of the more commonly used parameters are as follows:

- Corrected tip resistance (q_t) modified for pore pressure:

Due to action of water pressure on the unequal areas of the exposed surface behind the cone, the measured q_c does not represent the true total stress as depicted in Fig. 3.7. To overcome the difficulty, particularly in soft clays, q_c must be corrected using the following relationship (Baligh et al., 1981; Campanella et al., 1982):

$$q_t = q_c + u_2(1 - a) \tag{3.1}$$

where

q_t = corrected total cone tip resistance,
q_c = measured total cone tip resistance,
u = measured pore pressure behind the cone tip,
a = net area ratio = A_N/A_T,
$\boldsymbol{A_N}$ = load transfer area behind the cone tip (Fig. 3.7),
$\boldsymbol{A_T}$ = cross-sectional area at the base of cone tip (Fig. 3.7).

- Corrected sleeve friction (f_t) modified pore pressure:
  This parameter is valid only if the pore pressure is measured at both ends of the friction rod.

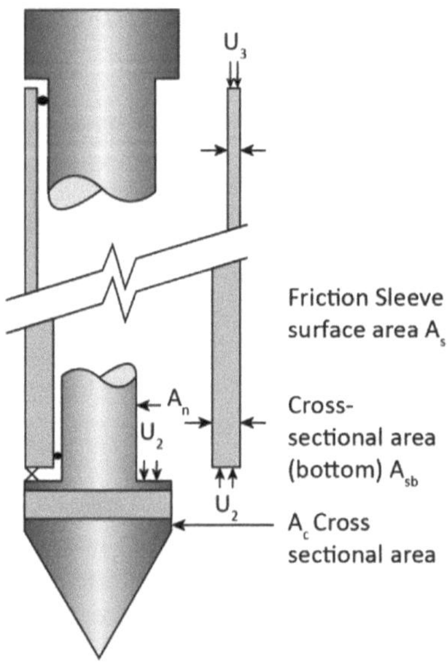

Figure 3.7 Effect of water pressure on the unequal areas of the exposed surface behind the cone.

- Friction ratio (R_f):
  It is calculated by Eq. (3.2) and is introduced as a percent.

$$R_f = \frac{f_s}{q_c} \tag{3.2}$$

- Pore pressure coefficient (B_q):

$$B_q = \frac{\Delta u}{q_t - \sigma_{v0}} \tag{3.3}$$

where
Δu = excess pore pressure ($u - u_0$),
u_0 = hydrostatic pore pressure,
σ_{v0} = vertical overburden total stress.

3.5.3 Graphical presentation

After processing the obtained CPT data, they are usually presented graphically, including q_c, f_s, R_f, and u versus depth as shown for a few cases in Fig. 3.8.

Fig. 3.8A is a CPTu profile recorded at North Western University, which consists of two main layers of sand from 0 to 7.4 m and clay from 7.4 to 20 m. As can be seen, higher values of cone tip and skin friction are recorded in sands, while these values are significantly less in clayey soil. On the other hand, in clayey soils, more excess pore pressure is observed. Moreover, the larger the friction ratio, the softer the soil is. Fig. 3.8B depicts the CPT profile for Shekar Novin Project site in Iran, which consists of a clay layer underlain by sandy soil. Finally, the CPT profile for University of British Columbia (UBC) research site in Vancouver delta is illustrated in Fig. 3.8C, which indicates 2 m of fill at the surface underlain by 13 m of clay, and 13 m of sand underlain by a clayey silt layer.

The graphical presentation of CPT result provides an accurate and useful indication of subsurface conditions at the site of the test. Also, graphical presentation provides a quick and convenient way for defining the bearing strata, and estimate the likely type and suitable depth for the required foundation.

3.6 Factors influencing CPT measurements and interpretation

There are several factors influencing the CPT and CPTu measurements, which can be partly attributed to the equipment, partly to the use of it. The factors that can affect significantly the CPT measurement and interpretation are as follows:

1. Equipment design and appropriate selection for a specific soil condition
2. Rate of penetration
3. In situ stress
4. Compressibility
5. Temperature

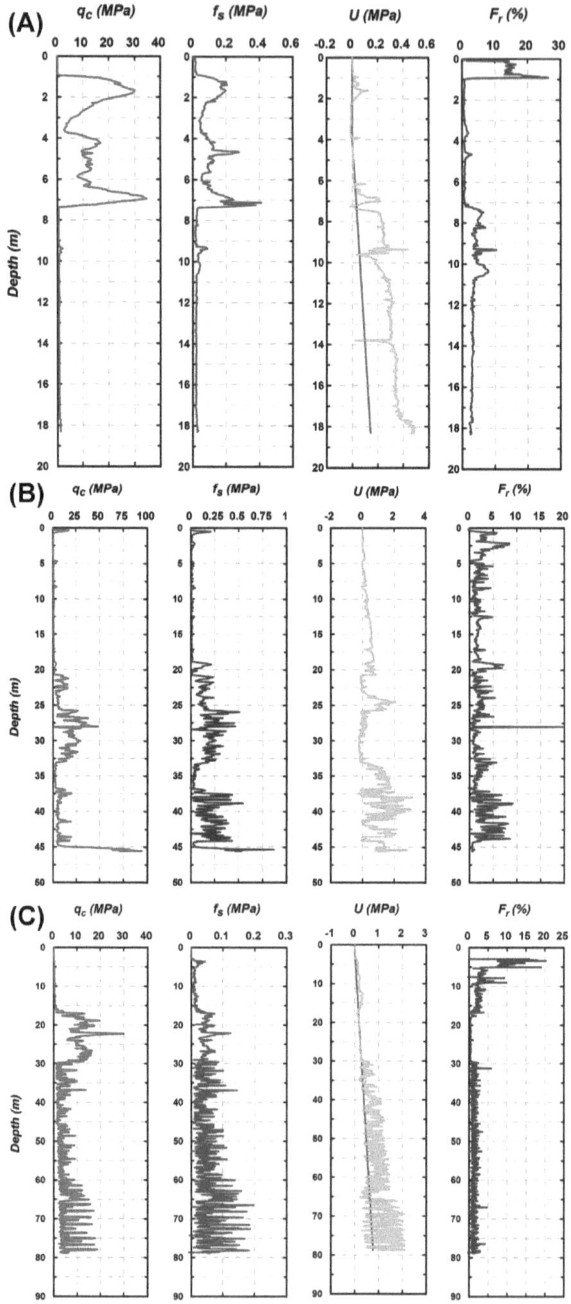

Figure 3.8 Typical piezocone profiles: (A) North Western University, Evanston, USA, (Eslami, 1996), (B) Shekar Novin project, Iran, (Hosseinzadeh Attar, 2013), (C) UBC, Vancouver, Canada, (Eslami, 1996).
(A) Eslami, A., 1996 Bearing Capacity of Piles from Cone Penetrometer Test Data (Ph.D. thesis). University of Ottawa, Department of Civil Engineering, 516pp; (B) (Hosseinzadeh Attar, 2013); (C) Eslami, A., 1996 Bearing Capacity of Piles from Cone Penetrometer Test Data (Ph.D. thesis). University of Ottawa, Department of Civil Engineering, 516pp.

6. Lack of maintenance
7. Lack of qualified operator and wrong use of methods
8. Penetrometer geometry

Despite performing the test according to available standards, recorded data from CPTu can still vary significantly depending on which equipment is used (Sandven, 2010). The three major areas of cone design that influence the acquired data, and subsequent interpretations, are the piezometer locations, accuracy of measurements (tolerances), and unequal end area effects (probe geometry). The first was explained previously in Section 3.4. The accuracy of load measurements mainly depends on the design of the load cells that may involve a calibration error and/or zero shift error.

Regarding the site condition, in granular soils where the measurement of pore pressure is less important, the choice of probe configuration has a marginal effect (Sandven, 2010). However, in fine-grained soils, where measurement of pore pressure is highly crucial, it is important to have a sufficiently saturated pore pressure measurement system.

According to Campanella et al. (1982), small variations in penetration rate relative to the standard rate of 20 mm/s have no significant influence on the cone resistance, q_c. Therefore, the rate variation of $\pm 25\%$ allowed by the ASTM D3448 seems acceptable. The mobilized pore pressures during cone penetration (u_1 and u_2) are more dependent on the rate of penetration than q_c and f_s.

Theoretical approaches and calibration chamber studies have shown that the in situ radial stress has a dominant effect on q_c and f_s. Baldi et al. (1982) have shown that for a given density of sand, the q_c depends not only on vertical effective stress but also on the in situ horizontal effective stress. Also, the resulting increase of the q_c and particularly f_s after soil improvement techniques such as dynamic compaction and vibrio flotation is due to the increase of σ'. Therefore, the in situ stress history is important in CPT interpretation.

The compressibility of sand can significantly influence q_c and f_s. Highly compressible sands tend to have low q_c and high friction ratio values. The compressibility of sand is a function of uniformity of grading, angularity of grains, as well as mineralogical type, such as mica and feldspar as opposed to quartz.

There is a temperature difference between the site environment and the soil in depth. Thus, as the cone penetrates the ground, it experiences a temperature gradient, and as a result, the readings might be affected till the temperature stabilizes. In addition, passing through a dense and noncohesive layer may result in an increase in cone temperature. This will not affect the readings of the dense layer. However, if the dense layer is underlain by a soft clay layer with limited resistance, the increased temperature will affect the readings in the clay layer (Sandven, 2010).

3.7 Special cones

Some of significant recent developments in CPT apparatus and testing are described in the following section.

3.7.1 Dual range penetrometer, Brecone

When performing CPT in clays, the values of cone tip resistance are relatively low, maximum 5 MPa; however, the cone resistance in sands may be 40 MPa or more. Therefore, the highly sensitive load cell designed for use in clays is likely to be overloaded when used in sands. On the other hand, a load cell designed for use in sands may not be sensitive enough when used in clays. A dual range cone called Brecone is designed to measure on a high-sensitivity, low-capacity load cell up to a certain limit of cone resistance and then on a high-capacity, low-sensitivity load cell for higher cone resistance.

3.7.2 Acoustic cone

Cones have been equipped with a microphone and data acquisition system to monitor acoustic response of the soil during cone penetrating (Muromachi, 1974; Tringle and Mitchell, 1982; Massarsch, 1986). Noise level, spectrum, and frequency are functions of soil type and density. It was also determined by visual inspections that sand and silt layers as thin as 1 mm could be identified using this apparatus, whereas such thin layers cannot be detectible by conventional CPT (Mitchell, 1988).

3.7.3 Lateral stress cone

The in situ horizontal stress is one of the important parameters required for many geotechnical analyses. However, its accurate determination involves difficulty. Several attempts were made for measuring normal stress at the location of the friction sleeve in order to estimate the in situ lateral stress prior to penetration (Huntsmann, 1985; Mitchell, 1988). Masood and Mitchell (1993) proposed a correlation between sleeve friction, overconsolidation ratio, and coefficient of earth pressure at rest. Also, a lateral stress sensing cone penetrometer (LSSCP) was developed at the University of Berkeley (Huntsmann and Mitchell, 1986). The simultaneous measurement of lateral stress and other CPT data can lead to an improved analysis of cone penetrometer data for determining soil parameters.

3.7.4 Resistivity cone

The resistivity piezocone (RCPTu) consists of a resistivity module which is added to a standard piezocone. The incorporation of the resistivity module enables continuous measuring of bulk resistivity of soil and pore fluid, in addition to all other CPTu measurements. The interpretation is based on resistivity contrast between remolded sand (at the vicinity of the interface) and undisturbed sand (relatively far away from the interface) induced during cone penetration. Soil electrical resistance depends on soil type, physical properties such as porosity, water content, etc., which can be employed for soil profiling in combination with other information derived from in situ tests. In addition, soil resistivity, relies on soil permeability, existence of ions and clay minerals. Thus, significant variations in electrical resistance indicate presence of chemicals

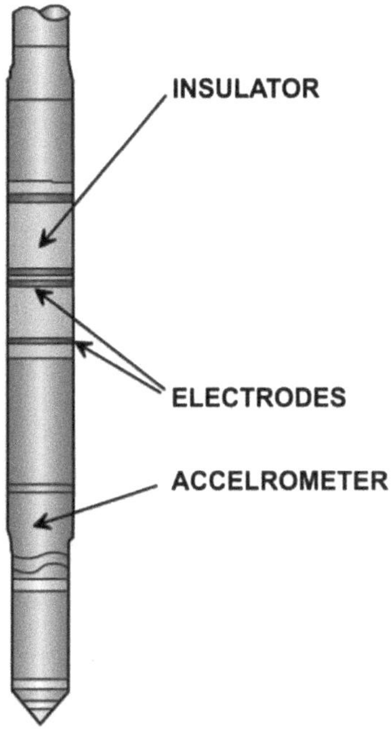

Figure 3.9 Resistivity cone penetrometer and electrodes arrangement named Wenner (Eslami and Sekhavatian, 2014).

or contaminations in soils. Accordingly, one of the main applications of the RCPTu measurements is evaluation of groundwater contamination and contaminant transportation (Campanella and Weemees, 1990).

Aside from estimating soil parameters and environmental applications of RCPTu, some other applications of RCPTu include assessment of dilatancy in saturated sand (Campanella and Kokan, 1993), ground improvement quality control (Daniel et al., 1999). An example of RCPTu as well as its results is presented in Figs. 3.9 and 3.10.

3.7.5 Piezocone with additional sensors

There are some supplementary sensors which can be incorporated along with the piezocone (CPTu) to record additional measurements.

- Inclinometer: it allows the user to monitor the deviation of the cone from vertical alignment, which can occur by possible large stones or local dense layers during penetration. It can prevent probable damages to the apparatus or incorrect measurements. The location of installing the inclinometer on the cone is depicted in Fig. 3.11.
- Temperature sensor: It measures the soil temperature at each depth while penetrating.
- Microphone or noise sensor: The level of the noise caused by penetration can be measured by means of microphones. The range and frequency of the noise depend on soil type as well as densification.

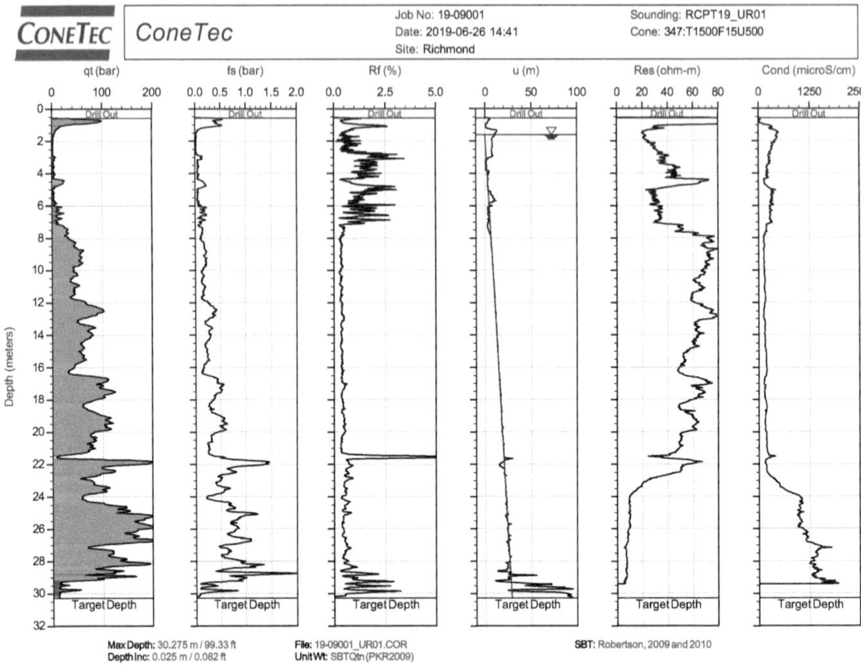

Figure 3.10 Example of resistivity piezocone profiles (ConeTec, 2019).

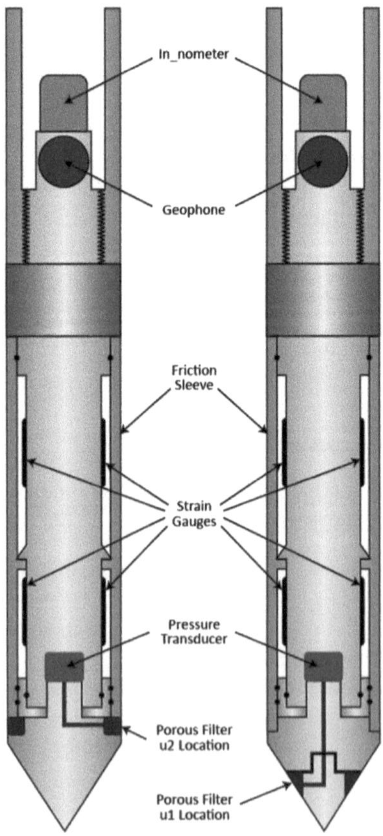

Figure 3.11 Location of inclinometer.

- Radioactive source: By installing a source of gamma rays close to the cone tip, the diffusion of the emitted gamma ray can be measured. Since magnitude of this diffusion is a function of soil density, it can represent a qualitative estimate of soil unit density.

3.7.6 Seismic cone

A seismic cone or SCPTu has been developed by incorporating small velocity seismometers and accelerometers into the conventional electric cone (Campanella et al., 1986). The introduction of seismic measurements into the CPT procedure permits the specific determination of shear wave velocity and therefore dynamic shear modulus (G_{max}). Downhole shear wave velocity measurements can be made during brief pauses in the penetrating of the cone. The seismic cone offers an advantage relative to the crosshole method for shear wave velocity measurement that only one hole is needed, and the test can be performed during a regular cone penetration test.

The incorporated sensors in this apparatus measure shear and compression wave velocities in depth. Consequently, based on the velocity variations with depth, the soil stratification, consolidation, and cracking can be determined with no need for sampling. A schematic view of the testing procedure is demonstrated in Fig. 3.12. In addition, a sample result obtained from SCPTu is presented in Fig. 3.13.

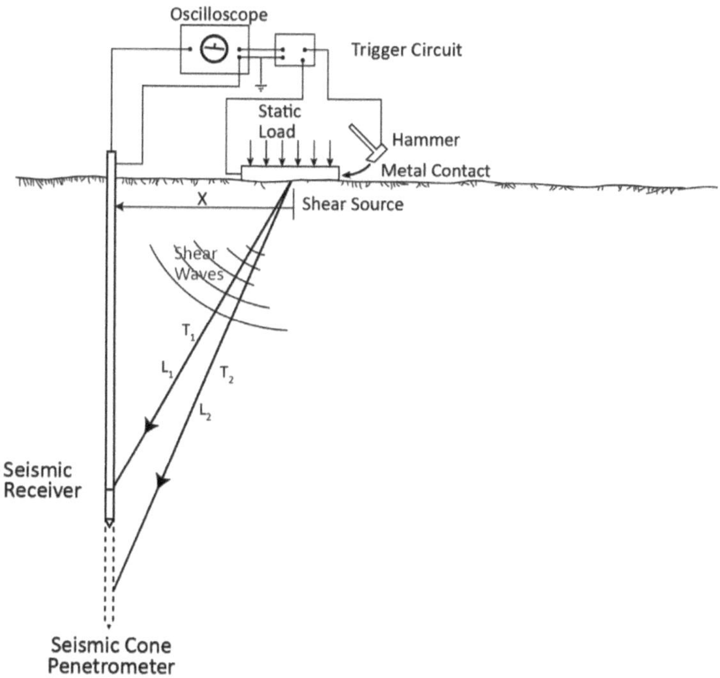

Figure 3.12 Schematic view of seismic penetrometer (Campanella and Stewart, 1992).

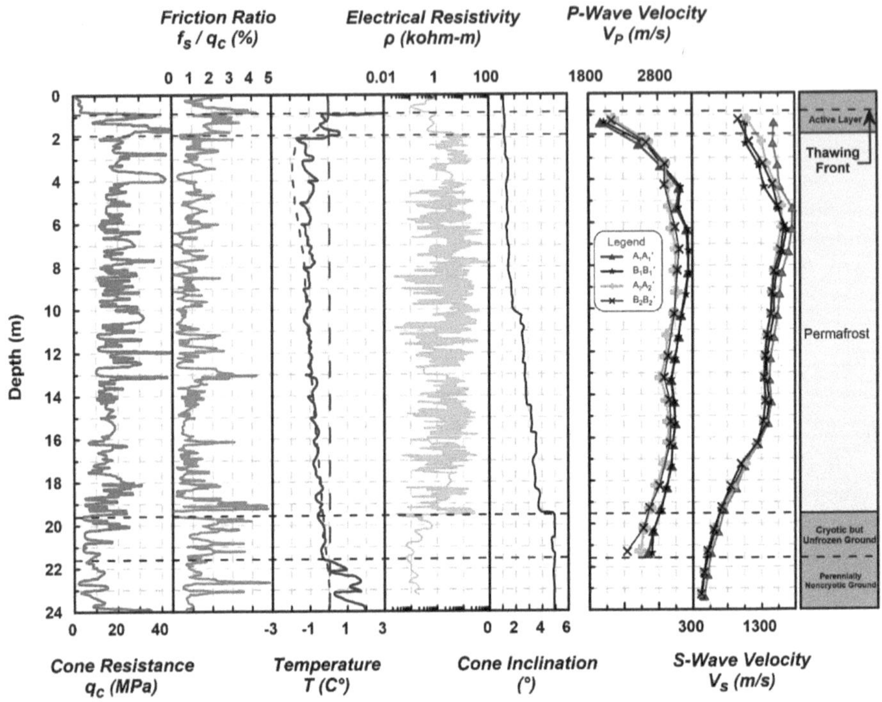

Figure 3.13 Example of seismic cone records and soil profiling.
Data from Eslami, A., Sekhavatian, A., 2014 Geotechnical Engineering: Principles, Investigations and Interpretations, Amirkabir University of Technology Publication, (In Persian).

3.7.7 Cone pressuremeter

The cone pressuremeter is an in situ testing device which combines the electric piezocone with the pressuremeter (Withers et al., 1989). The cone pressuremeter is a particular form of the full-displacement type, with a self-boring pressuremeter mounted behind a solid conical point with diameter of 43.7 mm. The combination of the profiling ability of piezocone together with stress–strain measurements from the pressuremeter, in the form of cone pressuremeter, provides an appropriate in situ measurements for geotechnical design.

3.7.8 Piezovibrocone and Vibrocone

Piezovibrocone or Vibrocone is a device which enables to directly measure the soil liquefaction potential by measuring the excess pore pressure in the vicinity of the cone as well as the dynamic load applied during penetration. It was first constructed by Sasaki et al. (1984) in order to apply a periodic load and consequently measuring the resisting load at the tip. A schematic view of the device is illustrated in Fig. 3.14.

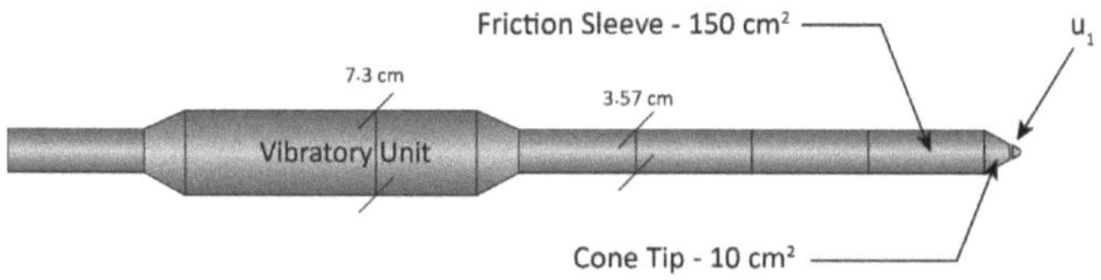

Figure 3.14 A schematic view of a vibrocone (Mayne and Kulhawy, 1990).

The vibrational force in the vibrocone can be applied in several different ways according to the soil type, the required information of the site, and site condition. This vibration can be applied by a rotating cam, solenoid, eccentric counterweight, or a cylinder. The values of cone tip resistance measured by a vibrocone would be less in liquefiable soil compared to those measured by a static penetrometer, while they would be equal in nonliquefiable soils.

3.7.9 Ultraviolet-induced fluorescence CPT

The combination of ultraviolet-induced fluorescence and CPTu provides a powerful tool for geo-environmental assessment of soil. The ultraviolet light is emitted through a special lens into the soil mass in a way that its reflection to fluorescence sheet located at the center of the cone demonstrates the amount of hydrocarbonated contaminants in soil and underground water. It should be mentioned that there are also several other types of cones employed for geo-environmental purposes which are beyond the scope of this book.

3.8 Mini-cone and CPT in physical modeling

The mini-CPT has been vastly used in the characterization of the soil properties of near-surface seafloors during installation of seafloor cable systems in sands (Tufenkjian and Thompson, 2005). Also, CPT with various diameters is applicable in marine works. The major advantages of the mini-CPT are the following:

1. Cones with smaller dimensions need smaller downward force to drive penetrometer into the soil (Tummay et al., 1998).
2. Smaller CPT could be installed in smaller vehicles, so it has better mobility and site accessibility (Tufenkjian and Thompson, 2005).
3. According to Schmertmann (1978), a penetrometer with smaller diameter is more sensitive to display the soil layers specifically, very thin lenses compared to a larger piezocone (Fig. 3.15).

A miniature CPT system is commonly used for superficial penetration and laboratory studies of CPT sounding particularly, in calibration chambers, where the full-size penetration tests are not practical (Mitchell, 1988), to reduce the error caused by the boundary conditions.

There has been a wide range of in situ and experimental studies for investigating the effect of CPT scale on the measured results reporting a variety of observations. Some

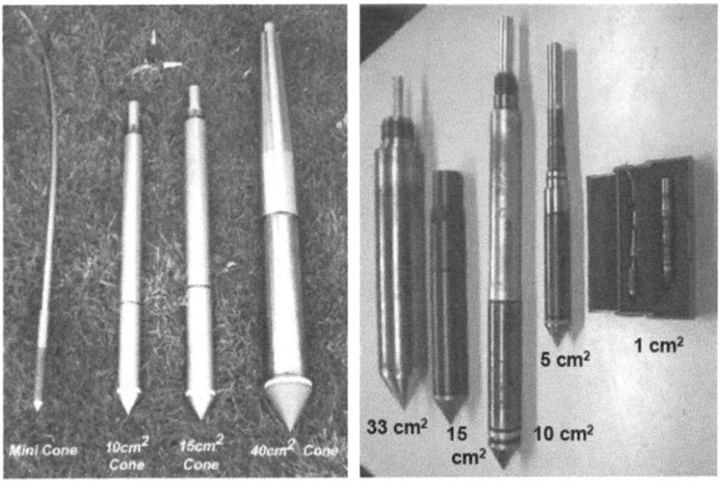

Figure 3.15 Different sizes of cone penetrometer (Robertson and Cabal, 2010).

research studies indicate that there has been no sign of size or scale effect for various penetrometers with different diameters and that different CPTs with various sizes result in the same outcome (Schmertmann, 1978). In this regard, Vesić (1967) and De Beer (1963) defined a critical depth at which the scale effect starts to dissipate, and for the depths less than the critical, different cone diameters result in different cone resistances as depicted in Fig. 3.16 (Vesić, 1967; De Beer, 1963).

Additionally, experiments carried out in the calibration chamber show that the cone resistance values obtained by mini-CPT are more than those measured by the standard CPT and the increase in the cone diameter will result in a reduction in the cone resistance values. Also, this scale effect increases proportionally as the soil density and vertical effective stress increase (Las, 1984; Ghionna and Jamiolkowski, 1991; Fioravante et al., 1991).

On the contrary, Eid (1987), using CPTs with the diameters of 43.7, 35.7, and 23.2 mm, concluded that as the cone diameter increases, the cone resistance and sleeve friction values both decrease and the amount of their reduction increases as the soil relative density and vertical effective stress increase (Eid, 1987).

More detailed conclusions were derived by Tufenkjian et al. (2010) who employed two CPTs, one with the standard cone area (10 cm^2) and a mini-CPT with the base area of 2 cm^2 to investigate the size effect on the resistance of CPT tip in sandy soils. According to their measurements, in fairly loose sands with the void ratio being higher than critical void ratio, the mini penetrometer resulted in lower cone resistance in comparison to the standard cone penetrometer. However, in soils with medium density, q_c values for both CPT and mini-CPT were almost equal. In loose sands, it was quite opposite, and the CPT with smaller dimensions had a higher cone resistance than CPT with the standard dimensions (Tufenkjian et al., 2010).

An example of assessing the mini-CPT in frustum confining vessel (FCV) is presented below. Two CPTs with different diameters, as shown in Fig. 3.17, were used, the properties of which are given in Table 3.1.

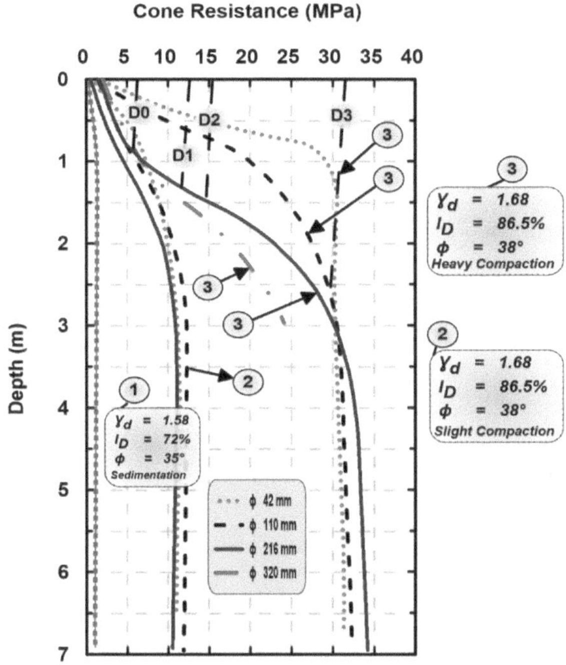

Figure 3.16 Comparison of the cone tip resistance with respect to cone diameters (De Beer, 1963).

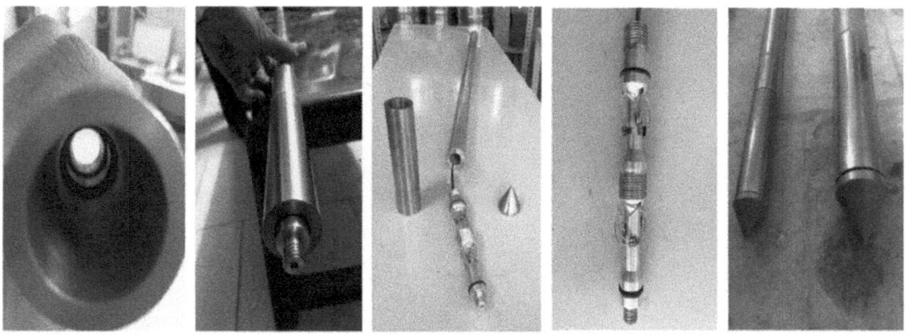

Figure 3.17 Various cone geometry and apparatus.

Strain gauge sensors were used at two different elevations to record cone resistance and sleeve friction separately. The cone resistance and sleeve friction values were simultaneously recorded at 2 s intervals (in each period 0.6 cm is read). An example of performing the penetrometer test via minicone is depicted in Fig. 3.18. Figs. 3.19 and Fig. 3.20 illustrate the cone resistance and sleeve friction in soils with various relative densities and effective vertical stresses for the standard and the mini-CPT, respectively. According to the results, it is clearly shown that as the bottom pressure in FCV increases, both the cone resistance and sleeve friction remarkably increase. The

Table 3.1 Properties of the cone penetrometers used in frustum confining vessel.

Apparatus	Cone Diameter (cm)	Tip Area (cm^2)	Apex Angle at Tip (degree)	Sleeve Friction Length (cm)	Sleeve Surface Area (cm^2)
Cone penetration test (CPT)	3.57	10	60	13.38	150
Mini-CPT	2.4	4.52	60	9.00	67.86

Figure 3.18 Performance cone penetrometer test via minicone in frustum confining vessel.

augmented amount of strength for all the three soil conditions (loose, medium dense, and dense) is appreciable. This trend is in complete compliance with the study done by Robertson and Campenella which indicates the increase in the cone resistance with an increase in equivalent overburden pressure (Robertson and Campanella, 1989) (it depends on either friction angle or soil density). It can be inferred from the results that this mobilized strength significantly increased with the increase in soil density. The abovementioned issues can be adduced for both CPT and mini-CPT.

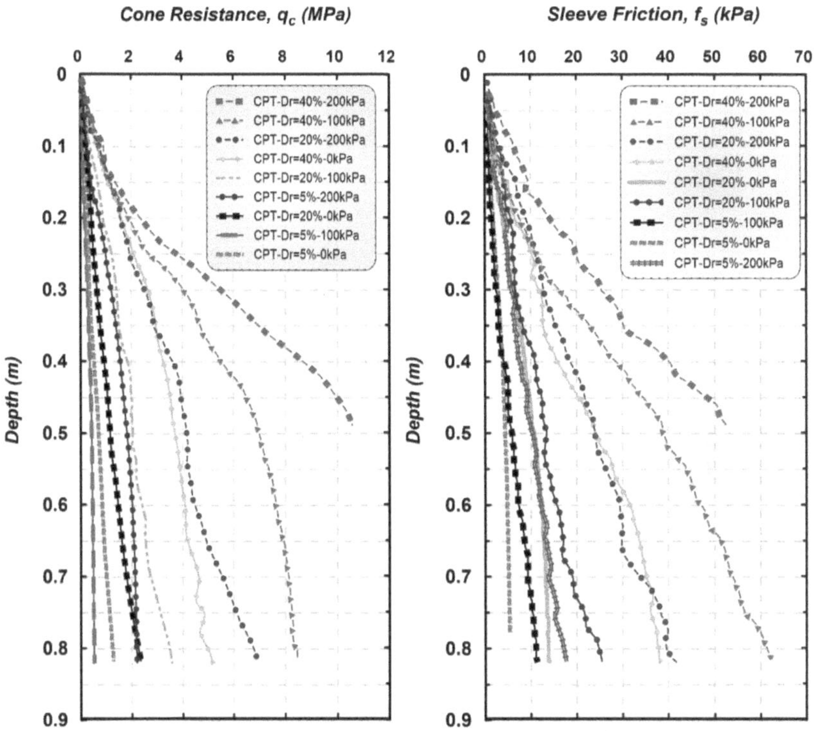

Figure 3.19 Cone resistance and sleeve friction of cone penetration test (CPT) in different relative densities and bottom pressures.

3.9 Codes, standards and software

In geotechnical engineering, in order to facilitate employment of CPT and CPTu data in analysis and design, several computer programs have been developed, some of which are as follows:

• Edison—SGI	• Unipile (www.unisoftltd.com)
• CPTPlot	• SPAS (geologismiki.gr)
• Almeid (UWA)	• PlotCPT (GeoMil Equipment)
• Go! (van den Berg)	• GEO5 Pile CPT (gintsoftware)
• PCL-pro (Geotech)	• INSITU (www.geo&soft.com)
• SCPT-DAA (Baziw)	• GeoExplorer (Gouda-Geo Equip)
• Georit (Geotech AB)	• CPTGL (Geotech AB reduction)
• Cliq (geologismiki.gr)	• CPTLiq (Geosoftwaresolutions)

Continued

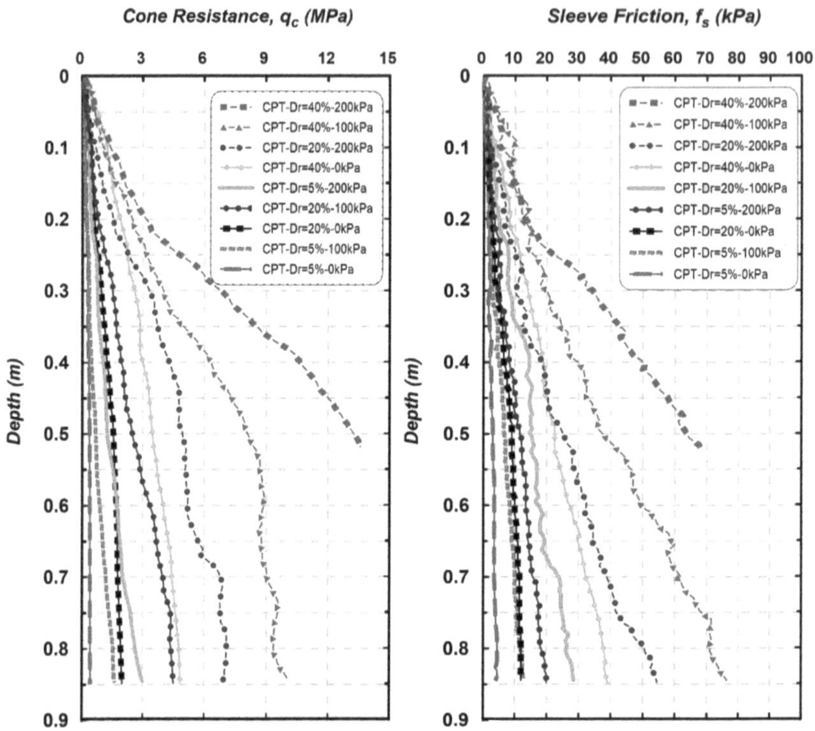

Figure 3.20 Cone resistance and sleeve friction of mini-CPT in different relative density and bottom pressures.

- LCPC360A (ConeTec)
- ScreenZw (ConeTec)
- CPT-Log (Geotech.se)
- MFoundation GeoDelft)
- GLAMCPT (Omer et al., 2006)
- CPeT-IT (geologismiki.gr)
- CPT7 (GeoMil Equipment)
- FB-Deep (University of Florida)
- Gorilla (A.P. van den Berg)
- CPTINT (Dick Campanella)
- DCCONE (dc-software.com)
- RapidCPT (dataforensics.net)
- CPT-Pro (www.geosoft.com.pl)
- OpenSounding (sourceforge.net)
- Static Penetrometer (Alpes-Geo)
- PL-aid (McTrans at Univ. Florida)
- Static Probing (www.geostru.com)
- Conrad (Swedish Geotech Institute)
- Shake2000 (CPT for soil liquefaction)
- Unicone (Fellenius, 2002)
- NonoCPT 1.2 (novosoftware.com)
- LPD-CPT (Louisana State Universitiy)
- P-Class (LSU and LTRC, Louisiana)

Moreover, some of the standards with respect to cone penetration testing are listed as follows:

- International: IRTP for CPT and CPTU (ISSMGE, 1998)
- Europe: CEN standard EN ISO 22476-12 (2009) for mechanical cone
- Europe: CEN standard EN ISO 22476-1 (2007) for electrical piezocone
- Austria: DIN 4094-1
- France: NF P94-113 CPT and NF P94-119 CPTu
- Germany: DIN 4094-1
- Netherlands: NEN 5140
- United Kingdom: BS1377, BS5930
- Bulgarian Standard (use German DIN 4094-90)
- Czech Standard STN 721033
- Slovakia Standard CSN 721033
- Estonia Reference LBN 207-01
- Latvia Reference LVS 437
- Lithuania Reference LBN 005-99
- Hungary Standard MI 15000/2 (1989)
- Poland: Polish Standard PN-B-04452 (2002)
- Romanian Standard NP 074-2007
- Russia Standards GOST 19912-2001 Code SP 11 105 97
- Egypt: Use British Standards: BS 1377—Part 9 on In Situ Testing
- USA and Canada: ASTM D 3441 (1998) (mechanical cone); ASTM D 5778 (2012) (electric piezocone)
- Norway: Norwegian Geotechnical Society Guideline 5 (1995)
- Sweden: Swedish Geotechnical Society SGF Report 1:93E
- Spain: Reference is the UNE 103-804-93
- Australian National Standard: AS1289.6.5.1-1999
- New Zealand: National Standard NZS 4402.6.5.3:1988

References

American Society for Testing and Materials D3441, 1998. Standard Test Method for Mechanical Cone Penetration Tests of Soil. ASTM Standard, West Conshohocken, PA, 19428-2959, United States.

American Society for Testing and Materials D5778, 2012. Standard Test Method for Electronic Friction Cone and Piezocone Penetration Testing of Soils. ASTM Standard, West Conshohocken, PA, 19428-2959, United States.

Baldi, G., Bellotti, R., Ghionna, V., Jamiolkowski, M., Pasqualini, E., 1982. Design parameters for sands from CPT. In: Proceedings of the 2nd European Symposium on Penetration Testing, vol. 2, pp. 425–436. ESOPT-2, Amsterdam.

Baligh, M.M., Azzouz, A.S., Wissa, A.Z.E., Martin, R.T., Morrison, M.J., 1981. The Piezocone Penetrometer. Symposium on Cone Penetration Testing and Experience. ASCE, Geotechnical Engineering Division, St. Louis, pp. 247–263.

Barentsen, P., 1936. Short description of a field testing method with a cone shaped sounding apparatus. In: Proceedings of 1st International Conference on Soil Mechanics and Foundation Engineering, vol. 2. ICSMFE, Boston, p. 10.

Begemann, H.K., 1953. Improved method of determining resistance to adhesion by sounding through a loose sleeve placed behind the cone. In: Proceedings of 3rd International Conference on Soil Mechanics and Foundation Engineering, vol. 1. ICSMFE, Zurich, pp. 213–217.

Begemann, H.K.S., 1965. The maximum pulling force on a single tension pile calculated on the basis of results of adhesion jacket cone. In: Proceedings of 6th International Conference on Soil Mechanics and Foundation Engineering, vol. 2. ICSMFE, Montreal, pp. 229–232.

Broms, B.B., 1988. History of soil penetration testing. In: Proceedings of the 1st International Symposium on Penetration Testing, vol. 1, pp. 157–220. ISOPT-1, Disney world.

Campanella, R.G., Gillespie, D., Robertson, P.K., 1982. Pore pressures during cone penetration testing. In: Proceedings of the 2nd European Symposium on Penetration Testing, vol. 2, pp. 507–512. ESOPT-2, Amsterdam.

Campanella, R., Kokan, M., 1993. A new approach to measuring dilatancy in saturated sands. Geotechnical Testing Journal 16 (4), 485–495.

Campanella, R.G., Robertson, P.K., 1988. Current status of the piezocone test. In: Proceedings of the 1st International Symposium on Penetration Testing, vol. 1, pp. 93–116. ISOPT-1, Orlando, March 22 - 24.

Campanella, R.G., Robertson, P.K., Gillespie, D., 1986. Seismic cone penetration test. In: Proceedings of In-Situ 86, ASCE Specialty Conference, Blacksburg, Virginia, pp. 116–130.

Campanella, R.G., Stewart, W.P., 1992. Seismic cone analysis using digital signal processing for dynamic site characterisation. Canadian Geotechnical Journal 29 (3), 477–486.

Campanella, R.G., Weemees, I., 1990. Development and use of an electrical resistivity cone for groundwater contamination studies. Canadian Geotechnical Journal 27, 557–567.

Daniel, C.R., Giacheti, H.L., Howie, J.A., Campanella, R.G., 1999. Resistivity piezocone (RCPTU) data interpretation and potential applications. In: Proc., XI Pan-American Conference on Geotechnical Engineering, Iguasu Falls, Brazil: Int. Soc. For Soil Mechanics and Geotechnical Engineering, pp. 1–8.

De Beer, E., 1963. The scale effect in the transposition of the results of deep-sounding tests on the ultimate bearing capacity of piles and caisson foundations. Géotechnique 13, 39–75. https://doi.org/10.1680/geot.1963.13.1.39.

De Ruiter, J., 1971. Electric penetrometer for site investigations. Journal of the Soil Mechanics and Foundations Division 97 (2), 457–472.

Eid, W.K., 1987. Scaling Effect in Cone Penetration Testing in Sand. Ph.D. thesis. Virginia Polytechnic In stitu te and State University, Virginia.

Eslami, A., 1996. Bearing Capacity of Piles from Cone Penetrometer Test Data. Ph.D. Thesis. University of Ottawa, Department of Civil Engineering, 516pp.

Eslami, A., Sekhavatian, A., 2014. Geotechnical Engineering: Principles, Investigations and Interpretations. Amirkabir University of Technology Publication (In Persian).
Fellenius, B.H., 2002. Excerpt from Chapter 6 of the red book: direct methods for estimating pile capacity. In: Background to UniCone. http://www.fellenius.net.
Fioravante, V., Jamiolkowski, M., Tanizawa, F., Tatsuok, F., 1991. Results of CPT's in Toyoura quartz sand. In: Calibration Chamber Testing. Elsevier, pp. 13–39.
Ghionna, V.N., Jamiolkowski, M., 1991. A critical appraisal of calibration chamber testing of sands. In: Proceedings of the 1st International Symposium on Calibration Chamber Testing (ISOCCTI). Elsevier, Potsdam, NY, pp. 13–39.
Holden, J.C., 1971. Research on Performance of Soil Penetrometers. Country Roads Board of Victoria Internal Report. CE-SM-71-1 (as referenced by Chapman, 1974).
Hosseinzadeh Attar, I, 2013. Influence of soil setup on shaft resistance variations of driven piles: case study. PhD Dissertation, Amirkabir University of Technology.
Huntsmann, S.R., 1985. Determination of In-Situ Lateral Pressure of Cohesionless Soils by Cone Penetrometer. Ph. D. Thesis. University of California, Berkeley (as referenced by Mitchell, 1988).
Huntsmann, S.R., Mitchell, J.K., 1986. Lateral stress measurement during cone penetration. In: Proceedings of In-Situ 86, ASCE Specialty Conference, Blacksburg, pp. 617–634.
International Society for Soil Mechanics and Geotechnical Engineering (ISSMGE, former ISSMFE), 1998. Report of the ISSMFE Technical Committee on Penetration Testing of Soils (TC 16) with Reference Test Procedure. Swedish Geotechnical Society/Swedish Geotechnical Institute (SGI), Linköping, Sweden. Information No. 7.
Las, N., 1984. Seminar on Cone Penetration Testing in the Laboratory. Held at Southampton University, Department of Civil Engineering. November 15-16.
Masood, T., Mitchell, J.K., 1993. Estimation of in-situ lateral stresses in soil by CPT. ASCE Journal of Geotechnical Engineering 119 (10), 1624–1639.
Massarsch, K.R., 1986. Acoustic penetration tests. In: Proceedings of 4th International Geotechnical Seminar, Field Instrumentation and In-Situ Measurements, Nanyang (as referenced by Mitchell, 1988).
Mayne, P.W., Kulhawy, F.H., 1990. Observations on the development of pore water stresses during piezocone penetration in clays. Canadian Geotechnical Journal 27, 418–428.
Mitchell, J.K., 1988. New developments in penetration tests and equipment. In: Proceedings of the 1st International Symposium on Penetration Testing, vol. 1, pp. 245–261. ISOPT-1, Disney world.
Muromachi, T., 1974. Phono-sounding apparatus. In: Proceedings of the 1st European Symposium on Penetration Testing, vol. 2 (1), pp. 110–112. ESOPT-1, Stockholm.
Omer, J.R., Delpak, R., Robinson, R.B., 2006. A new computer program for pile capacity prediction using CPT data. Geotechnical & Geological Engineering 24, 399–426.
Robertson, PK, Campanella, RG, 1983. Interpretation of cone penetration tests. Part I: Sand. Canadian Geotechnical Journal 20 (4), 718–733.
Robertson, P.K., Campanella, R.G., 1989. Guidelines for Geotechnical Design Using the Cone Penetrometer Test and CPT with Pore Pressure Measurement. Soil Mechanics Series, No.105, fourth ed. UBC, Dept. of Civil Engineering and Hogentogler and Company, Inc. 193 pp.
Robertson, P.K., Cabal, K.L., 2010. Guide to Cone Penetration Testing for Geotechnical Engineering. Gregg Drilling & Testing.
Sandven, R., 2010. Influence of test equipment and procedures on obtained accuracy in CPTU. In: 2nd International Symposium on Cone Penetration Testing, Huntington Beach (USA).
Sanglerat, G., 1972. The Penetrometer and Soil Exploration. Elsevier, Amsterdam, 464 pp.

Sasaki, Y., Itoh, Y., Shimazu, T., 1984. A study on the relationship between the results of vibratory cone penetration tests and earthquake-induced settlement of embankments. In: Proceedings of the 19th Annual Meeting of JSSMFE, Tokyo, Japan.

Schmertmann, J.H., 1978. Guidelines for Cone Test, Performance and Design. Federal Highway Administration, Washington. Report FHWA-TS-78209, 145 pp.

Torstensson, B.A., 1975. Pore pressure sounding instrument. In: Proceedings of ASCE Specialty Conference on In-Situ Measurement of Soil Properties, Raleigh, vol. 2, pp. 48–54.

Torstensson, B.A., 1982. A combined pore pressure and point resistance probe. In: Proceedings of the 2nd European Symposium on Penetration Testing, vol. 2, pp. 903–908. ESOPT-2, Amsterdam.

Tringle, P.T., Mitchell, J.K., 1982. An acoustic cone penetrometer for site investigation. In: Proceedings of the 2nd European Symposium on Penetration Testing, vol. 2, pp. 909–914. ESOPT-2, Amsterdam.

Tufenkjian, M., Thompson, D., 2005. Shallow penetration resistance of a minicone in sand. In: Proceedings of the International Conference on Soil Mechanics and Geotechnical Engineering, p. 1789.

Tufenkjian, M., Yee, E., Thompson, D., 2010. Comparison of cone and minicone penetration resistance for sand at shallow depth. In: 2nd International Symposium on Cone Penetration Testing. California, pp. 305–311.

Tummay, M., Kurup, P., Boggess, R., 1998. A continuous intrusion electronic miniature cone penetration test system for site characterization. In: Proceedings, International Conference on Site Characterization, pp. 1183–1188.

Vermeiden, J., 1948. Improved soundings apparatus, as developed in Holland since 1936. In: Proceedings of 2nd International Conference on Soil Mechanics and Foundation Engineering, vol. 1. ICSMFE, Amsterdam, pp. 280–287.

Vesić, A.S., 1967. A Study of Bearing Capacity of Deep Foundations. Final Report Project B 189. Georgia Institute of Technology, Engineering Experiment Station, Atlanta.

Vlasblom, A., 1985. The Electrical Penetrometer, a Historical Account of its Development. Delft Soil Mechanics Laboratory, Delft, 51 pp. (LGM mededelingen: 1985, No. 92).

Wissa, A.E.Z., Martin, R.T., Garlanger, J.E., 1975. The piezometer probe. In: Proceedings of ASCE Specialty Conference on In-Situ Measurement of Soil Properties, Raleigh, vol. 2, pp. 536–545.

Withers, N.J., Howie, J., Hughes, J.M.0, Robertson, P.K., 1989. Performance and analysis of cone pressuremeter tests in sands. Géotechnique, London, England 39 (3), 433–454.

Geotechnical parameters from CPT records

4

4.1 Introduction

Geomaterial behaviors are extremely complex particularly in their load–displacement and time-dependent relationships which can be referred to the soil nonlinear stress–strain behavior, different loading pattern, elastoplastic performance, and effects of stress history based on unloading and physiochemical alterations. For any earthen structure, foundation (a transition element) is applied to carry the loads from super structure to substructure or naturally deposits. Major aspects in geotechnical and foundation engineering practices such as bearing capacity, settlement, and structural design requires parameters that can be directly obtained from in situ investigations including plate load test for shallow and CPT for deep foundations.

Owing to difficulties in providing sample confirmed stress and scale effect from conventional lab testing, in situ investigations are preferred to determine bearing capacity and settlement indirectly through common in situ tests that are performed in geotechnical engineering such as the standard penetration test (SPT) and cone penetration test (CPT).

The CPT is versatile and reliable compared with other in situ tests. The main advantageous of CPT can be counted as simple and rapid application, economical aspects, production of continuous and precise records, and reduction in operator influence.

Several methods have been suggested for prediction of geomaterial properties for using foundation design based on CPT at different circumstances, which can be considered into the following four main groups:

- Empirical methods based on synthesis of case histories
- Simplified analytical methods
- Numerical analyses
- Soft computing in data handing

All prediction methods, based on any of the abovementioned methods, require determining input soil properties. However, based on versatility, experimental correlations by statistical analysis and system identification techniques remain as a major practice. Furthermore, to consider the complex behavior of the geotechnical materials, soft computing techniques such as Group Method of Data Handing have been applied recently. In this chapter, it mainly focuses on the following geotechnical soil parameters being predicted using CPT records.

Piezocone and Cone Penetration Test (CPTu and CPT) Applications in Foundation Engineering
https://doi.org/10.1016/B978-0-08-102766-0.00004-3

4.1.1 Group I: most geotechnical parameters applied in foundation engineering

- Unit weight (γ)
- Relative density (D_r)
- Internal friction angle (φ)
- Undrained shear strength (S_u)
- Over consolidation ratio (OCR)
- Stiffness (E)
- Constrained modulus (m_v)

4.1.2 Group II: related soil parameters

- Shear wave velocity (V_S) and small strain shear modulus (G_0)
- Sensitivity (S_t)
- CPT correlation with SPT
- Permeability (k)
- Liquefaction potential approximation (Fs)

Given that, in other chapters, the workability of CPT and CPTu tests in foundation-bearing capacity (shallow and deep foundation), soil behavior classification, soil settlement, and the applications of CPT-based soft computing techniques has been reviewed and discussed in detail, this chapter does not encapsulate the aforementioned issues.

In geotechnical engineering, many correlations have been introduced for soil parameters associated with CPT records. Because the CPT has additional equipment and sensors (e.g., seismic: SCPT, pore pressure: CPTu, and resistivity cone: RCPTu), the applicability to estimate soil parameters varies from provided measure records. As perceived and presented in Chapter 1, the applicability of CPTu for deriving soil parameters is more than the other in situ tests. Following the brief review of soil parameters and methods of estimating are presented.

4.2 Unit weight

In soil mechanic engineering, unit weight (γ) is a basic property of a soil which is used for different earthwork applications. Unit weight known as soil density is usually approximated in the lab (the ratio of weight to volume of an undisturbed soil sample). Sand cone and rubber balloon or nuclear densitometer tests are candidates for in situ unit weight measuring.

The CPT parameters including q_c, f_s, and γ have been mainly investigated by Robertson and Cabal (2010) and Mayne et al. (2010), Mayne (2014) and Baginska (2016) who proposed correlations as presented in Table 4.1. Recently, Lengkeek et al. (2018) suggested a new chart that predicts the saturated soil unit weight (γ_{sat}) for wide range of soils, from highly organic soils to sands (Fig. 4.1).

Table 4.1 Unit weight and CPT parameters' relationships.

Reference	Correlations	Parameter unit
Robertson and Cabal (2010)	$\frac{\gamma}{\gamma_w} = 0.27[logR_f] + 0.36\left[\log\left(\frac{q_t}{P_a}\right)\right] + 1.236$	R_f(%)
Mayne et al. (2010)	$\gamma = 1.95\gamma_w\left(\frac{\sigma'_v}{P_a}\right)^{0.06}\left(\frac{f_s}{P_a}\right)^{0.06}$	–
Mayne (2014)	$\gamma = 26 - \frac{14}{1+[0.5 \log f_s+1]^2}$	f_s (kPa)
Baginska (2016)	$\gamma = 11 + 24 \ln(f_s + 0.7)$	f_s (MPa)
Lengkeek et al. (2018)	$\gamma_{sat} = 19 - 4.12\frac{\log\left(\frac{5}{q_t}\right)}{\log\left(\frac{30}{R_f}\right)}$	R_f(%), q_t (MPa)

Pa, Atmospheric pressure; q_t, corrected q_c; σ'_v, Effective vertical stress; γ (kN/m^3).

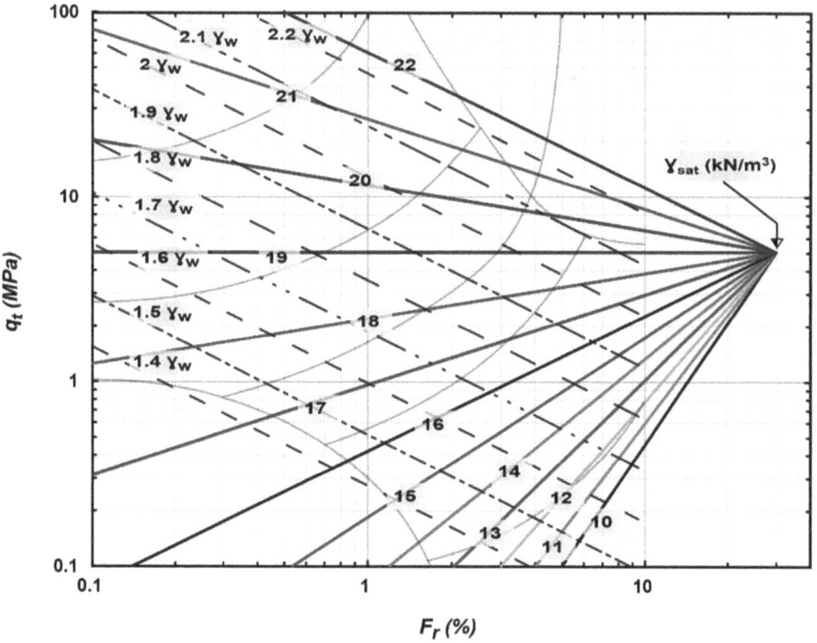

Figure 4.1 Normalized unit weight and R_f (%) by Robertson and Cabal (2010) (dotted line) and by Lengkeek et al. (2018) (continuous line).

4.3 Relative density

For coarse-grained soils, the in situ state of density is very important. This is because the major causes of settlement, compressibility, and liquefaction interacted from loose situation. Relative density (D_r), which is commonly used for sand deposits, is a factor to classify the level of cohesivèless material compaction and is defined by

$$Dr = \frac{e_{max} - e}{e_{max} - e_{min}} \tag{4.1}$$

where e_{max} *and* e_{min} are maximum and minimum void ratios, respectively, that is calculated by Eq. 4.2 and *e* is the current void ratio.

$$\begin{cases} e_{max} = \dfrac{G_s\gamma_w}{\gamma_{d\ min}} - 1 \\ e_{min} = \dfrac{G_s\gamma_w}{\gamma_{d\ max}} - 1 \end{cases} \tag{4.2}$$

where $\gamma_{d\ min}$ and $\gamma_{d\ max}$ are minimum and maximum dried unite weight, respectively, and G_s stands for specific gravity.

Various standards have detailed the laboratory procedures to obtain $\gamma_{d\ min}$ and $\gamma_{d\ max}$ can be found in geotechnical soil testing handbooks.

To describe the behavior of coarse-grained soils particularly sand deposits, density parameter (D_r) is classified as very loose, loose, medium dense, dense, and very dense which is provided in Table 4.2.

Because of limitations of using in situ density test (can only be measured at shallow depth), the use of *CPT* records to assess D_r based on empirical correlations has been highlighted. Many researchers noted that the D_r affected by q_c and σ_v' as presented firstly in Schmertmann (1970), Schmertmann (1978), and Villet and Mitchell (1981) publications (Fig. 4.2). In this regard, some of other important correlations have been listed in Table 4.3.

4.3.1 Internal friction angle

The shear strength of coarse-grained soils is typically stated in the form of friction angle, φ. Several researchers have suggested φ from CPT records in sands based on one of the following three theories:

- Bearing capacity
- Cavity expansion
- Empirical, based on calibration chamber tests

Table 4.2 Density classification of coarse-grained soils in term of D_r

D_r (%)	0–15	15–35	35–65	35–80	80–100
Density	Very loose	Loose	Medium dense	Dense	Very dense

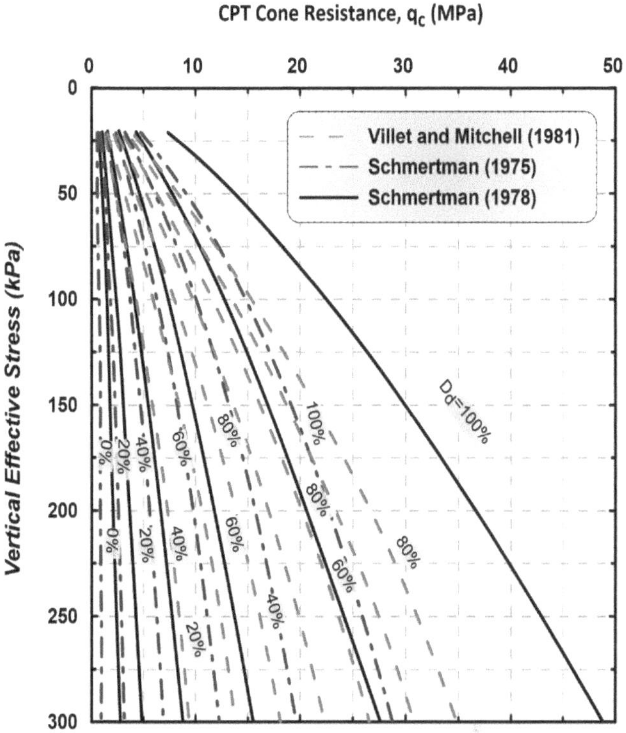

Figure 4.2 Comparison of relative density−qc relationships by Schmertmann (1970, 1975) and Villet and Mitchell (1981).

Numerous attempts have been put into interpreting ϕ of sand from CPT, mainly q_c. For CPT failure mechanism, it is realized that q_c is corresponding to ultimate load of a deep circular foundation in a subsoil which leads the soil mass to fail. Whereas failure assumption in bearing capacity and cavity expansion theory is based on the required pressure to form a deep hole in an elastic-plastic environment which is tailored to the pressure needed for creation of the cavity expansion in the same volume under identical conditions. Based on mentioned theories, numerous investigators have examined the prediction of ϕ from CPT and CPTu.

A review of calibration chamber test results was made by Robertson and Campanella (1983) to investigate the $\varphi - q_c$ relationship. Of a note, Robertson and Campanella (1983) used drained triaxial compression tests to measure φ. Fig. 4.3 shows the recommended correlations.

Other researchers [e.g., Meyerhof (1974), Robertson et al. (1986), Kulhawy and Mayne (1990), Mayne (2007), and Uzielli et al. (2013)] used the same method and introduced a couple of equations to predict φ as summarized in Table 4.4. Moreover, Mayne (2014) proposed correlations to predict friction angle which are based on empirical correlation and case histories.

In the following simplified analytical methods to predict φ and cohesion (C) by Motaghedi and Eslami (2014) is presented in details.

Table 4.3 Correlations predicting D_r from CPT records.

Reference	Proposed correlation	Remarks
Baldi et al. (1986)	$D_r = \frac{1}{C_2} \ln\left(\frac{q_c}{C_0(\sigma_v')^{0.55}}\right)$	C_0 and C_2: soil constants, $C_0 = 157$ and $C_2 = 2.41$ normally consolidated q_c and σ_v' are in kPa unit
Jamiolkowski et al. (2001)	$D_r = 26.8 \ln \frac{q_c/P_a}{(\sigma'_v/P_a)^{0.5}} - b_x$	$\begin{cases} b_x = 52.5 & \text{high compressibility sands} \\ b_x = 67.5 & \text{medium compressibility sands} \\ b_x = 82.5 & \text{low compressibility sands} \end{cases}$
Kulhawy and Mayne (1990)	$D_r^2 = \frac{Q_{cn}}{305 Q_c Q_{OCR}}$; $Q_{cn} = \frac{q_t/P_a}{\sqrt{\sigma'_v/P_a}}$	$Q_c =$ compressibility factor (0.91 for high, 1.0 for medium, and 1.09 for low). $Q_{OCR} =$ over consolidation factor $OCR^{0.18}$
Mayne (2007)	$D_r = 100\left[0.268 \ln\left(\frac{\frac{q_t}{P_a}}{\sqrt{\frac{\sigma'_v}{P_a}}}\right) - 0.675\right]$	

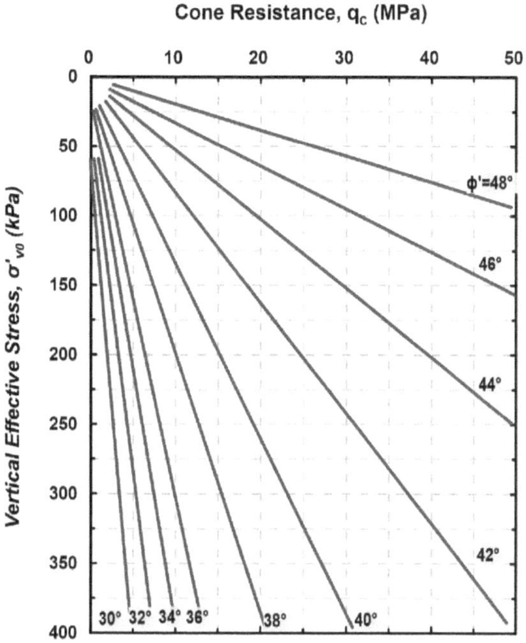

Figure 4.3 φ—qc and σ_v' relationship by Robertson and Campanella (1983).

Table 4.4 Proposed correlations for friction angle.

Reference	Correlations	Soil type and remarks
Meyerhof (1974)	$\varphi = tan^{-1}\left(\frac{q_c}{0.5N_q}\right)$	Sand; q_c (MPa)
Robertson et al. (1986)	$\varphi = tan^{-1}\left[0.1 + 0.38log\left(\frac{q_c}{\sigma'_v}\right)\right]$	Sand
Kulhawy and Mayne (1990)	$\varphi = 17.6 + 11\text{Log}\left(\frac{q_c}{\sqrt{100\sigma'_v}}\right)$	Sand q_c and σ_v' are in kPa unit
Uzielli et al. (2013)	$\varphi = 25\left(\frac{q_c}{\sqrt{100\sigma'_v}}\right)^{0.1}$	Sand q_c and σ_v' are in kPa unit
Mayne (2014)	$\varphi = 29.5B_q^{0.121}[0.256 + 0.33B_q + \log Q_t]$ $Q_t = \frac{q_t - \sigma_v}{\sigma'_v}$; $B_q = \frac{u_2 - u_0}{q_t - \sigma_v}$	Cohesive soils

By applying two basic equations on determination of the deep foundation—bearing capacity, one for corrected q_t (q_E) tip and the other for f_s, using the effective bearing capacity instead of total stress approach and extension of the relationships, a dual equation system with two unknowns can be achieved under static loading conditions as follows:

$$\begin{cases} CN_C + qN_q + 0.5\,\gamma B N_\gamma = q_E = q_t - u_2 \\ C + \sigma'_{hc}\tan(\delta) = f_s \end{cases} \quad (4.3)$$

Where at first, several parameters should be calculated or replaced via relationships that are function of φ and after that the only parameters which remain to be achieved will be cohesion (C). Hence, q_E and f_s are determined from CPTu log at the certain depth and N_C, N_q, and N_γ should be replaced in Eq. (4.3) as follows:

$$N_q = k_p \exp[(\pi - 2\beta)\tan\varphi] \quad (4.4)$$

$$N_c = (N_q - 1)\cot\varphi \quad (4.5)$$

$$N_\gamma = 2(N_q + 1)\tan\varphi \quad (4.6)$$

where, in Eq. (4.4), $k_p = \frac{1+\sin\varphi}{1-\sin\varphi}$ and β is the angles of plasticization which defines the failure zone around the penetrometer tip ($-40<\beta<+30°$). To solve the second part of Eq. (4.3), σ'_{hc} is needed to be evaluated too. A correlation has been presented in Eq. (4.7) for σ'_{hc} in the form of horizontal effective stress (σ'_h) and in situ mean stresses (σ_{mean}).

$$\frac{\sigma'_{hc}}{\sigma'_h} = 7.89 \times 10^{-4}\left[\frac{q_c - \sigma_{mean}}{\sigma'_{mean}}\right]^{1.44} \quad (4.7)$$

where

$$\sigma_{mean} = \frac{\sigma_v + 2\sigma_h}{3} = \frac{\sigma_v(1+2k_0)}{3} \quad (4.8)$$

$$\sigma'_{meam} = \frac{\sigma'_v + 2\sigma'_h}{3} = \frac{\sigma'_v(1+2k_0)}{3} \quad (4.9)$$

$$k_0 = k_{0(nc)} OCR^{1-\sin\phi} \quad (4.10)$$

$$k_{0(nc)} = 1 - \sin\phi \quad (4.11)$$

Substituting σ'_{hc}, N_q, N_c, and N_γ in Eq. (4.3), dual form can be presented as Eq. (4.12).

$$\begin{cases} q_E = \left(tan^2\left(\frac{\pi}{4}+\frac{\phi}{2}\right)e^{(\pi)tan\phi} - 1\right)c.cot\phi + \sigma'_v tan^2\left(\frac{\pi}{4}+\frac{\phi}{2}\right)e^{(\pi)tan\phi} + \\ 0.018\gamma\left(tan^2\left(\frac{\pi}{4}+\frac{\phi}{2}\right)e^{(\pi)tan\phi} + 1\right)tan\phi \\ \mathrm{f_s} = c + 0.000789\left[\dfrac{q_c - \left(\dfrac{\sigma_v + 2\sigma_h}{3}\right)}{\left(\dfrac{\sigma'_v + 2\sigma'_h}{3}\right)}\right] tan\left(\frac{2}{3}\phi\right)\sigma'_h \end{cases} \tag{4.12}$$

In the same research, Eslami and Mohammadi (2016) have presented approaches to predict φ and C of marine deposits from CPTu records based on bearing capacity and cavity expansion theories. They summarized soil classification chart in five categorized areas based on CPTu test results. For specified q_t and $F_r \left(= \frac{f_s}{q_c} \times 100\right)$, i.e., friction ratio, soil type can be determined in every depth interval. However, variation range of C and φ using the proposed method, at 200 kPa< σ'_v <300 kPa into five categorized soil type, is presented and also for 300 kPa< σ'_v <400 kPa in Fig. 4.4.

It should be noted that, in some cases, intersection of two lines which is the answer of the sets of equation occurs in two points which obtained cohesion (C) in one of them, probably is negative, and this amount is not reasonable in fact. Hence, positive obtained C will be the true answer and the other one would not be valid. Moreover, in some case studies, neither the current methods nor the proposed approach can obtain reasonable answers, and the results are deleted from data base.

Combining bearing capacity theories and direct shear modes of failure considering the pore water pressure, a set of equations is derived. By inputting CPTu at a certain depth, soil shear strength parameters can be calculated simultaneously as presented in Fig. 4.5.

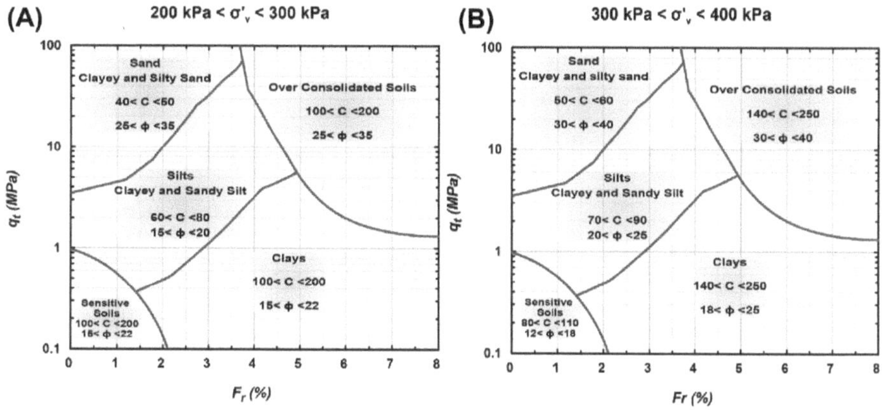

Figure 4.4 Variation range for C (kPa) and φ (Degree) by Eslami and Mohammadi (2016). (A) 200 kPa< σ'_v <300 kPa and (B) 300 kPa<σ'_v<400 kPa.

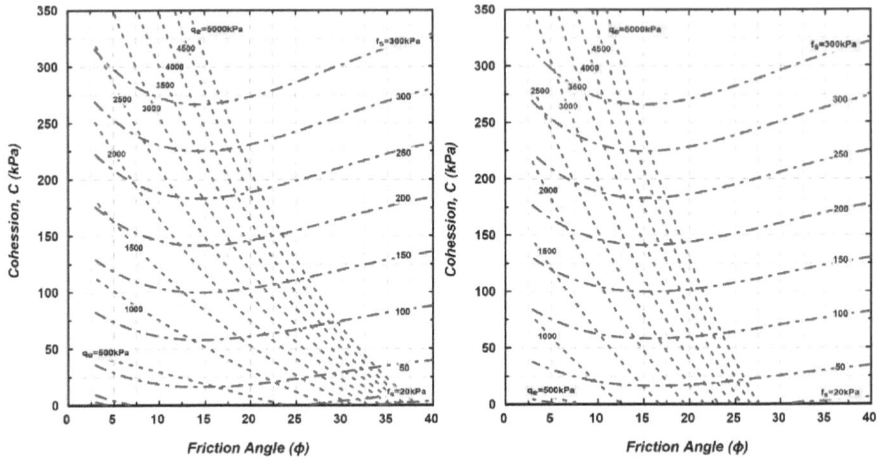

Figure 4.5 Graphic estimation of C and ϕ, values by Eslami and Mohammadi (2016) ($\sigma_v^{'}$ = 50kPa – left graph, $\sigma_v^{'}$ = 200kPa – right graph).

4.4 Undrained shear strength

Undrained shear strength (S_u) of a clayey soil, apart from classification of the soil profile, is one of the most common fine grained soil's properties derived from CPT. S_u cannot be directly measured by the CPT test, yet it can be derived by the aid of empirical correlations and needs to be calibrated by laboratory tests such as direct shear or triaxial compression tests. Therefore, several studies have investigated the S_u–CPT parameters relationship. Some important correlations have been enlisted in Table 4.5.

Table 4.5 Proposed correlations for cohesion of soils.

Reference	Correlations	Remarks
Lunne et al. (1997)	$S_u = \frac{(q_c - \sigma_v)}{N_c}$	Nc is called the cone factor
Kulhawy and Mayne (1990)	$S_u = \frac{\Delta u}{N_{\Delta u}}$	$\Delta u = u_2 - u_0$ = excess pore pressure measured at u_2 position and $N_{\Delta u}$ = pore pressure cone factor $N_{\Delta u} = Bq\ N_{kt}$ $N_{\Delta u}$ varies between 4 and 10
Naeini and Moayed (2007)	$\frac{S_u}{\sigma'_v} = 0.107 + 0.111 q_{cn1}$	q_{cn1}: normalized cone tip resistance; FC<30%
Rémai (2013)	The same as Kulhawy and Mayne (1990) method	$N_{\Delta u} = 24.3$ Bq

4.5 Over consolidation ratio

In geotechnical practice, to state the stress history of a soil deposit, OCR is introduced as follows:

$$Over\ consolidation\ ratio(OCR) = \frac{maximum\ past\ effective\ vertical\ stress}{current\ effective\ vertical\ stress} = \frac{\sigma'_p}{\sigma'_v} \tag{4.13}$$

Value of σ'_p can be determined from odometer test. However, conducting odometer test especially the sophisticated ones with stress path approach is time consuming, and the test results can be influenced by several factors such as sample disturbance and interpretation method employed for determining the maximum past consolidation pressure. Hence, CPTu is suggested to be carried out as an economic and efficient in situ testing method.

There are several existing methods for evaluating the OCR from CPT (Lunne et al., 1997; Mayne, 1991; Trevor and Mayne, 2004), most of which involve some empirical parameters. Table 4.6 summarizes some these important correlations.

4.6 Stiffness

Soil modulus is a complex parameter in foundation engineering due to different involved conditions. Young's modulus (E) parameter is usually introduced as the deformability parameter for soil relates to linear elastic behavior for geomaterial.

Table 4.6 Proposed correlations for OCR.

Reference	Correlations	Remarks
Mayne and Kemper (1988)	$OCR = 0.37\left(\frac{q_c-\sigma_v}{\sigma'_v}\right)^{1.01}$	
Trevor and Mayne (2004)	$OCR = 2(0.029 + 0.409M)\left[\frac{1}{1.95M+1}\left(\frac{q_t-u_2}{\sigma'_v}\right)\right]^{\frac{1}{\theta}}$	$\theta = 0.8 - 0.9M = \frac{6 \sin \varphi'}{3-\sin \varphi'}$
Mayne (2007)	$\sigma'_p = k(q_t - \sigma_v)$	k: Pre consolidation cone factor with an expected range of 0.2–0.5
Robertson (2009)	$OCR = 0.25Q_t^{1.25}$	
Robertson (2012)	$OCR = (2.625 + 1.75logFr)^{-1.25}Q_t^{1.25}$	
Chanmee et al. (2017)	$OCR = k\left(\frac{q_t-\sigma_v}{OCR\sigma'_v}\right)$	Under consolidated deposits k = 0.14-0.4

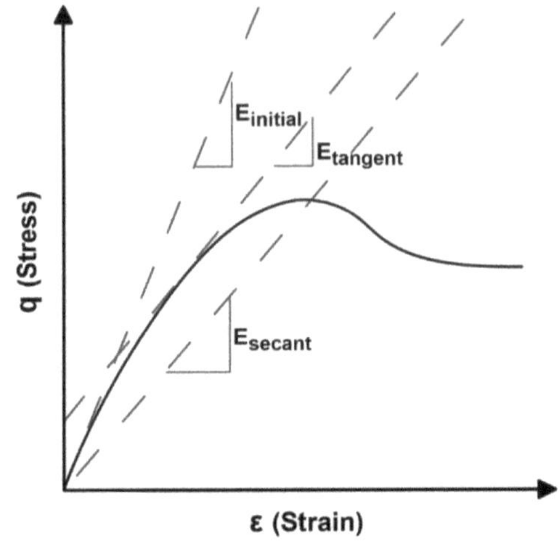

Figure 4.6 q−ε curve acquired from a triaxial test.

Because geomaterial is not an ideal linear elastic material, the soil modulus is not equal to the slope of the stress−strain (q−ε) curve. This indicates that the soil modulus is stress dependent. As presented in Fig. 4.6, q−ε behavior in a triaxial test, many soil moduli can be obtained.

When drawing a slope from the origin to an arbitrary point in Fig. 4.7, a secant slope is secant modulus *Es*. This is the modulus that can be used to predict the settlement of a

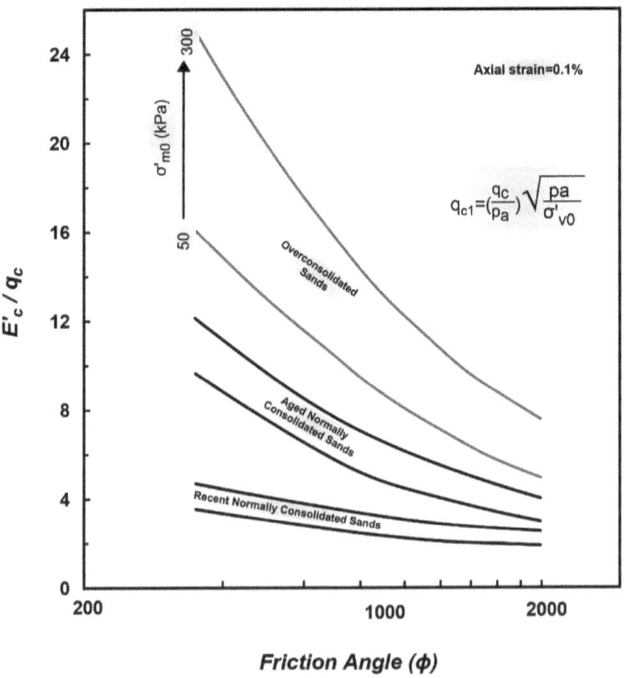

Figure 4.7 Drained Young's modulus from CPT for silica sands (Bellotti et al., 1989).

Table 4.7 Drained elastic modulus for soils using CPT records (Eslami and Sekhavatian, 2014)

Soil		q_c (MPa)	E_t From CPT
Sand	Loose	2.5–5	$(2-4)q_c$
	Medium dense	5–10	$(3-7)q_c$
	Dense	10–20	$(6-10)q_c$
Clay	Soft	2–5	$(1-2)q_c$
	medium	5–10	$(2-4)q_c$
	Hard	10–20	$(3-8)q_c$

footing which is loaded for the first time on a normally consolidated deposit. The tangent slope E_t relates to the tangent modulus and can be used to determine incremental movements when an incremental load is applied. Approximate relationships between *CPT* and elastic modulus for different soils are also enlisted in Table 4.7.

As Baldi et al. (1989) stated, although the fine grained soil's stiffness varies by many features including the grain size, mineralogy, particles shape, grain fabric, stress–strain behavior, mean effective stress, drainage conditions, etc., in a given soil, q_c is primarily controlled by the void ratio/relative density and the state of the effective stress. Based on a large number of carried out in situ and in calibration chamber tests, proposed correlations have been shown in Fig. 4.7 to obtain the drained Young's modulus of silica sands regarding q_c (Bellotti et al., 1989).

Robertson (2009) provided the following relations for the Young's modulus based on the shear modulus prediction procedure and specific Poisson's ratio and stress level assumption.

$$E = \alpha_E(q_t - \sigma_v)$$

$$\alpha_E = \frac{K_E}{Q_t} \tag{4.14}$$

where K_E and α_E are number and the modulus factor that can be defined by Fig. 4.8.

Engels (2016) has presented a correlation based on Robertson (2009) assumptions to calculate E as following:

$$E = 0.047\left[1 - \left(\frac{q}{q_{ult}}\right)^{0.3}\right]\left[10^{(0.55I_C+1.68)}\right](q_t - \sigma'_v) \tag{4.15}$$

$$I_c = \left[(3.47 - logq_t)^2 + (logFr + 1.22)^2\right]^2 \tag{4.16}$$

where q_{ult} stands for ultimate bearing capacity, and Ic is soil behavior type index which can be defined by Eq. (4.16).

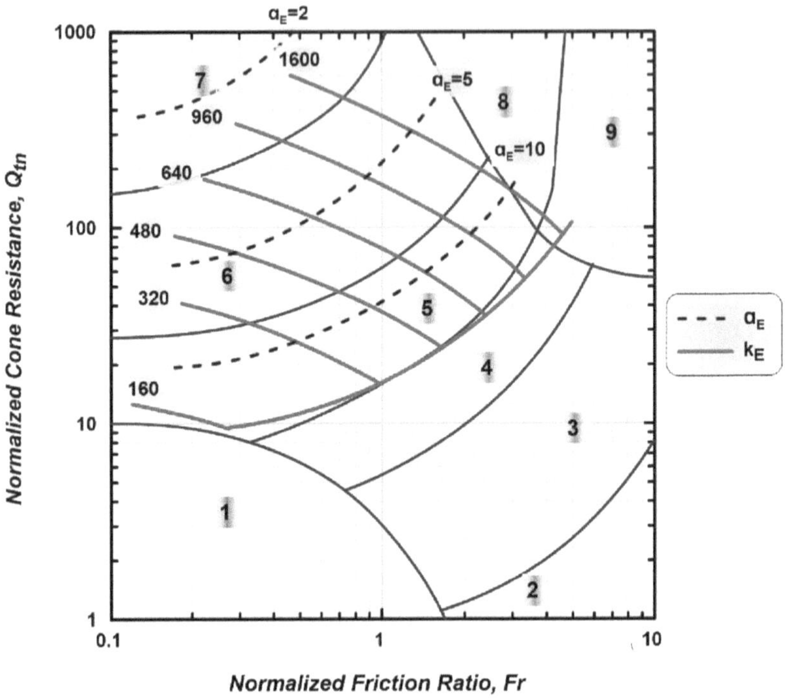

Figure 4.8 Contours of α_E and K_E on the SBC chart (Robertson, 2009).

4.7 Constrained modulus

Lunne et al. (1997) have pointed out that the most correlations between penetrometer test results and the drained constrained modulus refer to the tangent modulus as found from oedometer tests. The correlations are typically represented as follows:

$$M = \alpha q_c \tag{4.17}$$

where M is the constrained modulus and α is the constrained modulus factor, a constant Robertson and Campanella (1983), summarizing the work of Lunne and Kleven (1981) who reviewed calibration chamber results of different authors, commented that the results indicate an α of 3 should provide the most conservative estimate of 1-D settlement and that the choice of α value depends on judgment and experience(for normal and over consolidated soils, $\alpha = 5$, CH, MH, and ML; $\alpha = 6$, CL and OL; $\alpha = 1.5$, over consolidated OH with water content more than 100% $\alpha = 4$, CH, MH, CL, and ML; $\alpha = 2$, CL and ML with $q_c > 2$).

To calculate modulus for higher stress ranges, Lunne and Christophersen (1983) recommended Janbu's (1963) formulation:

$$M = M_0 \sqrt{\frac{\sigma'_v + \frac{\Delta\sigma_v}{2}}{\sigma'_v}} \tag{4.18}$$

where $\Delta\sigma_v$ is the additional stress above the initial stress

Robertson and Cabal (2014) suggested the following formulation:

$$M = \alpha_M(q_t - \sigma_v) \tag{4.19}$$

When Ic < 2.2 (coarse-grained soils):

$$\alpha_M = 0.0188\left[10^{(0.55I_C+1.68)}\right] \tag{4.20}$$

Constrained modulus of soft to firm clays is commonly assessed using the consolidation test:

$$M = {}^{1}/_{m_v} \tag{4.21}$$

where m_v is the coefficient of volume compressibility.

4.8 Shear wave velocity and shear modulus at small strain

Shear wave velocity (V_s) induced shear modulus known as a key geotechnical property corresponded to small strain which is important in earthquake investigations. Because of limitations in obtaining undisturbed sample especially in granular soils, in situ seismic tests instead of laboratory measurements are directly ideal tests to achieve the V_s. Performing surface wave velocity assessing techniques as well as down-hole and cross-hole techniques, a V_s profile can be established.

Several correlations were suggested for the estimation of the V_s based on CPT records for loose sands, silts, clays, and other soil types. The main variables as input parameters are q_c, f_s, and σ'_V in the correlation functions linear, logarithm, and power form. Some of the important correlations are presented in Table 4.8.

Small strain (<0.001%) shear modulus, G_0, is generally recognized the most appropriate way of assessing G_0 determined by shear wave velocity ($G_0 = \rho V_s^2$). However, it is generally the practice to use penetration tests to assess G_0 using empirical correlations unless project-specific detailed investigations are carried out. In Table 4.9, some of the CPT-based correlations for G_0 prediction are enlisted.

4.9 Sensitivity

Sensitivity (S_t) can be defined as the ratio of peak undisturbed shear strength to remolded shear strength. Many researchers have found that skin friction, f_s, has impact on shear strength of remolded samples.

Table 4.8 Empirical correlations between V_s and CPT data.

Reference	Proposed correlation (ms-1)	Soil type	Units of parameters	
			q_c	f_s
Hegazy and Mayne (1995)	$V_s = 12.02(q_c)^{0.319}(f_s)^{-0.0466}$	Sand	kPa	kPa
Mayne and Rix (1995)	$V_s = 1.75(q_c)0.627$	Clay	kPa	kPa
Iyisan (1996)	(1) $V_s = 55.3(q_c)^{0.377}$	Clay	kgf/cm2	
	(2) $V_s = 218 + 0.70\ q_c$	Sand	kgf/cm2	
Tun (2003)	$V_s = 109.29 + 52.674\text{Ln}(q_c)$	All	MPa	MPa
Madiai and Simoni (2004)	(1) $V_s = 211(q_c)^{0.23}$ (2) $V_s = 155(q_c)^{0.29}(f_s)^{-0.10}$	All	MPa	MPa
Mayne (2006)	$V_s = 18.5 + 118.8\log(f_s)$	All		kPa
Paoletti et al. (2010)	$V_s = 50[(q_c/pa)0.43-3]$	Sand	kPa	
MolaAbasi et al. (2015)	$V_s = 100[1.36 - 0.35f_s + 0.15q_c - 0.05f_s^2 - 0.018q_c^2 + 0.39f_s\ q_c]$	Clay	MPa	MPa
	$V_s = 100[1.73 + 2.74f_s + 0.03q_c - 4.015f_s^2 - 0.00026q_c^2 + 0.007f_s q_c]$	Sand	MPa	MPa
	$V_s = 100[1.47 + 2.07f_s + 0.10q_c + 9.50f_s^2 - 0.0023q_c^2 - 0.034f_s\ q_c]$	Mixed	MPa	MPa
	$V_s = 100[1.40 + 1.59f_s + 0.09q_c - 1.33f_s^2 - 0.002q_c^2 + 0.05f_s\ q_c]$	All	MPa	MPa

Table 4.9 Empirical correlations for G_0 based on q_c.

Reference	Correlations	Remarks
Mayne and Rix (1995)	$G_{max} = 99.5pa^{0.305}(qc)^{\frac{0.695}{e_0^{1.13}}}$	Cohesive soils, e_0 = initial void ratio
Eslaamizaad and Robertson (1997)	$\left(\frac{G_0}{q_c}\right) = 1634\left(\frac{q_c}{\sqrt{\sigma'_v}}\right)^{-0.75}$	Cohesionless soils
Schnaid (2009)	$G_0 = b(q_c\sigma'_v Pa)^{0.3}$	Cohesionless soils, b = 280 and 110 for upper and lower bond.

Units (kPa).

$$S_t = \frac{S_u}{S_{u(remolded)}} = \frac{q_t - \sigma_v}{N_{kt}}\left(\frac{1}{fs}\right) \tag{4.22}$$

Schmertmann (1978) proposal was based on this and could be simply written as follows:

$$S_t = \frac{N_s}{R_f} \tag{4.23}$$

where *Ns* is a constant, and R_f is the friction ratio.

Although Schmertmann (1978) suggested values for *Ns* which were based on a mechanical cone, they are not applicable to electric cones including piezocones. While several proposals have been put forward, Lunne et al.'s (1997) suggestion of an average value of 5 with 6–9 likely range appears practical and reasonable. Fig. 4.9 shows the sensitivities of clay versus depth related to Urmiyeh Lake Causeway piling projects.

4.10 *CPT* correlations with *SPT*

The simplicity and common usage of *SPT* around the world and the accuracy and continuity of *CPT* led the researchers to correlate the results of these two tests to be used in geotechnical designs. Relationship between *CPT* and *SPT* is generally

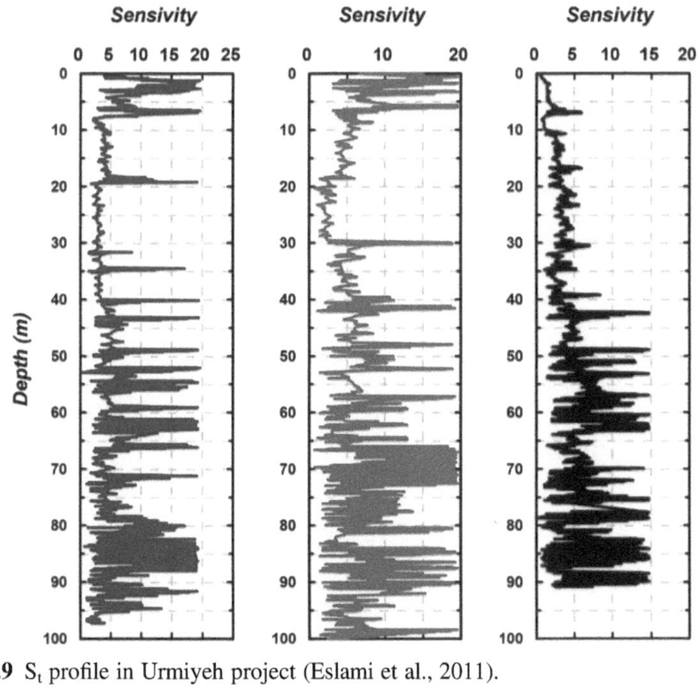

Figure 4.9 S_t profile in Urmiyeh project (Eslami et al., 2011).

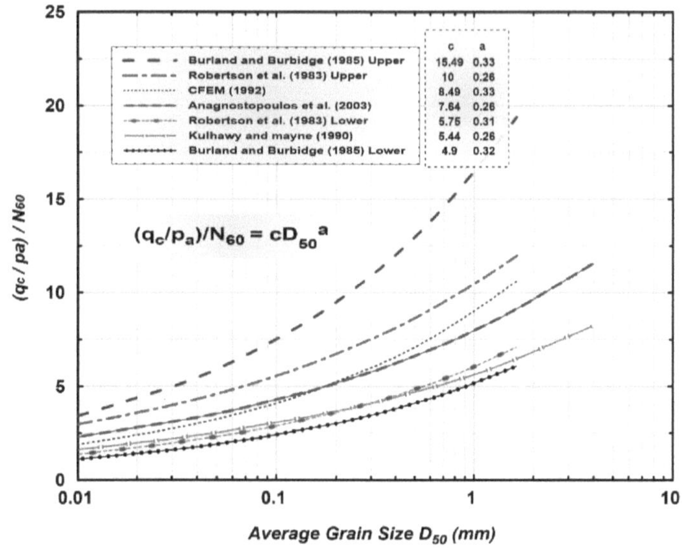

Figure 4.10 The graph determining the relation between q_c and N_{60} based on mean grain size of soil (D_{50}) (Robertson and Campanella, 1983).

specified by q_c/N_{SPT} ratio (q_c: cone tip resistance; N_{SPT}: *SPT blow counts*). The relationship between q_c/N_{SPT} ratio along mean grain size of soil particle D_{50}(mm) in power form as Eq. (4.24). In Eq. 4.24, a and c coefficients are constant and several researchers proposed these coefficients as illustrated in Fig. 4.10. For instance, Kulhawy and Mayne (1990) equation has the coefficients c = 5.44 and a = 0.26.

$$q_c / N_{SPT} = cD_{50}{}^{a} \tag{4.24}$$

To use the above figure, the median grain size (D_{50}) evaluation from grain size distribution test is necessary. Lunne et al. (1997) proposed a method for the estimation of N_{60} values directly from CPTu results without measuring D_{50}. They proposed the use of Ic to correlate CPT and SPT as the following relationship.

$$\frac{qc/pa}{N_{60}} = 8.5\left(1 - \frac{I_c}{4.6}\right) \tag{4.25}$$

4.11 Permeability

Robertson (1990) provided a correlation between permeability (k) and the soil behavior type for CPT as shown in Table 4.10. Robertson and Cabal (2010) presented the average permeability (k_{ave}) as shown in Eq. (4.26):

$$\begin{cases} k = 10^{(0.952-3.04I_c} & 1 < Ic \leq 3.27 \\ k = 10^{(-4.52-1.37I_c} & 3.27 < Ic \leq 4 \end{cases} \tag{4.26}$$

Table 4.10 Permeability from CPT results (Robertson, 1990).

Soil behavior type	Soil permeability (m/s)
Sensitive fine grained	3×10^{-9} to 3×10^{-8}
Organic soils–peats	1×10^{-8} to 1×10^{-6}
Clays—clay to silty clay	1×10^{-10} to 1×10^{-7}
Silt mixtures; clayey silt to silty clay	3×10^{-9} to 1×10^{-7}
Sand mixtures; silty sand to sandy silt	1×10^{-7} to 1×10^{-5}
Sands; clean sands to silty sands	1×10^{-5} to 1×10^{-3}
Gravelly sand to sand	1×10^{-3} to 1
[a] Very stiff sand to clayey sand	1×10^{-8} to 1×10^{-6}
[a] Very stiff fine grained	1×10^{-9} to 1×10^{-7}

4.12 Liquefaction

In the field of liquefaction potential approximation, experimental models based on case histories have remained as most popular methods in the past decades. Fig. 4.11 signifies the experimental approaches for liquefaction prediction.

In the context of the field test data analyses, the liquefaction potential prediction obtained by empirical correlations can be generally categorized as deterministic and probabilistic approaches. Recently, innovative soft computing techniques including artificial intelligence and metaheuristic algorithms have used to determine

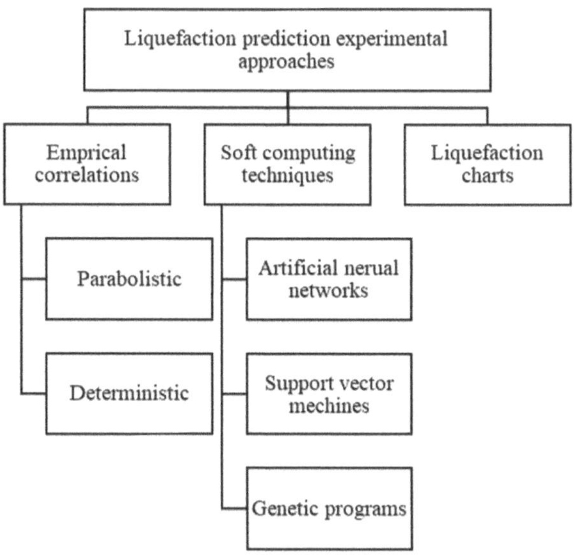

Figure 4.11 Experimental approaches for liquefaction prediction.

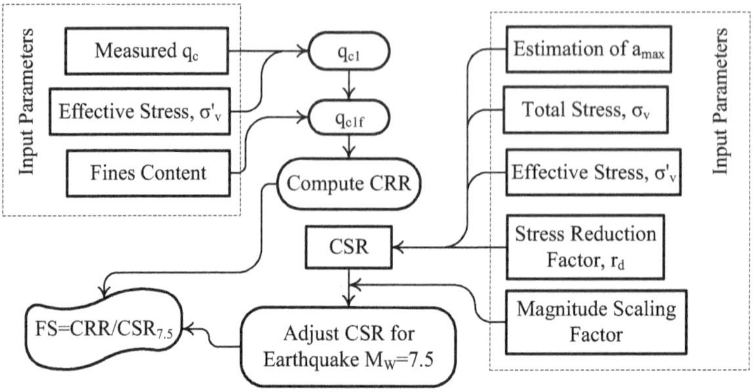

Figure 4.12 General procedure of simplified approaches based on CPT data (Eslami et al., 2014).

liquefaction-induced lateral spreading. The support vector machine, which combines the standards of structural risk minimization and the statistical learning theory, has been applied in solving liquefaction classification. In the latter approach namely "liquefaction charts," the possibility of liquefaction occurrence is distinguished considering parameters on chart sides. This method is mostly practical for CPT data. Robertson and Wride (1998) and Eslami and Fellenius (2004), Robertson (2010), and MolaAbasi et al. (2018) have presented standardized charts for determination of liquefaction susceptibility from CPT data. These charts, which are borne out of CPT data of liquefied and nonliquefied case histories, delineate different soil types regarding liquefaction potential.

The Seed and Idriss (1971) procedure is viewed as basic deterministic method that follows the classical format of liquefaction analyses. As conceptually, this main method correlates the safety factor (i.e., FS) for liquefaction susceptibility from input parameters through a stepwise procedure as illustrated in Fig. 4.12. If the value of FS for a particular case is less than 1, the occurrence of liquefaction is possible, otherwise it is considered as a nonliquefiable case.

The FS procedure to approximate liquefaction potential could be enlisted as shown in the following steps.

4.12.1 Step 1. CSR—Cyclic Stress Ratio computations

As exemplified in Figure 9.2, Fs is defined as $CRR/CSR_{7.5}$, where $CSR_{7.5}$ stands for cyclic stress ratio (CSR) adjusted for earthquake with $M_W = 7.5$ ($M_W =$ earthquake moment magnitude) and CRR is cyclic resistance ratio. Magnitude scaling factor (MSF) is implemented to determine $CSR_{7.5}$ from CSR calculated as follows:

$$F_s = \frac{CRR}{CSR_{7.5}}$$

$$CSR_{Mw=7.5;K_\sigma=1atm} = \frac{CSR}{MSF.K_\sigma} \tag{4.27}$$

$$CSR = 0.65\left(\frac{a_{\max}}{g}\right) \times \left(\frac{\sigma_v}{\sigma'_v}\right) \times r_d \tag{4.28}$$

a_{max} = maximum ground surface acceleration, σ_v = total vertical stress, σ'_v = effective vertical stress, $K\sigma$ = overburden factor, and r_d = stress reduction factor.

4.12.2 Step 2 overburden stress tip resistance correction

CPT penetration resistance, q_c, is initially corrected for overburden stress effects using an equivalent effective vertical stress (σ'_v) of 1 atmosphere (100 kPa) and an overburden correction factor, Cq, as part of the semiempirical procedure The measured CPT tip resistance has to be corrected for overburden pressure. This is done as follows:

$$q_{CN} = C_N q_C \tag{4.29}$$

4.12.3 Step 3. Fines Content correction of tip resistance

The $CRR_{7.5}$ liquefaction curves for CPT are based on clean sand. Therefore, the tip resistance values of soil containing fines have to be increased to take into account the higher liquefaction resistance. In this regard, recommended design (RD) is used for correction of q_{c1} for fines content, FC, by using this formula:

$$q_{cNf} = q_{cN} + \Delta q_{cN} \tag{4.30}$$

4.12.4 Step 4. CRR—Cyclic Resistance Ratio

Calculate the CRR at $\sigma'_v = 1$ atm

In Table 4.11, parameters r_d, MSF, C_N, Δq_{cN}, and CRR involving in liquefaction prediction, which are suggested by renowned reference namely Idriss and Boulanger (2006), have been presented.

Example 4.1. Calculating soil parameters for the CPT profile shown in Fig. 4.13, calculate soil parameters.

As can be realized from Fig. 4.13, the CPT profile includes two separated soil profile, a sandy soil from surface down to the depth of 7.34 m underlain by fine grained soil.

4.12.5 Section one. Sandy soil

Step 1. As presented by Eslami and Fellenius (1997), to filter the CPTu records, geometric average of records (q_c, q_t, and f_s) are considered (Table 4.12)

Step 2. To define soil parameters σ_v and σ'_v are needed. Based on previous empirical correlations γ and γ_{sat} parameters are calculated as presented in Table 4.13 and Eq. (4.31) demonstrates the σ_v and σ'_v determination process

Table 4.11 Calculation of correction factors: r_d, MSF, C_N, Δq_{c1}, and CRR (Idriss and Boulanger, 2006).

	Factor	Idriss and Boulanger (2006)
1	r_d	$r_d = e^{(\alpha_{(z)} + (\beta_{(z)} M)}$ $\Delta\alpha_{(Z)} = -1.012 - 1.126 \sin\left(\frac{Z}{11.73} + 5.133\right)$ $\beta_{(Z)} = 0.106 - 0.118 \sin\left(\frac{Z}{11.28} + 5.142\right)$
2	MSF	$MSF = 1 + (MSF_{max} - 1)\left[8.64 \exp\left(\frac{-M}{4}\right) - 1.325\right]$ $MSF_{max} = 1.09 + \left(\frac{q_{c1Ncs}}{180}\right)^3 \leq 2.2$
3	C_N	$C_N = \left(\frac{P_a}{\sigma'_v}\right)^{1.338 - 0.249(q_{cN})^{0.264}} \leq 1.7$ Note: Because q_{c1} is required to compute q_{c1} (on which C_N depends), iteration is required.
4	Δq_{cN}	Δq_{c1}
5	$K\sigma$	$K_\sigma = 1 - C_\sigma \ln\left(\frac{\sigma''_v}{P_a}\right) \leq 1.1$ $C_\sigma = \frac{1}{37.3 + 8.27(q_{cNf})^{0.264}} \leq 0.3$
	CRR	$CRR_{Mw=7.5, \sigma'_v = 1\ atm} =$ $\exp\left[\frac{q_{cNf}}{540} + \left(\frac{q_{cNf}}{67}\right)^2 - \left(\frac{q_{cNf}}{80}\right)^3 + \left(\frac{q_{cNcsf}}{114}\right)^4 - 3\right]$

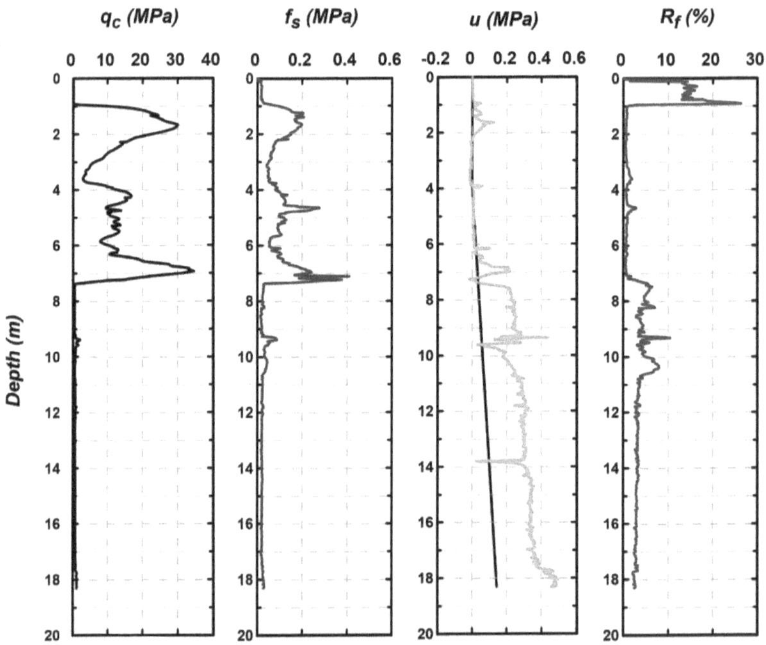

Figure 4.13 CPTu profile for Example 4.1.
Data from Eslami A (1996) Bearing Capacity of Piles from Cone Penetrometer Test Data, (Ph.D. thesis). University of Ottawa, Department of Civil Engineering, 516p.

Table 4.12 CPTu main parameters from Fig. 4.13 up to depth = 7.34m.

Parameters	value	Parameters	value
q_c (Geometric mean)	8.21 MPa	q_t (Geo mean)	8.23 MPa
f_s (Geometric mean)	23.8 kPa	R_f	1.12%
u_2 (Average)	23.8 kPa	u_{0}-—average	8.21 kPa

Table 4.13 γ and γ_{sat} calculations using information in Table 4.12

Parameters	Reference	Calculation
γ	Robertson and Cabal (2010)	$\frac{\gamma}{\gamma_w} = 0.27[log1.12] + 0.36\left[\log\left(\frac{8230}{100}\right)\right] + 1.236;$ $\gamma = 19\ kN/m^3$
γ_{sat}	Lengkeek et al. (2018)	$\gamma_{sat} = 19 - 4.12\frac{\log\left(\frac{5}{8.23}\right)}{\log\left(\frac{30}{1.12}\right)} = 19.6\ kN/m^3$

$$Dw = 4m; \begin{cases} \sigma_{v-Ave} = \dfrac{(4 \times 19 + 3.34 \times 19.6)}{2} = 70.8\ kPa \\ \sigma'_{v-Ave} = \dfrac{(4 \times 19 + 3.34 \times (19.6 - 9.8)}{2} = 54.4\ kPa \end{cases} \tag{4.31}$$

Step3. In this step, two main CPT parameters to measure soil properties are defined as follows

$$Q_t = \frac{q_t - \sigma_v}{\sigma'_v} = \frac{8230 - 70.8}{54.4} = 150\ kPa \tag{4.32}$$

$$B_q = \frac{u_2 - u_0}{q_t - \sigma_v} = \frac{23.8 - 8.21}{8230 - 70.8} = 0.0019 \tag{4.33}$$

Step 4. Geotechnical soil parameter related to foundation engineering is calculated in Table 4.14

4.12.6 Section two: fine grained soil

Step 1. Filtering CPT records

Step 2. Calculating V_s, σ_v, and σ'_v:

Table 4.14 Geotechnical soil parameters.

Parameters	Reference	Calculation
Dr	Mayne (2007)	$Dr = 100\left[0.268\ \ln\left(\frac{\frac{8230}{100}}{\sqrt{\frac{54.4}{100}}}\right) - 0.675\right] = 58.9\%$
φ'	Uzielli et al., 2013	$\varphi = 25\left(\frac{8210}{\sqrt{100\times 54.4}}\right)^{0.1} = 40$ degree
E	Robertson 2009	Based on Figure 4.8 $K_E = 2$; $a_E = \frac{2}{0.15} = 4$; $E = 4(8230 - 70.4) = 32638\ kP_\alpha$
M	Lunne et al. (1997)	$E = 2 \times 8.23 = 16.46\ MP_\alpha$
V_s	Mola-Abasi et al., (2015) Sand	$V_s = 100[1.36\text{-}0.35 \times 0.024 + 0.15 \times 8.21 - 0.05\times 0.024^2 - 0.018 \times 8.21^2 + 0.39 \times 0.024 \times 8.21] = 183$ m/s

$$\begin{cases} \gamma_{sat} = 19 - 4.12\dfrac{\log\left(\dfrac{5}{0.96}\right)}{\log\left(\dfrac{30}{3.47}\right)} = 15.9\ kN/m^3 \\ \sigma_{v-Ave} = \dfrac{(141.6 + 10.96 \times 15.9)}{2} = 157.9\ kPa \\ \sigma'_{v-Ave} = \dfrac{(108.8 + 10.96 \times (15.9 - 9.8))}{2} = 87.8\ kPa \end{cases} \tag{4.34}$$

Step3. Calculating q_t and B_q

$$Q_t = \frac{q_t - \sigma_v}{\sigma'_v} = \frac{962 - 157.9}{87.8} = 9.16\ kPa \tag{4.35}$$

$$B_q = \frac{u_2 - u_0}{q_t - \sigma_v} = \frac{301 - 80.9}{962 - 157.9} = 0.27 \tag{4.36}$$

Step 4. Calculating geotechnical soil parameters in Table 4.16

Example 4.2. Liquefaction: for the CPTu profile presented in Fig. 4.14, calculate soil liquefaction potential for an earthquake with $M_w = 6.5$ and $a_{max/g} = 0.2$.

Table 4.15 CPTu main parameters from Fig. 4.13 for depths of 7.34 m below.

Parameters	value	Parameters	value
q_c (Geo mean)	0.67 MPa	q_t(Geo mean)	0.96 MPa
f_s (Geo mean)	23.1 kPa	R_f	3.47%
u_2(Average)	297 kPa	u_{0}-Average	89.2 kPa

Table 4.16 Geotechnical soil parameters.

Parameters	Reference	Calculation
φ'	Mayne (2014)	$\varphi' = 29.5 \times 0.27^{0.121}[0.256 + 0.33 \times 0.27 + \log 9.16] = 33$
S_u	Rémai (2013)	$S_u = \frac{301-80.9}{24.3\times0.27} = 33.1\ kP\alpha$
OCR	Mayne and Kemper 1988	$OCR = 0.37\left(\frac{960-158}{87.8}\right)^{1.01} = 3.47$
E	Robertson 2009	Based on Figure 4.9 $K_E = 2; \alpha_E = \frac{2}{0.09} = 22.2;$ $E = 22.2(960 - 157.9) = 17806\ kP\alpha$
M	Lunne et al. (1997)	$M = 1.5 \times 0.96 = 1.44\ MP\alpha$
V_s	MolaAbasi et al., (2015) clay	$V_s = 100[1.73 + 2.74 \times 0.023 + 0.03 \times 0.67 - 4.015 \times 0.023^2 - 0.00026 \times 0.67^2 + 0.007 \times 0.023 \times 0.67] = 145$m/s

According to Fig. 4.14, two layers can be distinguished. One down to the depth of 9 m representing a loose sand with liquefaction potential underlain by a dense sand layer.

In the first step, V_s, σ_v, σ'_v, and r_d are determined (Fig. 4.15).

Considering Eqs. (4.27) and (4.28), CSR and CSR $_{mw=7.5,K\sigma=1atm}$ are calculated in Fig. 4.16. And the final calculation of safety factor is presented in Fig. 4.17.

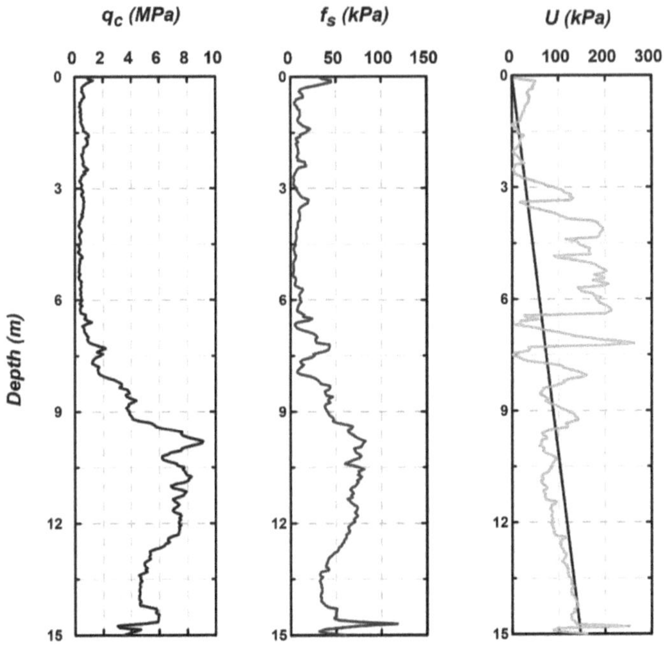

Figure 4.14 CPTu profile for Example 4.2.

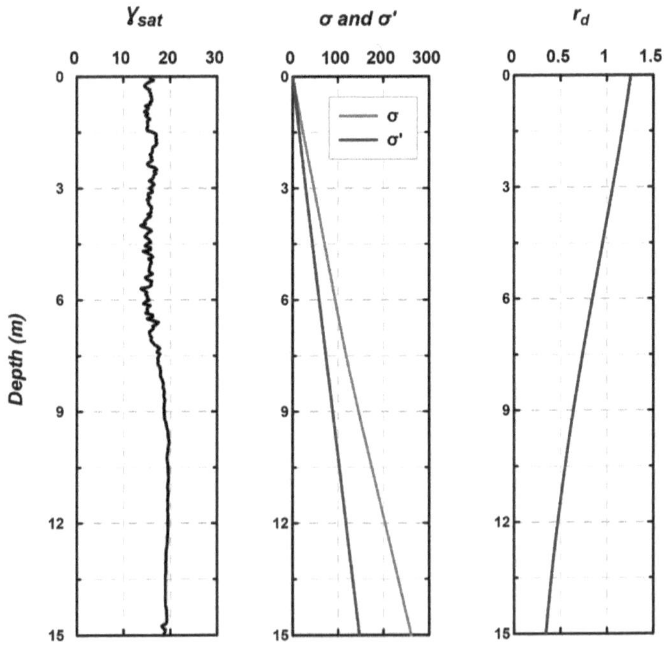

Figure 4.15 γ_{sat}, σ_v, σ'_v, and r_d parameters versus depth.

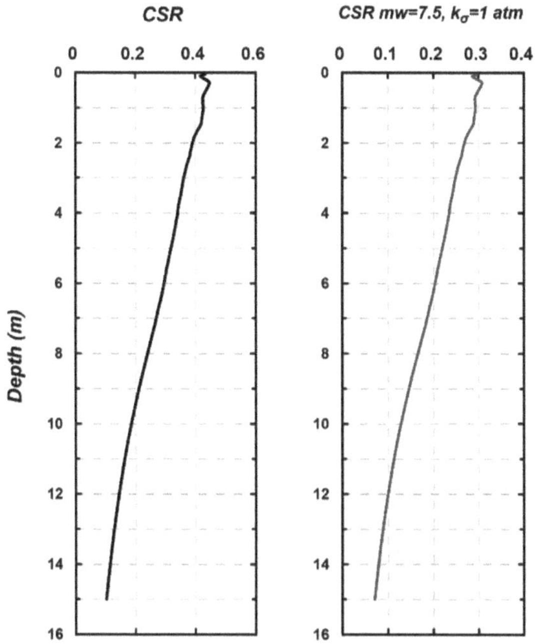

Figure 4.16 CSR and CSR $_{mw=7.5,K\sigma=1atm}$ parameters versus depth.

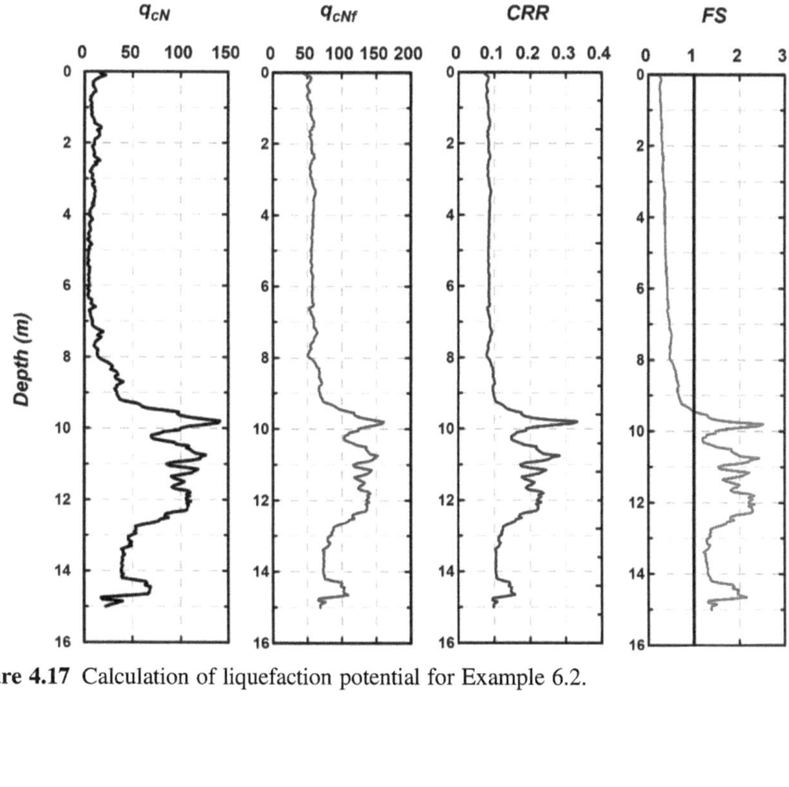

Figure 4.17 Calculation of liquefaction potential for Example 6.2.

References

Baldi, G., Bellotti, R., Ghionna, V., Jamiolkowski, M., Pasqualini, E., 1986. Interpretation of CPT's and CPTU's. 2nd Part: drained penetration. In: Proceedings 4th International Geotechnical Seminar, Singapore, pp. 143–156.

Baldi, G., Jamiolkowski, M., Lo Presti, D.C.F., Manfredini, G., Rix, G.J., 1989. Italian experience in assessing shear wave velocity from CPT and SPT. In: Proceedings of Discussion Session on Influence of Local Conditions on Seismic Response, vol. XII. ICSMFE, Rio De Janerio, pp. 157–168.

Bagińska, I., 2016. Estimating and verifying soil unit weight determined on the basis of SCPTu tests. Annals of Warsaw University of Life Sciences–SGGW. Land Reclamation 48 (3), 233–242.

Bellotti, R., Ghionna, V., Jamiolkowski, M., Robertson, P.K., Peterson, R.W., 1989. Interpretation of moduli from self-boring pressure meter tests in sand. Géotechnique 39 (2), 269–292.

Chanmee, N., Chai, J., Hino, T., Wang, J., 2017. Methods for evaluating overconsolidation ratio from piezocone sounding results. Underground Space 2 (3), 182–194.

Engels, S., 2016. Correlating CPT Data to Stiffness Parameters of Sand in FEM. Master Thesis. Delft University of Technology.

Eslami, A., 1996. Bearing Capacity of Piles from Cone Penetrometer Test Data. Ph.D. Thesis. University of Ottawa, Department of Civil Engineering, p. 516p.

Eslami, A., Fellenius, B.H., 2004. CPT and CPTu Data for Soil Profile Interpretation: Review of Methods and a Proposed New Approach.

Eslami, A., Aflaki, E., Hosseini, B., 2011. Evaluating CPT and CPTu based pile bearing capacity estimation methods using Urmiyeh Lake Causeway piling records. Scientia Iranica 18 (5), 1009–1019.

Eslami, A., Mola-Abasi, H., Shourijeh, P.T., 2014. A polynomial model for predicting liquefaction potential from cone penetration test data. Scientia Iranica. Transaction A, Civil Engineering 21 (1), 44.

Eslami, A., Mohammadi, A., 2016. Drained soil shear strength parameters from CPTu data for marine deposits by analytical model. Ships and Offshore Structures 11 (8), 913–925.

Eslaamizaad, S., Robertson, P.K., 1997. Evaluation of settlement of footings on sand from seismic in-situ tests. In: Proceedings of the 50th Canadian Geotechnical Conference, Ottawa, Ontario, vol. 2, pp. 755–764.

Hegazy, Y.A., Mayne, P.W., 1995. Statistical correlations between vs and cone penetration data for different soil types. In: Proceedings of the International Symposium on Cone Penetration Testing (CPT'95), Linkoping, Sweden, 4–5 October 1995, vol. 2. Swedish Geotechnical Society, pp. 173–178.

Idriss, I.M., Boulanger, R.W., 2006. Semi-empirical procedures for evaluating liquefaction potential during earthquakes. Soil dynamics and earthquake engineering 26 (2–4), 115–130.

Iyisan, R., 1996. Correlation between Shear Wave Velocity and In-Situ Penetration Test Results. Digest, pp. 371–374.

Jamiolkowski, M., Lo Presti, D.C.F., Manassero, M., 2001. Evaluation of relative density and shear strength of sands from CPT and DMT. In: CC Ladd Symposium. October, MIT, Cambridge, Mass.

Janbu, N., 1963. Soil compressibility as determined by oedometer and triaxial tests. In: European Conference on Soil Mechanics and Foundation Engineering, Wiesbaden, vol. 1, pp. 19–25, 2, pp. 17–21.

Kulhawy, F.H., Mayne, P.W., 1990. Manual on Estimating Soil Properties for Foundation Design. Report EL- 6800 submitted to Electric Power Research Institute, Palo Alto, California, 306 pp..

Lengkeek, H.J., de Greef, J., Joosten, S., Hicks, M.A., Pisano, F., Peuchen, J., 2018. CPT based unit weight estimation extended to soft organic soils and peat. In: Cone Penetration Testing 2018: Proceedings of the 4th International Symposium on Cone Penetration Testing (CPT'18), 21–22 June, 2018. CRC Press, Delft, The Netherlands, p. 389.

Lunne, T., Christoffersen, H.P., 1983. Interpretation of cone penetrometer data for offshore sands. In: Proceedings of 15th Annual OTC, OTC 4464, pp. 181–188.

Lunne, T., Kleven, A., 1981. Role of CPT in North sea foundation engineering. In: Cone Penetration Testing and Experience. ASCE, pp. 76–107.

Lunne, T., Robertson, P.K., Powell, J.J.M., 1997. Cone Penetration Testing in Geotechnical Practice. Blackie Academic & Professional/Chapman-Hall Publishers, London, 312 pp.

Madiai, C., Simoni, G., 2004. Shear wave velocity-penetration resistance correlation for Holocene and Pleistocene soils of an area in central Italy. In: Viana da Fonseca, Mayne (Eds.), Proceedings ISC-2 on Geotechnical and Geophysical Site Characterization. Millpress, Rotterdam, ISBN 90 5966 009 9, pp. 1687–1694.

Mayne, P.W., 1991. Determination of OCR in clays by piezocone tests using cavity expansion and critical state concepts. Soils and Foundations 31 (2), 65–76.

Mayne, P.W., 2006. Undisturbed sand strength from seismic cone tests. Geomechanics and GeoEngineering Taylor & Francis Group 1 (4), 239–257. London.

Mayne, P.W., 2007. Cone penetration testing State-of-Practice. In: NCHRP Synthesis. Transportation Research Board Report Project 20–05, 118 pp.

Mayne, P.W., 2014. Interpretation of geotechnical parameters from seismic piezocone tests. In: 3rd International Symposium on Cone Penetration Testing. Las Vegas, Nevada, USA, 2014.

Mayne, P.W., Kemper, J.B., 1988. Profiling OCR in stiff clays by CPT and SPT. Geotechnical Testing Journal 11 (2), 139–147.

Mayne, P.W., Rix, G.J., 1995. Correlations between shear wave velocity and cone tip Resistance in natural clays. Soils and Foundations 35 (2), 107–110.

Mayne, P.W., Peuchen, J., Bouwmeester, D., 2010, May. Soil unit weight estimation from CPTs. In: Proceedings of the 2nd International Symposium on Cone Penetration Testing (CPT'10), Huntington Beach, Calif, vol. 2, pp. 169–176.

Meyerhof, G.G., 1974. Ultimate bearing capacity of footings on sand layer overlying clay. Canadian Geotechnical Journal 11 (2), 223–229.

Mola-Abasi, H., Dikmen, U., Shooshpasha, I., 2015. Prediction of shear-wave velocity from CPT data at Eskisehir (Turkey), using a polynomial model. Near Surface Geophysics 13 (2), 155–167.

MolaAbasi, H., Kordtabar, B., Kordnaeij, A., 2018. Liquefaction prediction using CPT data by triangular chart identification. International Journal of Geotechnical Engineering 12 (4), 377–382.

Motaghedi, H., Eslami, A., 2014. Analytical approach for determination of soil shear strength parameters from CPT and CPTu data. Arabian Journal for Science and Engineering 39 (6), 4363–4376.

Naeini, S.A., Ziaei, M.R., 2007. Evaluation of undrained shear strength of loose silty sands using CPT results. International Journal of Civil Engineering 5 (2), 104–117.

Paoletti, L., Hegazy, Y., Monaco, S., Piva, R., 2010. Prediction of shear wave velocity for offshore sands using CPT data – Adriatic Sea. In: Second International Symposium on Cone Penetration Testing, Huntington Beach, CA, USA, pp. 1–8.

Rémai, Z., 2013. Correlation of undrained shear strength and CPT resistance. Periodica Polytechnica: Civil Engineering 57 (1), 39–44.

Robertson, P.K., 1990. Soil classification using the cone penetration test. Canadian Geotechnical Journal 27 (1), 151–158.

Robertson, P.K., 2009. Interpretation of cone penetration tests – a unified approach. Canadian Geotechnical Journal 46 (11), 1337–1355.

Robertson, P.K., Cabal, K.L., 2010. Estimating soil unit weight from CPT. In: 2nd International Symposium on Cone Penetration Testing, CPT'10, Hungtington Beach, California.

Robertson, P.K., Campanella, R.G., 1983. Interpretation of cone penetration tests. Part I: Sand. Canadian Geotechnical Journal 20 (4), 718–733.

Robertson, P.K., Campanella, R.G., Gillespie, D., Grieg, J., 1986. Use of piezometer cone data. In: Clemence, S., Blacksburg (Eds.), Proceedings of American society of civil engineers, ASCE, In-Situ 86 Specialty Conference, pp. 1263–1280. June 23-25, Geotechnical Special Publication GSP No. 6.

Robertson, P.K., Wride, C.E., 1998. Evaluating cyclic liquefaction potential using the cone penetration test. Canadian geotechnical journal 35 (3), 442–459.

Robertson, P.K., 2012. Interpretation of in-situ tests–some insights. Mitchell Lecture-ISC 4, 1–22.

Robertson, P.K., Robertson, K.C., 2014. Guide to cone penetration testing for Geotechnical Engineering, 6ta ed. Signal Hill, California: Gregg Drilling & Testing.

Schmertmann, J.H., 1970. Static cone to compute static settlement over sand. Journal of the Soil Mechanics and Foundations Division, ASCE 96 (SM3), 1011–1043.

Schmertmann, J.H., 1975. Measurement of in-situ shear strength. In: Proceedings of ASCE Specialty Conference on In-Situ Measurements of Soil Properties, Raleigh, NC, vol. 2, pp. 57–138.

Schmertmann, J.H., 1978. Guideline for Cone Penetration Test, Performance and Design. U.S. Department of Transportation, Washington, D. C., p. 145. Rep. No. FHWA-TS-78-209.

Schnaid, F., 2009. In: Situ Testing in Goemechanics. Taylor and Francis, London, 329 pp.

Seed, H.B., Idriss, I.M., 1971. Simplified procedure for evaluating soil liquefaction potential. Journal of Soil Mechanics and Foundations Division 97 (SM9), 1249–1273.

Trevor, F.A., Mayne, P.W., 2004. Undrained shear strength and OCR of marine clays from piezocone test results. In: Viana da Fonseca, Mayne (Eds.), Proceeding ISC-2 on Geotechnical and Geophysical Sitecharacterization. Millpress, Rotterdam, pp. 391–398.

Tun, M., 2003. Investigation of the Characteristics of Eskisehir Soils Due to Shear Wave Velocity and Determination of Their Fundamental Vibration Periods. Master Thesis. Anadolu University Institute of Science and Technology Department of Physics (in Turkish).

Uzielli, M., Mayne, P.W., Cassidy, M.J., 2013. Probabilistic Assessment of Design Strengths for Sands from In-Situ Testing Data. In: Modern Geotechnical Design Codes of Practice, pp. 214–227.

Villet, W.C.B., Mitchell, J.K., 1981. Cone resistance, relative density, and friction angle. In: ASCE National Convention on Cone Penetration Testing. ASCE. American Society of Civil Engineers, St. Louis, Missouri, pp. 178–208.

Soil behavior classification (SBC) using CPT and CPTu records

5

5.1 Introduction

Visual examinations and various laboratory tests are commonly used for recognition and classification of geotechnical deposits. However, visual identification is only applicable in preliminary investigation, and laboratory testing involves a few limitations, such as sample recovery, size effect, and actual stress modeling.

Early on, information about the soil type was approximated and the cone penetrometer was limited for determining the location of soil type boundaries and not much further details were provided. The soil type had to be confirmed from the results of conventional borings, with the exception of empirical interpretations limited to the geological area where they had been developed. The measured pore pressure, tip resistance (q_c), and sleeve friction (f_s) are functions of soil type and behavior. One of the major applications of the cone penetration test (CPT) is soil type identification profiling. Consequently, several charts were proposed in the literature to classify the soil from the CPT or CPTu data (Eslami and Fellenius, 2004; Ku et al., 2010; Robertson, 2010; Cai et al., 2015).

5.2 Brief survey of soil profiling methods

Begemann (1965) pioneered soil profiling from the CPT, showing that while coarse-grained soils generally demonstrate larger values of cone resistance, q_c, and sleeve friction, f_s, as opposed to fine-grained soils, the soil type is not a strict function of either cone resistance or sleeve friction, but of the combination of the these values.

The Begemann chart was derived from tests on Dutch soils with the mechanical cone. The chart is site-specific, i.e., directly applicable only to the specific geologic locality where it was developed. For example, cone tests in sand usually show a friction ratio smaller than 1%. However, the chart is of an important general qualitative value.

Sanglerat et al. (1974) proposed the chart presenting data from an 80-mm-diameter research penetrometer. The chart plots the cone resistance (logarithmic scale) versus the friction ratio (linear scale). This manner of plotting has the apparent advantage of combining the two important parameters, the cone resistance and the friction ratio. However, plotting the cone resistance versus the friction ratio implies, falsely, that the values are independent of each other; the friction ratio would be the independent variable and

Piezocone and Cone Penetration Test (CPTu and CPT) Applications in Foundation Engineering
https://doi.org/10.1016/B978-0-08-102766-0.00005-5

the cone resistance the dependent variable. The friction ratio is the inverse of the ordinate and the values are patently not independent—the cone resistance is plotted against its own inverse self, multiplied by a variable that ranges, normally, from a low of about 0.01 through a high of about 0.15.

Plotting of a variable against itself violates a fundamental principle of data representation. Simply, while both cone resistance and sleeve friction are important soil profiling parameters, plotting one as a function of the other distorts the information.

Schmertmann (1978) proposed the soil profiling chart shown in Fig. 5.1. The chart is based on results from mechanical cone data in "North Central Florida" and incorporates Begemann's CPT data and indicates envelopes of zones of common soil type. It also presents boundaries for density of sands and consistency (undrained shear strength) of clays and silts, which are imposed by definition and not related to the soil profile interpreted from the CPT results.

Also the Schmertmann (1978) chart (Fig. 5.1) presents the cone resistance as a plot against the friction ratio, that is, the data are plotted against their inverse self. Fig. 5.2 shows the Schmertmann chart converted to a Begemann-type graph (logarithmic scales), replotting Fig. 5.1 envelopes and boundaries as well as text information. When the plotting of the data against own inverse values is removed, a visual effect comes forth that is quite different from that of Fig. 5.1. Note also that the consistency boundaries do not appear very logical when seen in this type of presentation.

Schmertmann (1978) states that the correlations shown in Fig. 5.1 may be significantly different in areas of dissimilar geology. The chart is intended for typical reference and includes two warnings: "Local correlations are preferred" and "Friction ratio values decrease in accuracy with low values of q_c." Schmertmann also mentions that

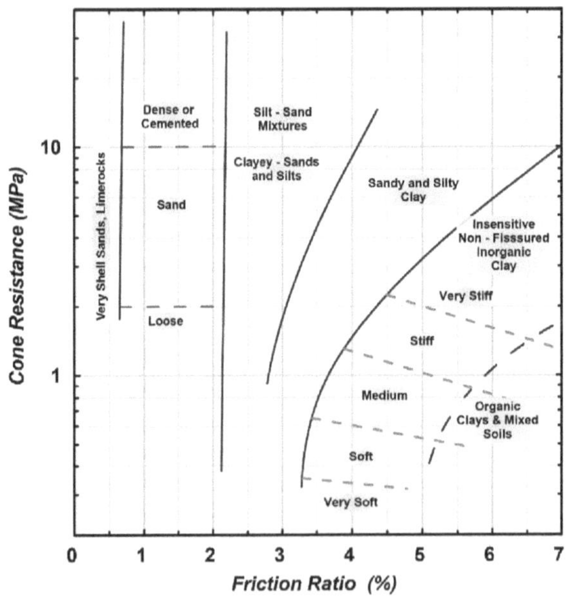

Figure 5.1 The Schmertmann profiling chart (Schmertmann, 1978).

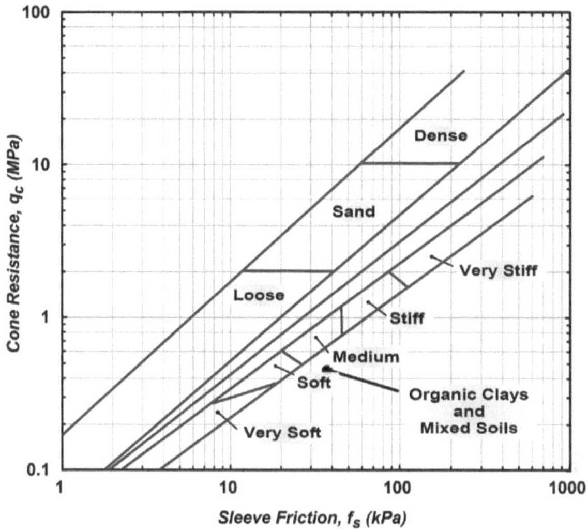

Figure 5.2 The Schmertmann profiling chart converted to a Begemann-type profiling chart (Eslami and Fellenius, 2004).

soil sensitivity, friction sleeve surface roughness, soil ductility, and pore pressure effects can influence the chart correlation.

Douglas and Olsen (1981) proposed a soil profiling chart based on tests with the electrical cone penetrometer. The chart, which is shown in Fig. 5.3, appends classification per the unified soil classification system to the soil type zones. The chart also indicates trends for liquidity index and earth pressure coefficient, as well as sensitive soils and "metastable sands." The Douglas and Olsen chart envelopes several zones using three upward curving lines representing increasing content of coarse-grained soil and four lines with equal sleeve friction. This way, the chart distinguishes an area (lower left corner of the chart) where soils are sensitive or "metastable."

Comparing Fig. 5.1 and 5.3, a difference emerges in implied soil type response; while in the Schmertmann chart the soil type envelopes curve downward, in the Douglas and Olsen chart they curve upward. Zones for sand and for clay are approximately the same in the two charts, however.

A comparison between the Schmertmann and Douglas and Olsen charts (Figs. 5.2 and 5.3) is more relevant if the charts are prepared per the Begemann-type of presentation. Thus, Fig. 5.4 shows the Douglas and Olsen chart converted to a Begemann-type graph. The figure includes the three curved envelopes and the four lines with equal sleeve friction and a heavy dashed line which identifies an approximate envelop of the zones indicated to represent "metastable" and "sensitive" soils.

Comparing the Begemann-type presentations of the Douglas and Olsen chart (Fig. 5.4) and Schmertmann (Fig. 5.2) chart, the former offers a smaller bandwidth for dense sands and sandy soils (q_c, larger than 10 MPa) and a larger bandwidth in the low range of cone resistance (q_c, smaller than 1 MPa).

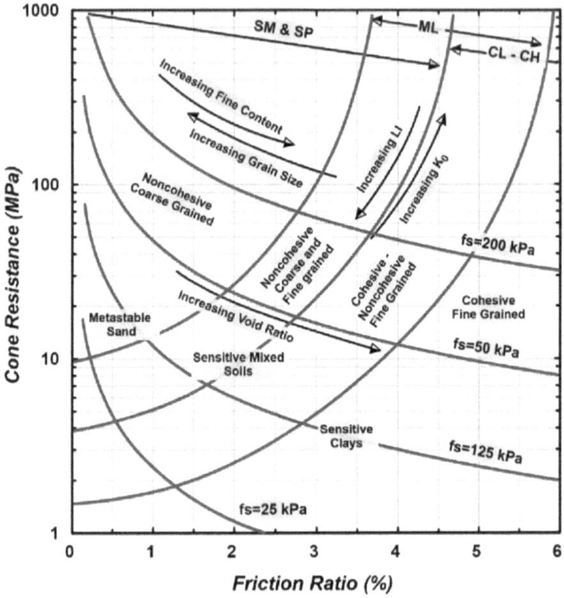

Figure 5.3 Profiling chart per Douglas and Olsen (1981).

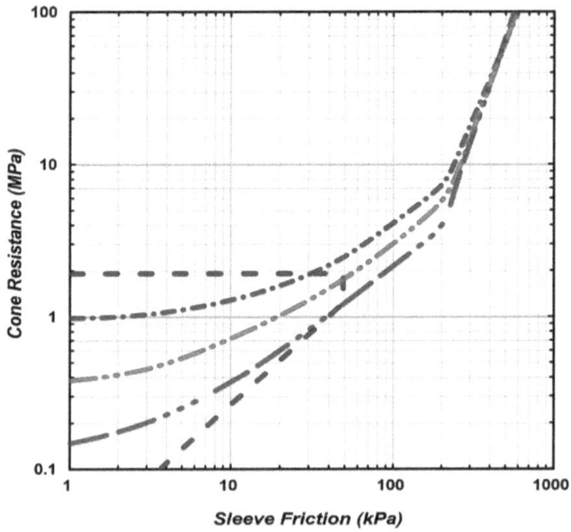

Figure 5.4 The Douglas and Olsen profiling chart converted to a Begemann-type chart (Eslami and Fellenius, 2004).

Vos (1982) suggested using the electrical cone penetrometer for Dutch soils to identify soil types from the friction ratio. The percentage values are similar but not identical to those recommended by Begemann (1965).

Campanella et al. (1982, 1985) presented a chart, which was the first chart to be based on the piezocone with the cone resistance corrected for pore pressure at the shoulder according to Eq. (5.1) as explained in Chapter 3.

$$q_t = q_c + u_2(1 - a) \quad (5.1)$$

where

q_t = cone resistance corrected for pore water pressure on shoulder,
q_c = measured cone resistance,
u_2 = pore pressure measured at cone shoulder,
a = ratio between shoulder area (cone base) unaffected by the pore water pressure to total shoulder area.

Robertson et al. (1986) proposed a dual profiling chart as illustrated in Fig. 5.5 and introduced a pore pressure ratio, B_q, defined by Eq. (5.2), as follows.

$$B_q = \frac{u_2 - u_0}{q_t - \sigma_v} \quad (5.2)$$

where

B_q = pore pressure ratio,
u_2 = pore pressure measured at cone shoulder,
u_0 = in situ pore pressure,
q_t = cone resistance corrected for pore water pressure on shoulder,
σ_v = total overburden stress.

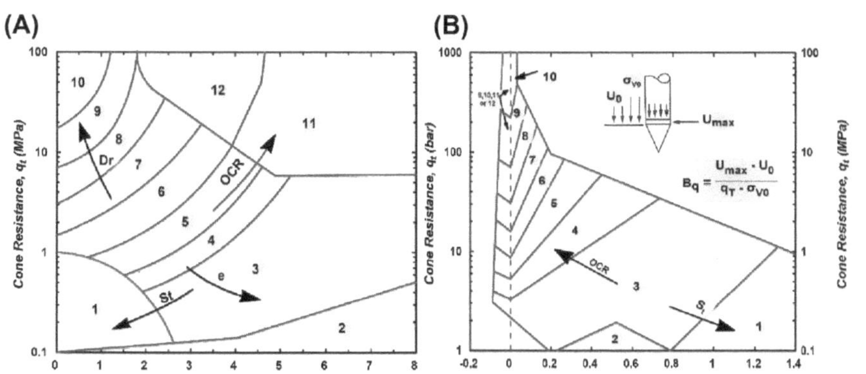

Figure 5.5 Profiling chart per Robertson et al. (1986).

Directly, the B_q-chart shows zones where the u_2 pore pressures become smaller than the initial pore pressures (u_0) in the soil during the advancement of the penetrometer, resulting in negative B_q-values. Otherwise, the B_q-chart appears to be an alternative rather than an auxiliary chart; one can use one or the other depending on preference. However, near the upper envelopes, a CPTu datum plotting in a particular soil-type zone in the friction ratio chart will not always appear in the same soil-type zone in the B_q-chart. Robertson et al. (1986) points out that "occasionally soils will fall within different zones on each chart" and recommends that the user studies the pore pressure rate of dissipation (if measured) to decide which zone applies to the questioned data.

The pore pressure ratio, B_q, is an inverse function of the cone resistance, q_t. Therefore, also the B_q-plot represents the data as a function of their own self-values.

Eslami and Fellenius (1996, 2004) proposed a pore pressure ratio, B_E, defined as follows:

$$B_E = \frac{u_2 - u_0}{u_0} \tag{5.3}$$

where B_E = "effective" pore pressure ratio.

A diagram showing q_t versus B_E provides a more perceptible picture of the pore pressure induced by the cone, and it does not violate the principles of plotting. According to Eslami and Fellenius (2004), research may show that the pore pressure ratio B_E will be useful for assessing liquefaction potential, degree of overconsolidation, and compressibility of sand and silt soils. It is also hypothesized that the B_E ratio may also show to be useful in predicting the magnitude of increase (setup) of capacity of driven piles between initial driving and after the soils have reconsolidated.

Robertson (1990) proposed a development of the Robertson et al. (1986) profiling chart, shown in Fig. 5.6, plotting a "normalized cone resistance," q_{cnrm}, against a "normalized friction ratio," R_{fnrm}, in a cone resistance chart. The accompanying pore pressure ratio chart plots the "normalized cone resistance" against the pore pressure ratio, B_q, defined by Eq. (5.2) applying the same B_q-limits as the previous chart (Zone 2 is not included in Fig. 5.6).

$$q_{cnrm'} = \frac{q_t - \sigma_v'}{\sigma_v} \tag{5.4}$$

where

$q_{cnrm'}$ = cone resistance normalized according to Robertson (1990),
q_t = cone resistance corrected for pore water pressure on shoulder,
σ_v = total overburden stress,
σ_v' = effective overburden stress,
$q_t - \sigma_v'$ = net cone resistance.

The normalized friction factor is defined as the sleeve friction over the net cone resistance, as follows:

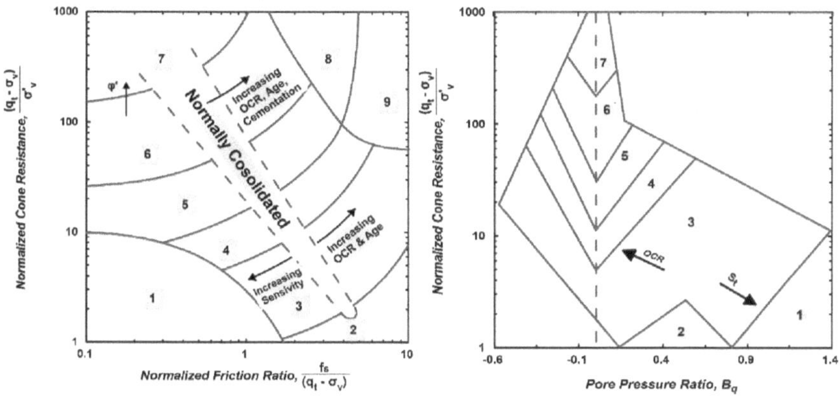

Figure 5.6 Profiling chart per Robertson (1990).

$$R_{fnrm'} = \frac{f_s}{q_t - \sigma_v} \tag{5.5}$$

where

f_s = sleeve friction,
q_t = cone resistance corrected for pore water pressure on shoulder,
σ_v = total overburden stress.

The numbered areas in the profiling chart separate the soil types in nine zones, as follows:

1. Sensitive, fine-grained soils
2. Organic soils and peat
3. Clays (clay to silty clay)
4. Silt mixtures (silty clay to clayey silt)
5. Sand mixtures (sandy silt to silty sand) or cemented soil
6. Sand (silty sand to clean sand)
7. Sand to gravelly sand
8. Sand (clayey sand to "very stiff" sand)
9. Very stiff, fine-grained, overconsolidated

The first two and last two soil types (i.e., zones 1, 2, 8, and 9) are the same as those used by Robertson et al. (1986) and Types 3 through 7 correspond to former Types 3 through 10. The Robertson (1990) normalized profiling chart has seen extensive use in engineering practice (as has the Robertson et al. (1986) chart).

The normalization is supposedly to compensate for that the cone resistance is influenced by the overburden stress. Therefore, when analyzing deep CPTu soundings (i.e., deeper than about 30 m), a profiling chart developed for more shallow soundings does not apply well to deeper sites. At very shallow depths, however, the proposed normalization will tend to lift the data in the chart and imply a coarser soil than reality. Moreover, the effective stress at depth is a function of the unit weight of the soil and, to a greater degree, of the pore pressure distribution with depth. Where soil types alternate between dense soils and light-weight soils (soil densities can range from 2100 kg/m^3 to 1400 kg/m^3, i.e., vary by 30%) and/or where upward or downward pore pressure gradients exist, the normalization is not straightforward. For these reasons, it would appear that the normalization merely exchanges one difficulty for another.

More importantly, the chart still includes the plotting of data against the inverse of own self. A chart with the same soil zones could just as well have been produced with normalized cone resistance against a normalized friction factor. Accepting the Robertson (1990) normalization, Fig. 5.6 shows the envelopes of the Robertson (1990) converted to a Begemann-type chart. The ordinate is the same and the abscissa is the multiplier of the normalized cone resistance and the normalized friction factor of the original chart (the normalized sleeve friction is the sleeve friction divided by the effective overburden stress). Where needed, the envelopes have been extended with a thin line to the frame of the diagram. As reference to Figs. 5.2 and 5.4, Fig. 5.7 presents the usual Begemann-type profiling chart converted from Fig. 5.6 under the assumption that the data apply to a depth of about 10 m at a site where the groundwater table lies about 2 m below the ground surface. This chart is approximately representative for a depth range of about 5−30 m. Comparing the "normalized" chart with the "as-measured" chart does not indicate that normalization would be advantageous.

Other early profiling charts were proposed by Searle (1979), Jones and Rust (1982), Olsen and Farr (1986), Olsen and Malone (1988), Erwig (1988). CPTu charts similar to that of Robertson (1990) were proposed by Larsson and Mulabdic (1991), Jefferies and Davies (1991, 1993), and Olsen and Mitchell (1995).

Jefferies and Davies (1993) introduced soil behavior index (I_{SBT}) which describes the boundary between different soil types and can be calculated by Eq. (5.6).

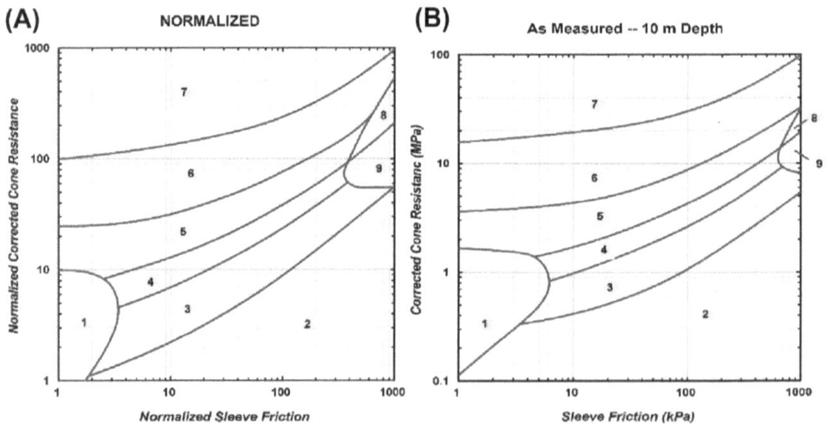

Figure 5.7 The Robertson (1990) profiling chart converted to Begemann-type charts: (A) Normalized corrected cone resistance versus normalized sleeve friction, (B) Corrected cone resistance versus sleeve friction (Eslami and Fellenius, 2004).

$$I_{SBT} = \left[\left(3.47 - \log \left(q_c / P_a \right) \right)^2 + (\log R_f + 1.22)^2 \right]^{0.5} \tag{5.6}$$

where

R_f = friction ratio,
q_c = cone resistance.

The values of I_c defined on the boundaries of various soil types on the Robertson (1990) classification chart is depicted in Fig. 5.8 (Table 5.1).

5.3 The Eslami–Fellenius CPTu profiling and soil type classification method

To investigate the use of cone penetrometer data in pile design, Eslami and Fellenius (1997) compiled a database consisting of CPT and CPTu data linked with results of boring, sampling, laboratory testing, and soil characteristics. The cases are from 18 sources reporting data from 20 sites in 5 countries. About half of the cases are from piezocone tests, CPTu, and include pore pressure measurements (u_2). Non-CPTu tests are from sand soils and were used with the assumption that the u_2 values are approximately equal to the hydrostatic pore pressure (u_0). The database values are separated on five main soil type categories as listed below.

1. Sensitive and collapsible clay and/or silt
2. Clay and/or silt

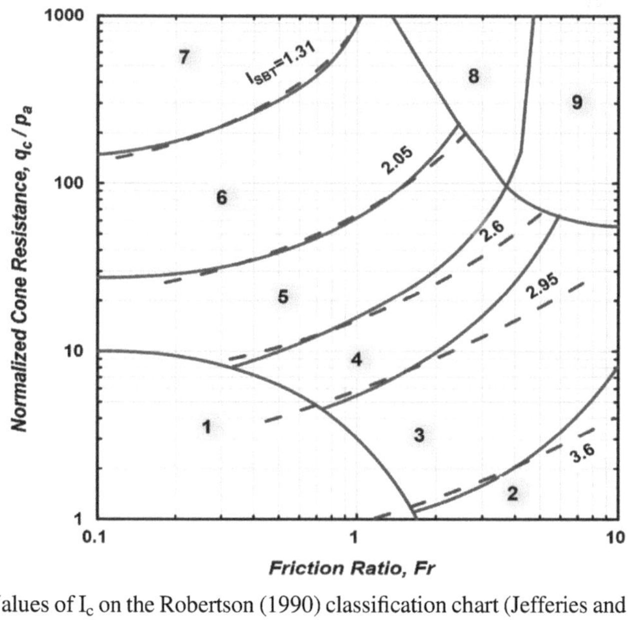

Figure 5.8 Values of I_c on the Robertson (1990) classification chart (Jefferies and Davies,1991).

Table 5.1 Soil classification based on I_{SBC} (Li et al., 2015).

Soil classification	Zone	CPTu index (I_{SBC})
Gravelly sands	7	$I_{SBC}<1.25$
Sands—clean sand to silty sand	6	$1.25<I_{SBC}<1.90$
Sand mixture—silty sand to sandy silt	5	$1.90<I_{SBC}<2.54$
Silt mixture—clayey silt to silty clay	4	$2.54<I_{SBC}<2.82$
Clays	3	$2.82<I_{SBC}<3.22$
Organic clays	2	$I_{SBC}>3.22$

3. Silty clay and/or clayey silt
4. Sandy silt and/or silty sand
5. Sand and/or sandy gravel

The data points were plotted in a Begemann-type profiling chart, and envelopes were drawn enclosing each of the five soil types. The envelopes are shown in Fig. 5.9. The database does not include cases with cemented soils or very stiff clays, and, for this reason, no envelopes for such soil types are included in the chart.

Plotting an "effective" cone resistance defined by Eq. (5.7) was found to provide a more consistent delineation of envelopes than a plot of only the cone resistance. (Note: subtracting the pore pressure does not make the value an effective stress. It was only found that the subtraction made the data collect in more easily delineated zones.)

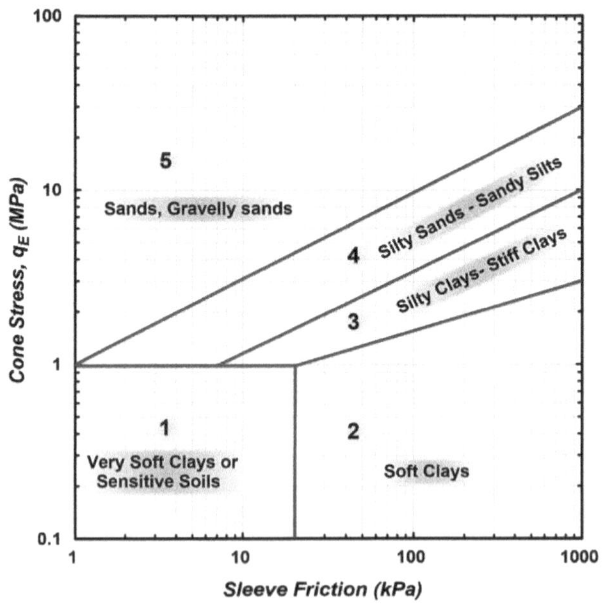

Figure 5.9 The Eslami–Fellenius profiling chart (Eslami, 1996; Eslami and Fellenius, 1997).

$$q_E = (q_t - u_2) \tag{5.7}$$

where

q_E = "effective" cone resistance,
q_t = cone resistance corrected for pore water pressure on shoulder (Eq. 5.1),
u_2 = pore pressure measured at cone shoulder,

The q_E-value was shown to be a consistent value for use in relation to soil responses, such as pile shaft and pile toe resistances (Eslami, 1996, Eslami and Fellenius, 1995, 1996; 1997). Notice again that, as mentioned by Robertson (1990), the measured pore water pressure is a function of where the pore pressure gage is located. Therefore, the q_E-value is by no means a measurement of effective stress in conventional sense. Because the sleeve friction is a rather approximate measurement, no similar benefit was found in producing an "effective" sleeve friction. In dense, coarse-grained soils, the q_E-value differs only marginally from the q_t value. In contrast, cone tests in fine-grained soils could generate substantial values of excess pore water pressure causing the q_E-value to be much smaller than the q_t value.

The Eslami–Fellenius chart is simple to use and requires no adjustment to estimated value of the overburden stresses. The chart is primarily intended for soil type (profiling) analysis of CPTu data. With regard to the boundaries between the main soil fractions (clay, silt, sand, and gravel), international and North American practices agree, but differences exist with regard to how soil-type names are modified according to the contents of other than the main soil fraction. The chart assumes the lower and upper boundaries for adjectives, such as clayey, silty, sandy to be 20% and 35%, assumes "some" to mean 10% through 20%, and assumes "trace" to mean smaller than 10% by weight as indicated in the Canadian Foundation Engineering Manual (CFEM, 2006).

The data from the CPT diagrams presented in Fig. 5.10A are presented in the chart shown in Fig. 5.10B. The three layers are presented with different symbols.

5.4 Comparison between the Eslami–Fellenius and Robertson (1990) methods

To provide a direct comparison between the Robertson (1990) profiling chart and the Eslami–Fellenius chart, three short series of CPTu data were compiled from sites with very different geologic origin where the soil profiles had been established independently of the CPTu. The borehole information provides soil description and water content of recovered samples. For one of the cases, grain size distribution data are also available. The CPTu diagrams from the site are shown in Fig. 5.11. The soil and CPTu information for the specific points are compiled in Table 5.2. The three sites are:

1. North Western University, Evanston, Illinois (Finno, 1989). CPTu data were obtained in a soil profile consisting of 7 m of sand deposited on normally consolidated silty clay.

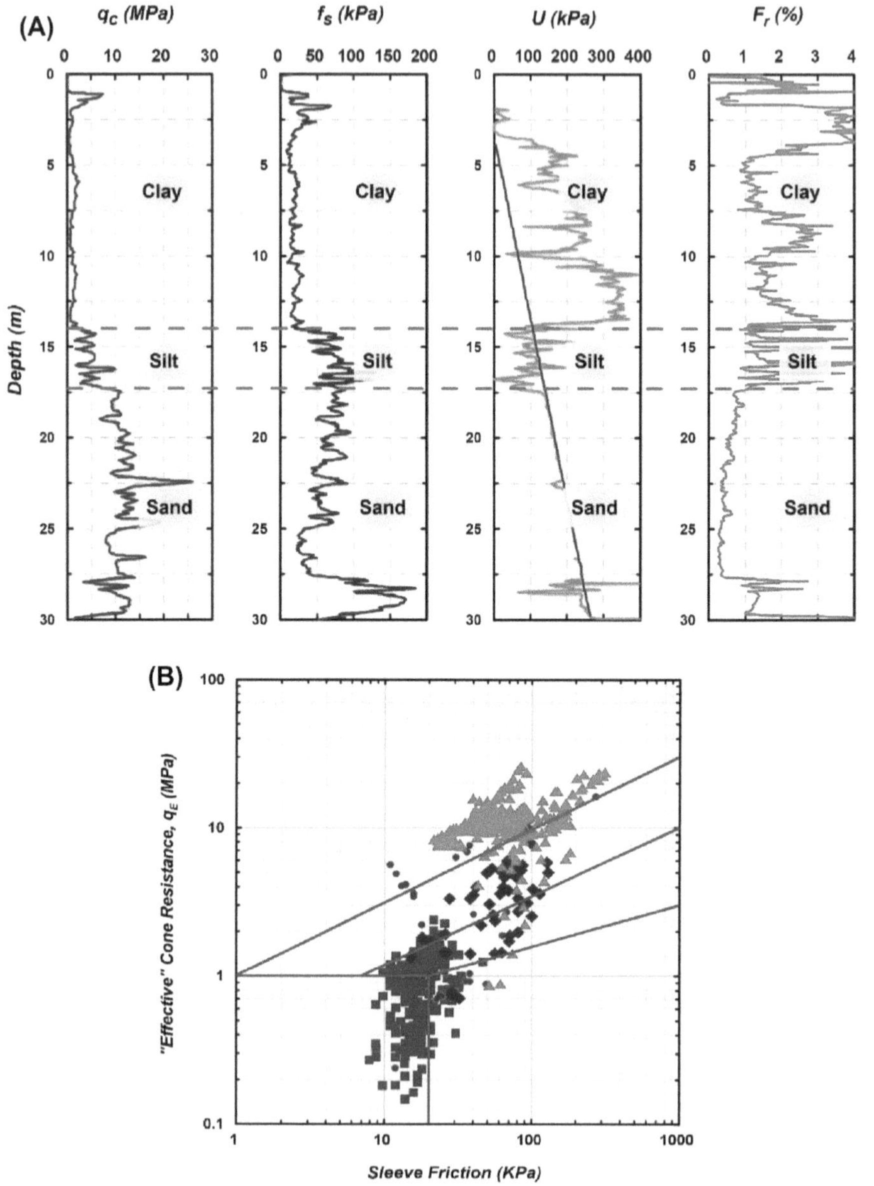

Figure 5.10 Example of soil profiling by Eslami–Fellenius (1997) chart: (A) Results from a piezocone, (B) Classification chart (Eslami and Fellenius, 2004).

The piezometer was attached to the cone face (u_1) and not behind the shoulder (u_2). The method of converting the pore pressure measurement to the u_2 value presented by Finno (1989) has been accepted here, although the conversion is disputed. For comments, see Mayne et al. (1990).

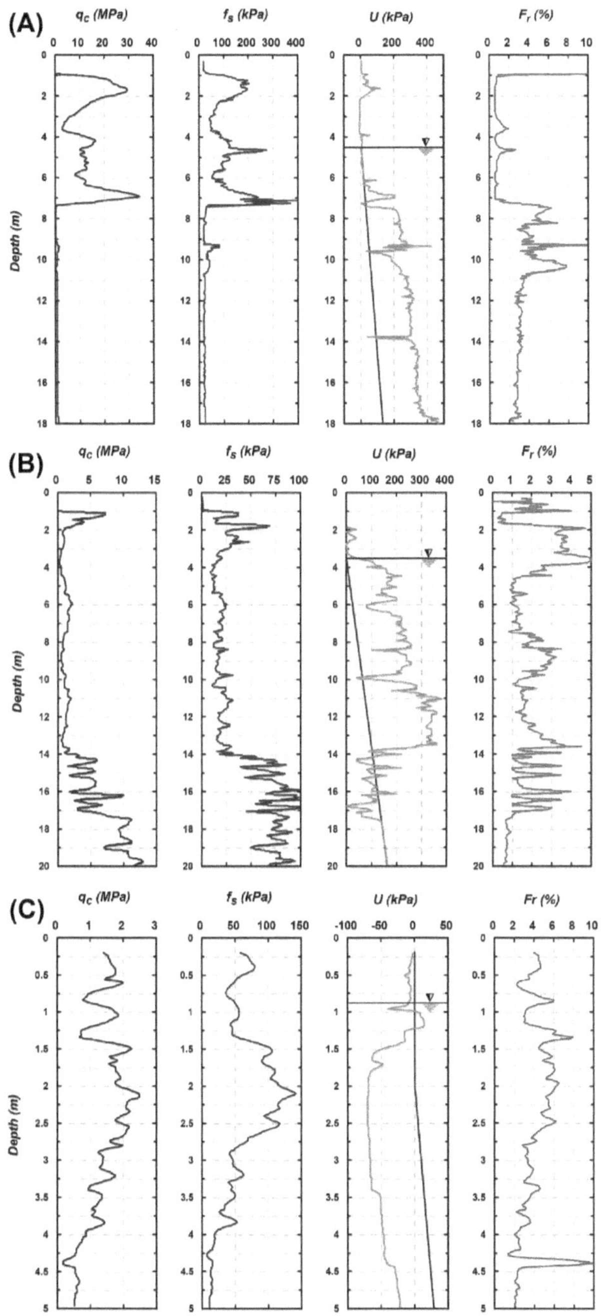

Figure 5.11 CPTu diagram from the three sites: (A) Evanston, (B) Vancouver, (C) Amherst (Eslami and Fellenius, 2004).

Table 5.2 Case records characteristics for comparing Eslami–Fellenius. Robertson (1990) charts (Eslami and Fellenius, 2004).

				Soil fractions (%)			CPTu data		
No.	Depth (m)	Description	Water content (%)	Clay	Silt	Sand	q_t (MPa)	f_s (kPa)	u_2 (kPa)
Evanston, IL (Groundwater table at 4.5 m)									
1	1.5	Sand, fine to medium, trace gravel	29				25.08	191.5	49.8
2	3.4	Sand, medium, trace gravel	16				3.48	47.9	−16.0
3	6.7	Sand, fine, trace silt, organics	26				32.03	162.8	111.7
4	8.5	Silty clay, trace sand	28				0.51	21.1	306.4
5	9.5	Silty clay, little gravel	22				0.99	57.5	39.6
6	12.8	Silty clay, little gravel	23				0.69	19.2	383.0
7	16.5	Silty clay, little gravel	24				0.77	17.2	427.1
Vancouver, BC (Groundwater table at 3.5 m)									
1	3.7	Clay to clayey silt	52				0.27	16.1	82.5
2	5.8	Clayey silt to silt	34				1.74	20.0	177.1
3	10.2	Silty clay	47				1.03	13.4	183.5
4	14.3	Silty clay	40				4.53	60.2	54.3
5	17.5	Silty sand	25				10.22	77.8	118.5
Amherst, MA (Groundwater table at 2.0 m)									
1	0.6	Sand and silt, trace clay	20	60	30	10	2.04	47.5	−9.4
2	1.5	Clayey silt, trace sand	28	10	67	23	2.29	103.3	−47.3

3	2.0	Clayey silt, trace sand	36		4	75	21	1.87	117.0	−69.5
4	2.5	Clayey silt, trace sand	29		2	65	33	1.86	117.0	−70.3
5	3.0	Clayey silt, trace sand	40		2	62	36	1.37	46.8	−66.3
6	3.5	Clayey silt, trace sand	53		2	58	40	1.38	48.9	−50.7
7	4.0	Clayey silt, trace sand	60			58	40	0.91	17.9	−46.9
8	4.5	Clayey silt	30		1	57	42	0.55	12.9	−29.3

2. Along the shore of Fraser River, Vancouver, British Columbia. A 20-m-thick mixed soil profile of deltaic deposits of clay, silt, and sand. The first four data points are essentially variations of silty clay or clayey silt. The fifth is a silty sand.
3. University of Massachusetts, Amherst, Massachusetts. A soil profile (Mayne, 2008) consisting of 5 m of a thick homogeneous overconsolidated clayey silt. This case includes also information on grain size distribution. The borehole records show the soil samples for data points 3 through 7 to be essentially identical. Notice that the u_2-measurements indicate substantial negative values, that is, the overconsolidated clay dilates as the cone is advanced.

For each case, the soil information in Table 5.2 is from depths where the CPTu data were consistent over a 0.5 m length. Then, the CPTu data from 150 mm above and below the middle of this depth range were averaged using geometric averaging, preferred over the arithmetic average as it is less subject to influence of unrepresentative spikes and troughs in the data (a redundant effort, however, as the records contain no such spikes and troughs).

The CPTu data were analyzed by Eslami–Fellenius (1996) and Robertson (1990) profiling methods and the results are shown in Fig. 5.12.

Evanston data: The first three samples are from a sand soil and both methods identify the CPTu data accordingly. The remaining data points (Nos. 4 through 7) given as silty clay in the borehole records are identified as clay/silt by the Eslami–Fellenius and as clay to silty clay by the Robertson method, that is, both methods agree with the independent soil classification.

Vancouver data: Both methods properly identify the first four data points to range from clayey silt to silty clay in agreement with the independent soil classification. The fifth sample (silty sand) is identified correctly by the Eslami–Fellenius method as a sand close to the boundary to silty sand and sandy silt. The Robertson method identifies the soil as a sandy silt to clayey silt, which is essentially correct, also.

Amherst data: Both methods identify the soils to be silt or clay or silt and clay mixtures. Moreover, both methods place Points 3 through 7 on the same soil type boundary line, that is, confirming the similarity between the soil samples. However, the spread of plotted points appear to be larger for the Robertson method; possibly due to that its profiling does not consider the pore pressures developed by the advancing penetrometer (but for the pore pressure on the shoulder, of course), while the Eslami–Fellenius method does account for the substantial negative pore pressures that developed.

5.5 Comments on current methods

I. The CPT methods (mechanical cones) do not correct for the pore pressure on the cone shoulder, and the profiling developed based on CPT data may not be relevant outside the local area where they were developed. The error due to omitting the pore water pressure correction is large in fine-grained soils and small in coarse-grained soils.
II. Except for the profiling chart by Begemann (1965) and Eslami-Fellenius (1997), all of the referenced soil profiling methods plot the cone resistance versus its own inverse value in

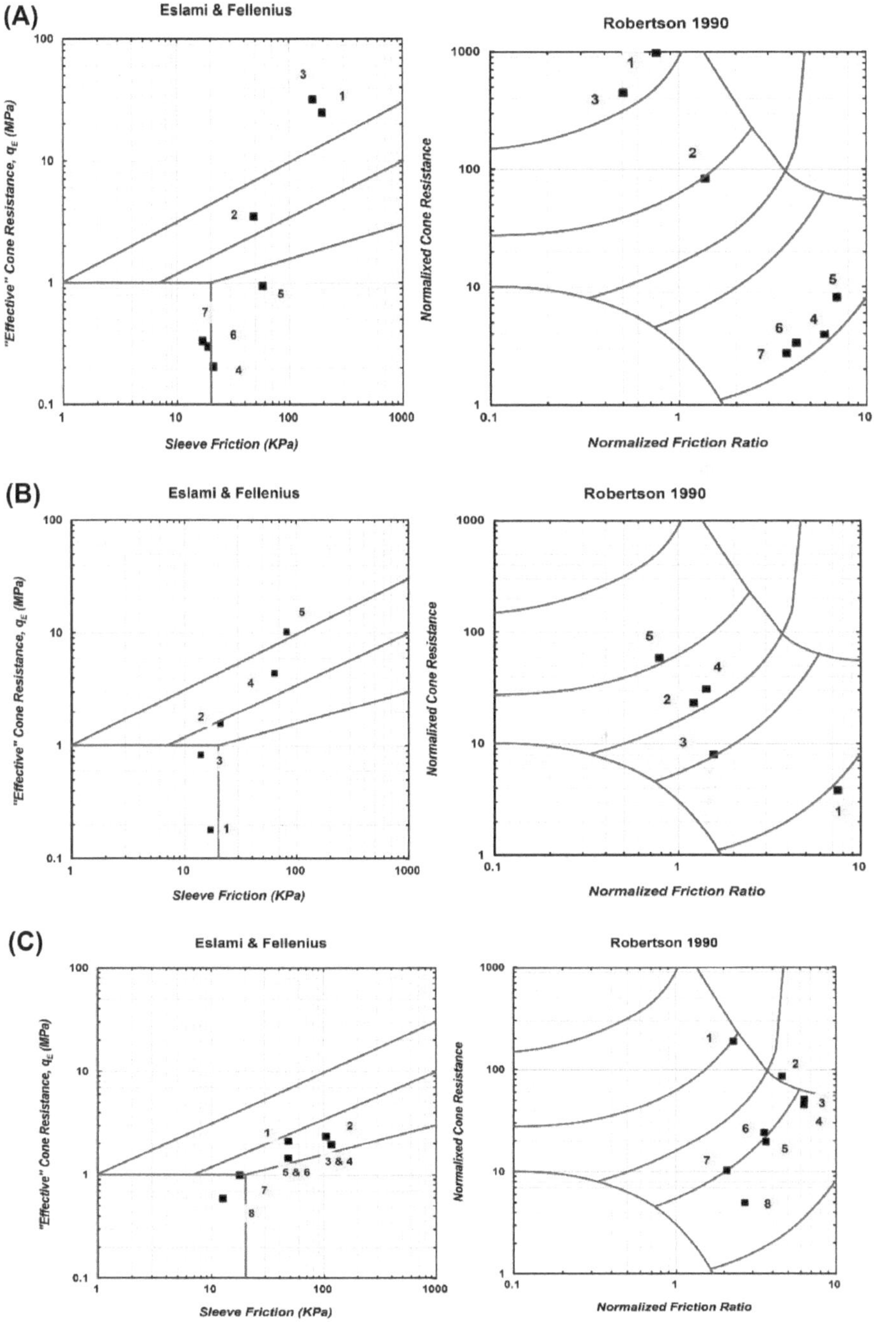

Figure 5.12 Comparison between Table 5.2 data plotted in Eslami–Fellenius (1997) and Robertson (1990) profiling charts: (A) Evanston, (B) Vancouver, (C) Amherst (Eslami and Fellenius, 2004).

one form of another. This generates data distortion and violates the rule that dependent and independent variables must be rigorously separated.

III. Some profiling methods, e.g., Robertson (1990), include normalization which requires complicated manipulation of the CPT data. For example, in a layered soil, should a guess-timated "typical" density value be used in determining the overburden stress or a value that accurately reflects the density? Moreover, regardless of soil layering, determining the effective overburden stress (needed for normalization) requires knowledge of the pore pressure distribution, which is not always hydrostatic but can have an upward or downward gradient, and this information is rarely available.

IV. The normalization by division with the effective overburden stress does not seem to always lead to correct classification. For example, the normalized values of fine-grained soils obtained at shallow depth (where the overburden stress is small) will often plot in zones for coarse-grained soil.

V. The Robertson (1990) and the Eslami–Fellenius (1997) CPTu methods of soil profiling were applied to data from three geographically apart sites having known soils of different types and geologic origins. Both methods identified the soil types accurately.

VI. Eslami–Fellenius (1996) method has the advantage over Robertson (1990) that it avoids the solecism of plotting data against their own inverted values and associated distortion of the data.

VII. Eslami–Fellenius (1997) method has the additional advantage over other referenced piezocone methods in that it allows the user to directly assess a value without first having to determine distribution of total and effective stress to use in a subtraction and multiplication effort in calculating a "normalized" set of values.

VIII. A soil profiling chart based on a Begemann-type plot, such as the Eslami–Fellenius (1996) method can easily be expanded by adding delineation of strength and consistency of fine-grained soils and relative density and friction angle of coarse-grained soils per the user-preferred definitions or per applicable standards.

IX. No doubt, CPTu sounding information from a specific area or site can be used to further detail a soil profiling chart and result in delineation of additional zones of interest. However, there is a possibility in producing a very detailed chart inasmuch as the resulting site dependency easily gets lost, leading an inexperienced user to apply the detailed distinctions beyond their geologic validity.

X. The CPTu is an excellent tool for the geotechnical engineers in developing a site profile. Naturally, it cannot serve as the exclusive site investigation tool, and soil sampling is still required. However, when the CPTu is used to govern the depths from where to recover soil samples for detailed laboratory study, fewer samples are needed, reducing the costs of site investigation while simultaneously increasing the quality of the information.

5.6 UniCone

UniCone is a program for processing and reporting results of cone penetrometer tests (CPT and CPTU) and for soil profiling and pile capacity analysis. It includes two methods of soil profiling, the Campanella–Robertson (1988) method and the Eslami–Fellenius method.

Rows of values can be edited or deleted, e.g., for reasons of being suspect or erroneous, noise in the records, etc. and data found affected by an error in calibration can be corrected. The records can be filtered by imposing a running average to smoothen out peaks and troughs and emphasize the dominant soil type in the profile.

Fig. 5.13 demonstrates an example of UniCone output for soil profiling, including digital data, soil types, and graphical presentation of CPT logs with soil behavior classification (SBC) charts.

5.7 Problematic soils evaluation via SBC

In seismically active regions, loose saturated sands, liquefiable soils are prone to liquefaction. Moderate to strong ground shaking can create large excess pore water pressures in these soils, which temporarily decrease the effective stress and shear strength, which leads to strength loss, excessive settlements, lateral spreading, and other damages.

The most common of sensitive clays are the soft, saturated mud and marine deposit clays often found near the delta area of rivers, along the perimeter of bays and wetlands. These soils are very weak and compressible and thus are subject to bearing capacity and settlement problems.

Peaty soils are composed of organic materials. The void ratio of peat tends to decrease with depth. Such high void ratios give rise to high water contents, and most of the peculiarity of peat soils is attributable to its moisture content. The shear strength of a peaty deposit appears to be influenced by its degree of humidification and its mineral content. Increasing moisture content decreases shear strength and the peaty soils under loading show high potential of primary and secondary settlements (Coduto, 2002).

Collapsible soils consist predominantly of sand and silt size particles arranged in a loose honeycomb structure. This loose texture is held together by small amounts of water-softening cementing agents, such as clay or calcium carbonate. As long as the soil remains dry, these cements produce a strong soil that is able to carry large loads. However, if the soil becomes inundated, under the pressure of surcharges, the cementing agents soften and the honeycomb structure collapses.

Expansive soils are sensitive to changes in moisture and will swell upon wetting and shrink upon drying. These soils have high plastic index and can cause cracks in the surface and deep due to the loss of moisture. Expansive soils are not as dramatic as hurricanes or earthquakes, but they cause only property damage, not loss of life. In addition, they act more slowly, and the damage is spread over wide areas rather than being concentrated in a small locality (Holtz et al., 2011).

5.8 Case records for problematic soils

Behavioral assessment of problematic soils using cone penetration test results and classification of diagrams are used for this context. A database has been compiled including 2500 points from 54 sites and based on 140 boreholes. The sources are

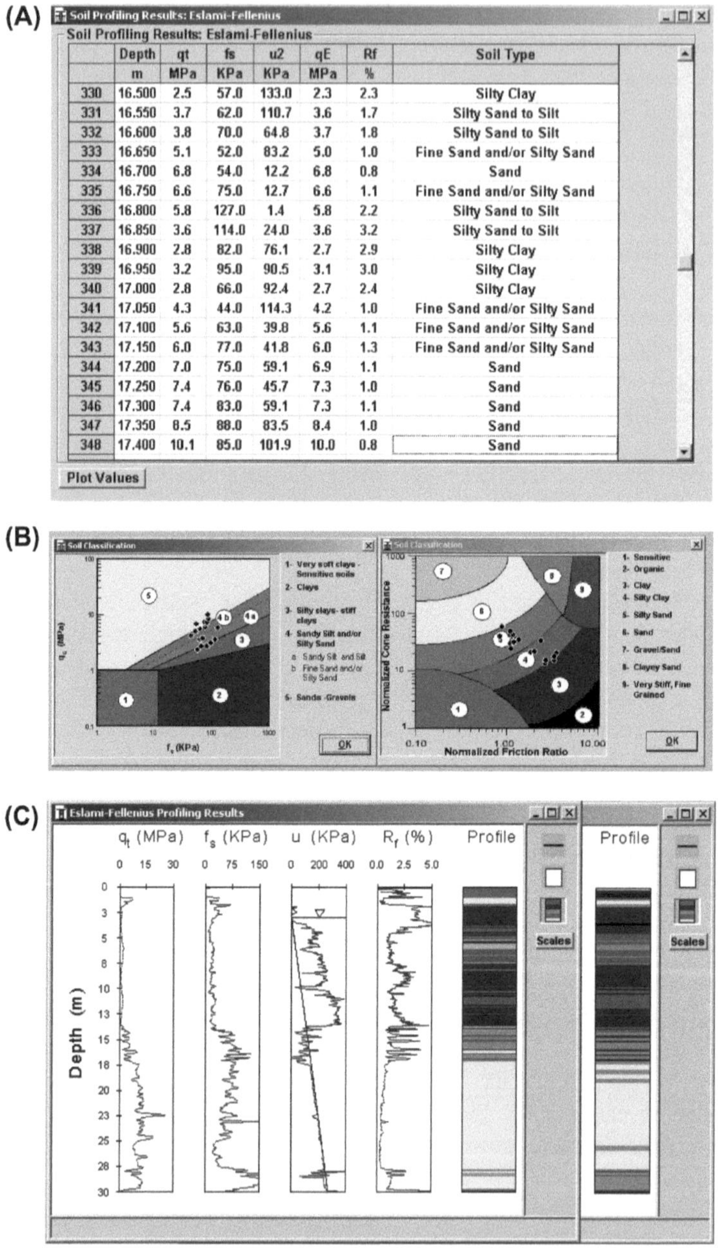

	Depth m	qt MPa	fs KPa	u2 KPa	qE MPa	Rf %	Soil Type
330	16.500	2.5	57.0	133.0	2.3	2.3	Silty Clay
331	16.550	3.7	62.0	110.7	3.6	1.7	Silty Sand to Silt
332	16.600	3.8	70.0	64.8	3.7	1.8	Silty Sand to Silt
333	16.650	5.1	52.0	83.2	5.0	1.0	Fine Sand and/or Silty Sand
334	16.700	6.8	54.0	12.2	6.8	0.8	Sand
335	16.750	6.6	75.0	12.7	6.6	1.1	Fine Sand and/or Silty Sand
336	16.800	5.8	127.0	1.4	5.8	2.2	Silty Sand to Silt
337	16.850	3.6	114.0	24.0	3.6	3.2	Silty Sand to Silt
338	16.900	2.8	82.0	76.1	2.7	2.9	Silty Clay
339	16.950	3.2	95.0	90.5	3.1	3.0	Silty Clay
340	17.000	2.8	66.0	92.4	2.7	2.4	Silty Clay
341	17.050	4.3	44.0	114.3	4.2	1.0	Fine Sand and/or Silty Sand
342	17.100	5.6	63.0	39.8	5.6	1.1	Fine Sand and/or Silty Sand
343	17.150	6.0	77.0	41.8	6.0	1.3	Fine Sand and/or Silty Sand
344	17.200	7.0	75.0	59.1	6.9	1.1	Sand
345	17.250	7.4	76.0	45.7	7.3	1.0	Sand
346	17.300	7.4	83.0	59.1	7.3	1.1	Sand
347	17.350	8.5	88.0	83.5	8.4	1.0	Sand
348	17.400	10.1	85.0	101.9	10.0	0.8	Sand

Figure 5.13 Example of UniCone output: (A) digital soil and pile data, (B) employed classification charts, (C) cone penetration test and soil profile.

Figure 5.14 Cases history site locations (Eslami et al., 2016).

from 15 countries such as the United States, the Netherlands, Ireland, Taiwan, Norway, Iran, Turkey, Canada, New Zealand, Egypt, Australia, England, Scotland, Poland, and China. Sixty-one boreholes of compiled records contain usual deposits (Farhadi and Eslami, 2012). In Fig. 5.14, the site locations are marked and illustrated.

5.9 Analysis and comparison

To characterize usual and problematic soils, five diagrams have been considered, including Campanella et al. (1985), Robertson (1990), Jefferies and Davies (1991), Eslami and Fellenius (1997) and Robertson (2010). After determining the depth and thickness of the desired deposits, arithmetic mean is taken for 300 mm sublayers. Accordingly, it represents the soil parameters such as cone tip resistance q_c, sleeve friction f_s, and pore pressure u_2.

Correct evaluation of total and effective overburden stresses is important to normalize the values of the cone tip resistance and sleeve friction for the charts. Equations proposed by Robertson (2010) were used to calculate the specific gravity of the soils.

For evaluation and recognition of unusual deposits, according to CPT records for the current database, as indicated in Fig. 5.15, a chart of Campanella et al. (1985) is partially able to capture the nature of the collapsible soils. Data shown are larger values due to normalizing in the charts of Jefferies and Davies (1991) and Robertson (1990). A couple of data points will be magnified specially in low overburden stress, located out of specified zone. Other diagrams showed scattered data points and presentations.

For expansive soils as denoted in Fig. 5.16, diagrams of Robertson (1990) and Jefferies and Davies (1991) differentiate this soil type with good accuracy, and the Eslami and Fellenius (1997) shows a specific range for expansive soils.

As illustrated in Fig. 5.17, the chart of Campanella et al. (1985) presented reasonable range for liquefiable soils. On the other hand, Robertson's (1990) diagram based on B_q-value can recognize the nature of the liquefiable soils, involving positive pore

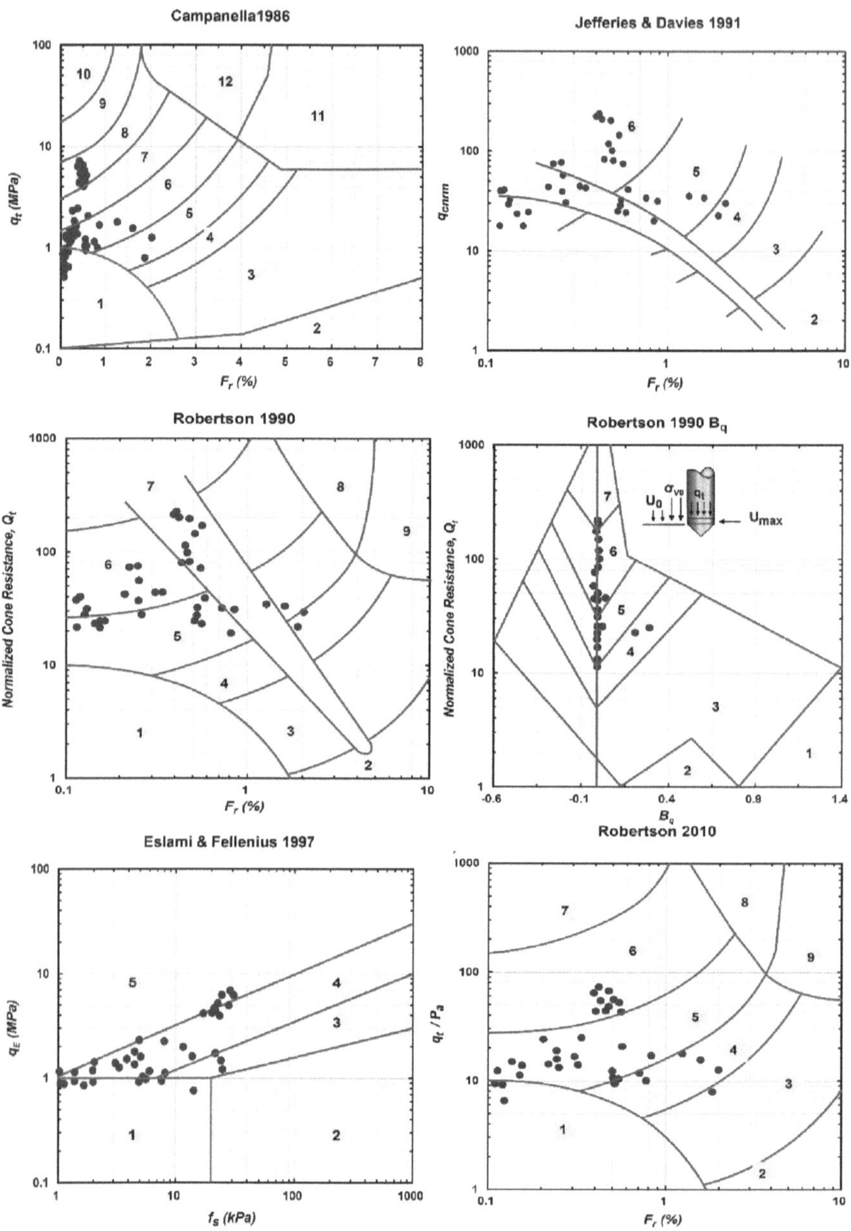

Figure 5.15 Collapsible soil boundaries in different charts (Eslami et al., 2016).

water pressure mobilization during penetration for the impose of shear force. But, the problem is the mismatching between the two charts. Jefferies and Davies's (1991) chart shows somehow overconsolidated soils for this category of soils, which involve incorrect interpretation.

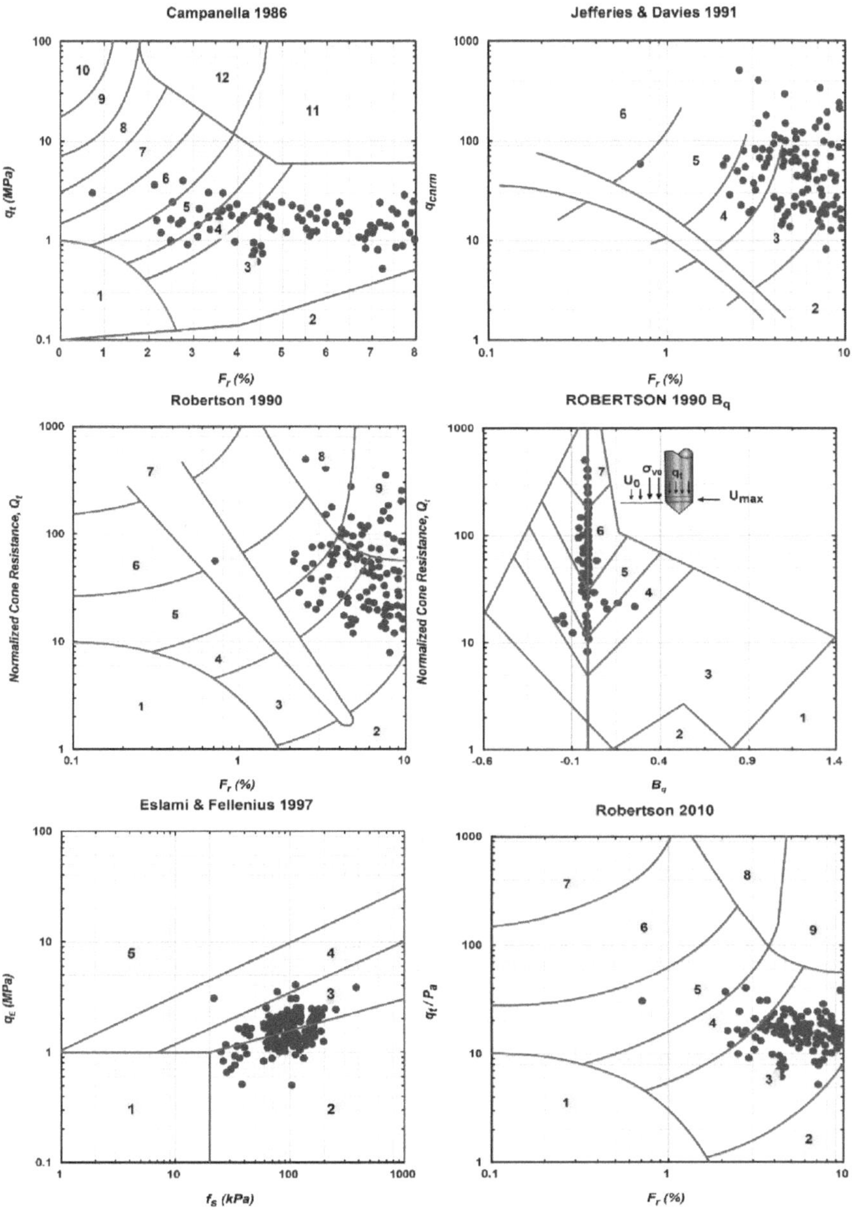

Figure 5.16 Expansive soil boundaries in different charts (Eslami et al., 2016).

Diagram of Eslami and Fellenius (1997) shows peaty soils pose within the boundary of clay and sensitive soils, this can be due to the weak nature of these deposits as indicated in low q_c and f_s values (Fig. 5.18). But, there is not a specified zone for peaty soils. Diagrams of Campanella et al. (1985) and Jefferies and Davies (1991) are able to

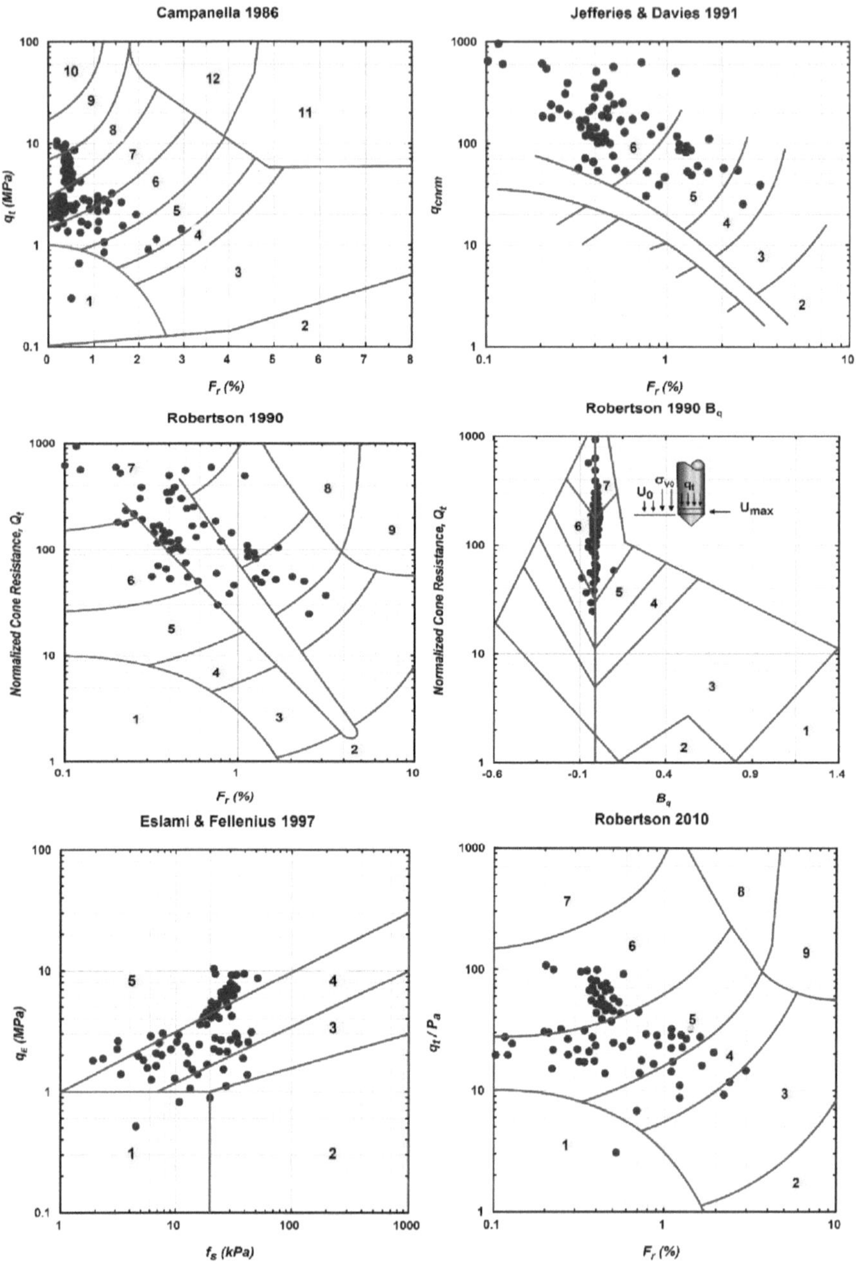

Figure 5.17 Liquefiable soils boundaries in different charts (Eslami et al., 2016).

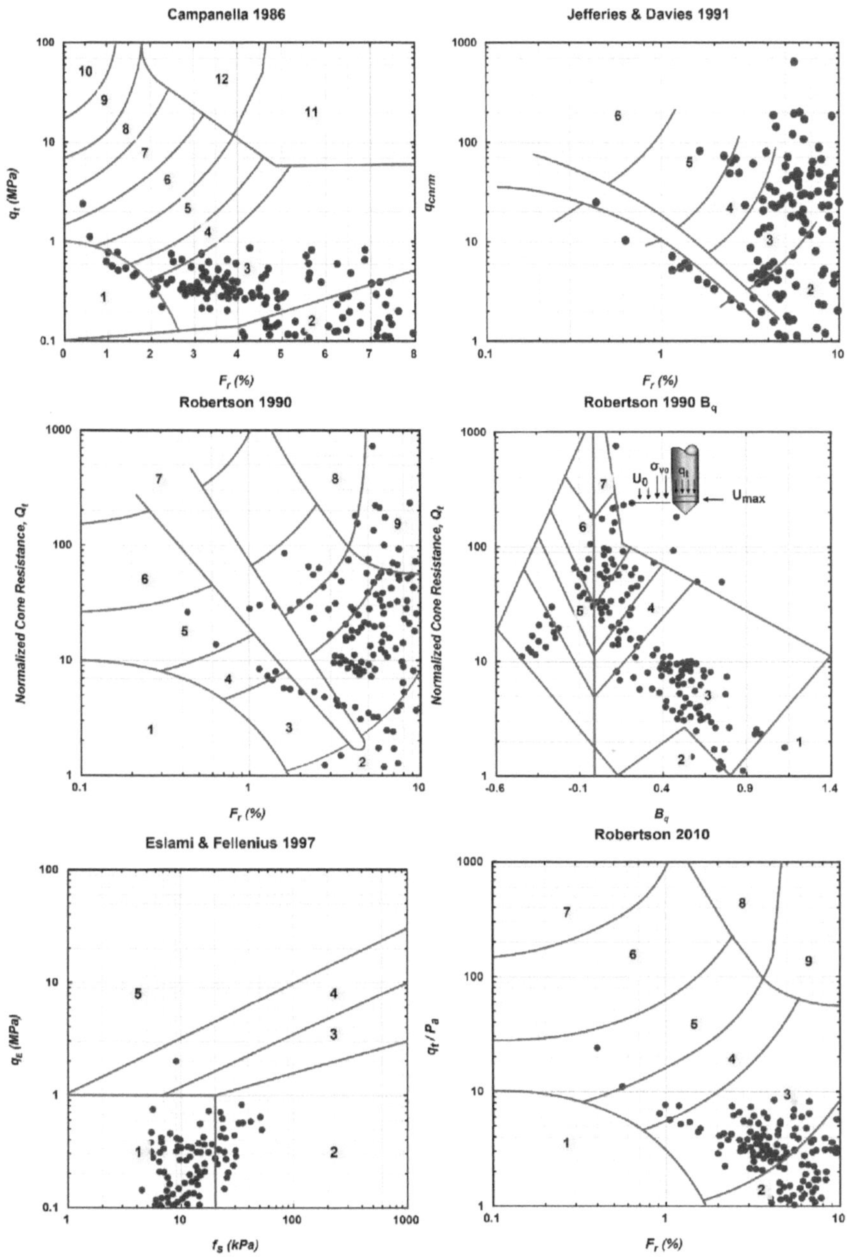

Figure 5.18 Peaty soil boundaries in different charts (Eslami et al., 2016).

capture a part of the organic soil, suitably. Diagrams of Jefferies and Davies (1991) and Robertson (1990) categorize this group of soils to overlap with overconsolidated soils which are not consistent to inherent properties of these deposits.

While diagrams of Campanella et al. (1985), Eslami and Fellenius (1997), and Robertson (1990) can detect sensitive soils in the specified zones according to Fig. 5.19, the other charts showed a wide range. However, the results of analysis for data have not been matched with this zone.

5.10 Triangular chart

Investigation and comparison using database indicate that the current methods for interpretation of usual and especially problematic soils involve a few limitations and difficulties. They are attributed to excluding one of the CPTu data in classification system, such as locally developed charts, manipulation of CPTu data, and stress normalization. Deficiency arises when using the chart with only two axes will be more pronounced because current methods attempt to present and interpret the soil boundaries via combination of CPTu data. Some charts represent the data as a function of their own self-values, i.e., y versus x/y ratio which variables mathematical rule (Eslami and Fellenius, 2004). Trends include effective cone resistance; stress normalization based on overburden stress estimation rather represent data as a function of each other or using two supplementary charts.

Therefore, due to shortcomings of the current soil behavioral classification charts, it seems necessary for a new approach to be developed to overcome a portion of difficulties and recognize most of soil deposits particularly deltaic deposits including compressible and weak soils.

It is necessary to examine the possibility of defining a new chart with different axes including q_t, f_s, and u_2 to illustrate more accurately the soil conditions, relying all of CPTu records.

After investigation of various charts for the soil behavior classification by collected data including 2500 points, it is required to define a new relatively accurate chart comparing all of the CPTu records, q_t, f_s, and u_2 in one diagram.

A three-axes chart was proposed by Eslami et al. (2016) to employ all three CPTu data, simultaneously. A triangular chart that is similar to soil classification chart recommended by United State Department of Agriculture (USDA) for classification of deltaic sands, silts and clays. In this chart, the sum of three parameters (soil content percentage) must be equal to 100. However, CPTu output parameters have no relation to each other and their sum cannot be realized equal to 100. Therefore by defining a new approach used for triangle chart, it is possible to include three CPTu records in one diagram.

Following are the step-by-step procedure for forming and using the triangular chart as shown in Fig. 5.20:

a. Draw a triangle by assigning each side for three CPTu records, i.e., q_t, f_s, and u_2. q_t and f_s are positive values due to negative and positive values for u_2. It is defined from minus to plus amount.

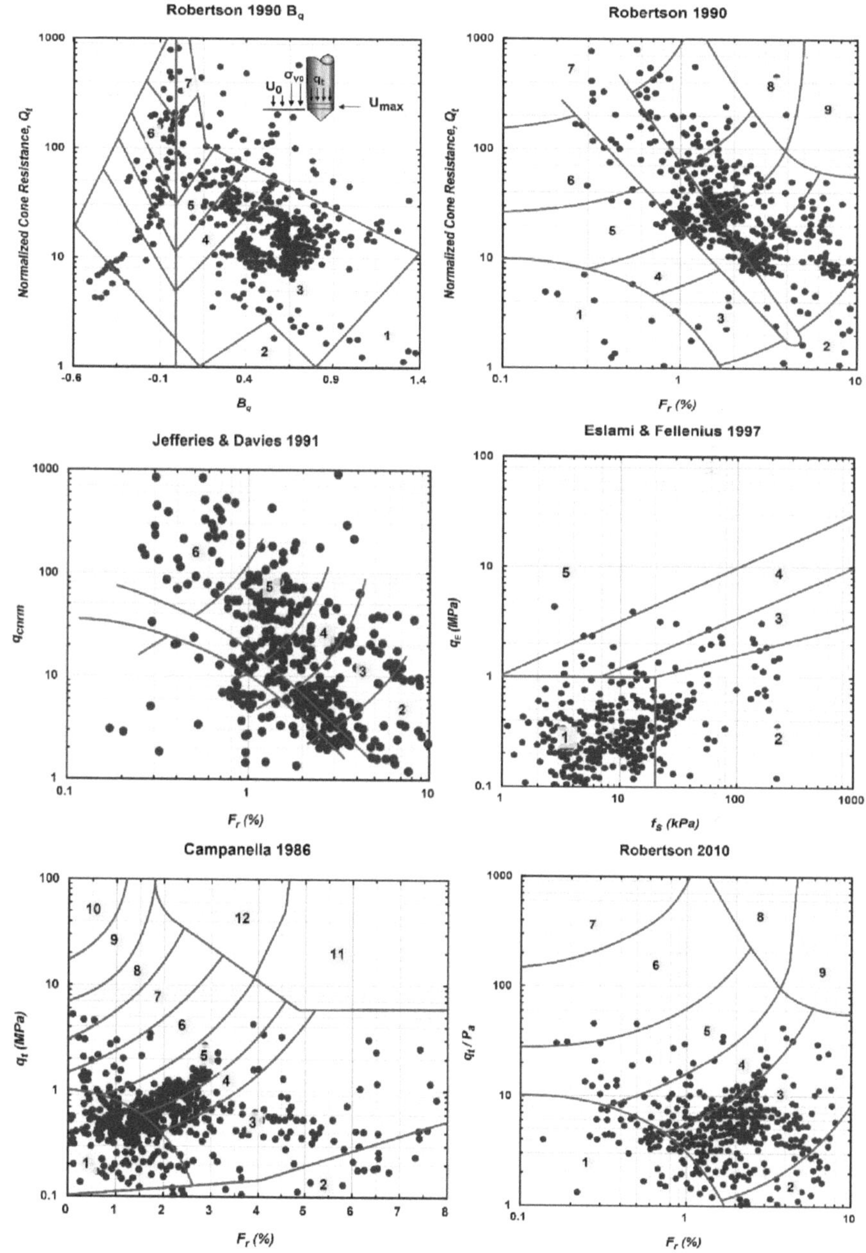

Figure 5.19 Sensitive soil boundaries in different charts (Eslami et al., 2016).

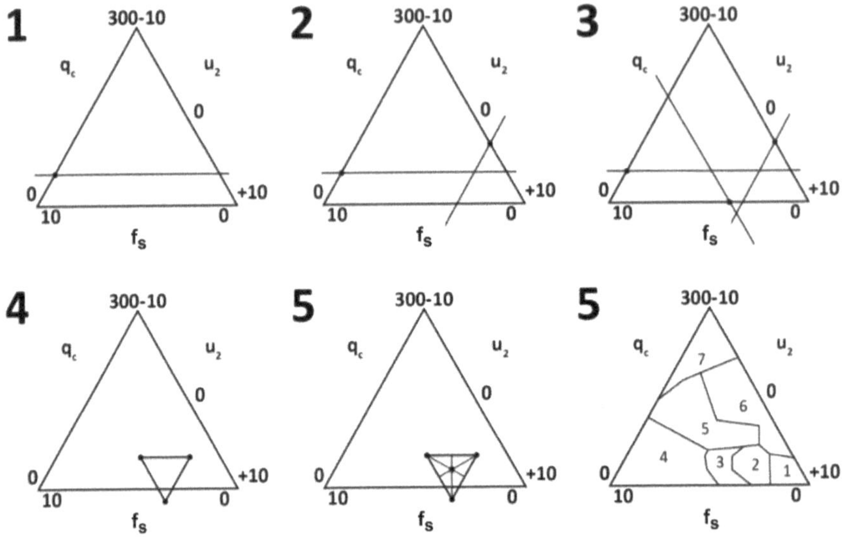

Figure 5.20 Step-by-step procedure for drawing triangular soil behavior classification chart (Eslami et al., 2016).

b. Data to be converted to normalized values by dividing to reference stress, i.e., ≈ 1 atm ≈ 1 kg/cm^2 ≈ 100 kPa ≈ 1 bar.
c. Find q_t value on the graph, and then draw a line parallel to f_s axis (step 1).
d. Find u_2 value on the graph, and then draw a line parallel to q_c axis (step 2).
e. Find f_s value on the graph, and then draw a line parallel to u_2 axis (step 3).
f. The intersection of these three lines, represent three points which form an isosceles triangle (steps 4 and 5).
g. The center of the triangle represents the soil condition, and its location can be assigned a typical zonation (step 6).

As shown in Fig. 5.21, the proposed chart shows collapsible soils in Zones 2 and 5. By considering the nature of this group that is including honeycomb structures, the results are acceptable. Most of the expansive soils are shown at Zone 2 which belongs to clay soils indicating the accuracy of the chart to determine the nature of these soils. Liquefiable soils are in the range of 6, 2, and 5 which due to their fine sand and silt contents. Hence, this type of problematic soil has been diagnosed very well. Peaty and sensitive soils are shown in Zones 1 and 2. Based on the size and nature of their low resistance, suitable characterization has been achieved. Therefore, the triangle chart is a basic chart with master zoning via using tree parameter simultaneously for deltaic soils. For more accuracy of the proposed chart, it is required to focus more on it to be divided into microzoning.

According to the points overlapping group, seven zones were defined on the chart for different types of soils as follows:

1 Sensitive soils
2 Clays

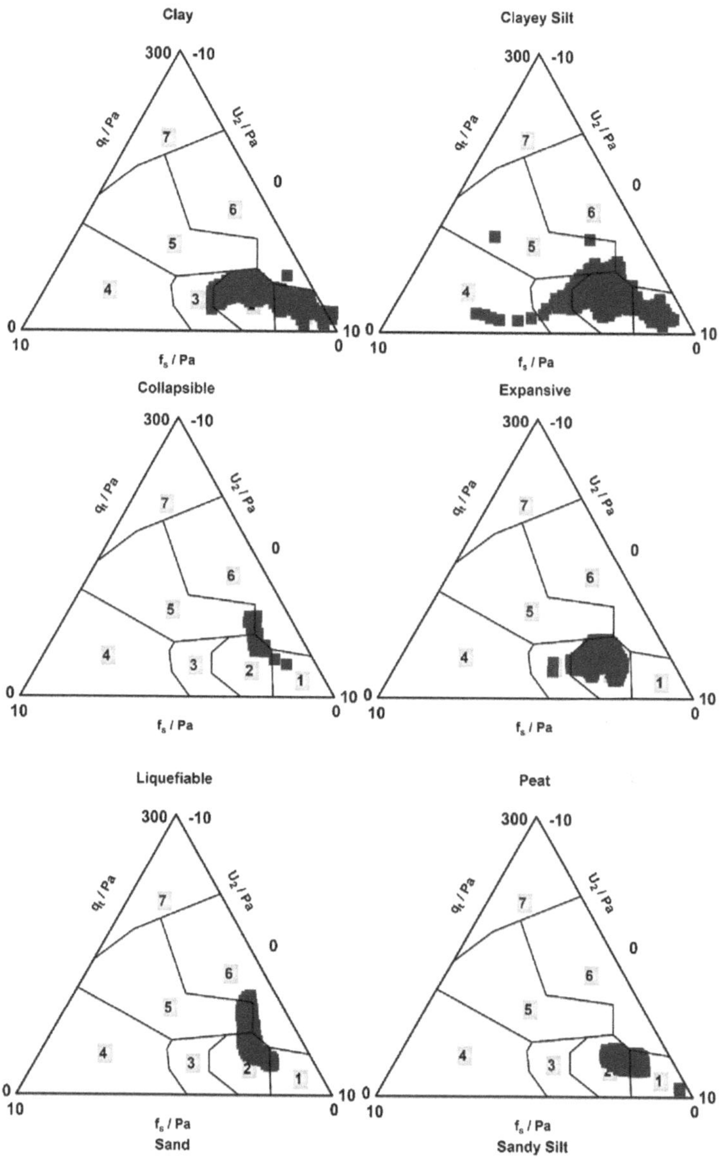

Figure 5.21 Illustration of deferent soil deposit zones using proposed Triangular chart (Eslami et al., 2016).

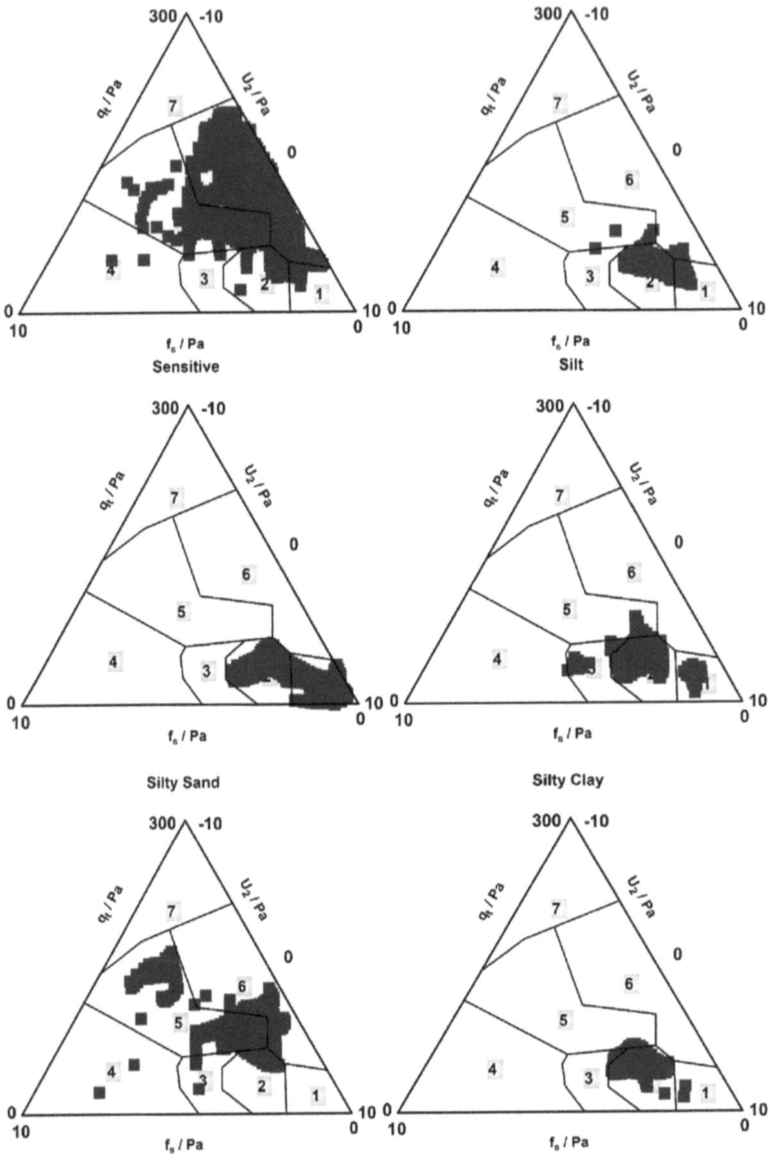

Figure 5.21 cont'd.

3 Silts
4 Overconsolidated clay
5 Mixed soils (clayey silt to sandy silt)
6 Sands
7 Very dense sand

The final zoning and boundaries are illustrated in Fig. 5.22.

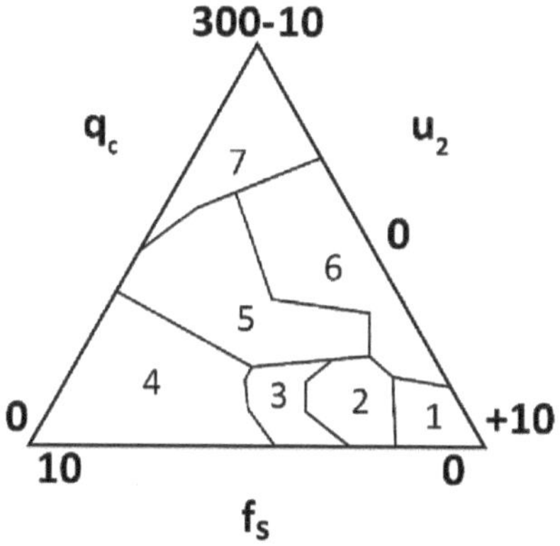

1	2	3	4	5	6	7
Sensitive Soil	Clays	Silts	Over Consolidated Clay	Mixed Soil (Clayey silt to sandy silt)	Sands	Very Dense Sand

Figure 5.22 The axes and zones for the triangular chart (Eslami et al., 2016).

Due to the increasing use of CPT significantly for behavioral identification of soils, this study examines the ability of the five most commonly used CPT-based classification charts. One hundred and forty CPT, CPTu sounding records have been collected from different countries around the world. The investigation focused on common deltaic soils such as clay, silt, and sand as well as five groups of problematic soils, including liquefiable, sensitive, peaty, collapsible, and expansive soils. Data analysis was performed by Campanella et al. (1985), Robertson (1990), Jefferies and Davies (1991), Eslami and Fellenius (1997), and Robertson (2010) (SBC) diagrams. Upon investigation and study of usual and problematic soil behavior classification using the current charts and compiled database, the following points are derived and concluded:

- Range of cone tip resistance and friction sleeve were presented for the five groups. Diagrams of Jefferies and Davies (1991), Campanella et al. (1985), and Robertson (1990) were unable to locate the soils to a specific zone for peaty or sensitive soils. The Eslami and Fellenius's (1997) chart has covered low zones, but, the accuracy and consistency in the type of categories are well. Robertson's (1990) chart based upon the B_q showed good results for liquefiable soils. The problem is the mismatch between the two supplementary charts. Liquefiable and sensitive soils can be recognized well by Campanella et al. (1985), Eslami and Fellenius (1997) charts. Diagrams of Robertson (1990), due to normalizing the data, have an inherent deviation for shallow boreholes. Charts of Campanella et al. (1985), Eslami and Fellenius (1997), and Robertson (2010) are more accurate than the others.
- Investigation and comparisons focused on database records for soil behavioral identification showed a few limitations and deficiencies for the current methods. The inadequacies when using the current diagrams with two axes will be more pronounced. This is due to that the methods try to present and interpret the soil boundaries via combination of CPTu data. Trends include effective cone resistance, stress normalization based on overburden stress estimation rather represent data as a function of each other or using two supplementary charts.

- To overcome a few of shortcomings, a new chart has been defined by the ability to use all three parameters simultaneously in a triangular format which is similar to the soil classification chart proposed by United State Department of Agriculture (USDA).
- Applying this procedure, it can be facilitated to use all three records of CPTu, concurrently including q_t, f_s and u_2 values by defining a new way for use of the triangle chart which consists of seven, mainly deltaic, deposits zones. Twelve groups of soils which are collected in the database have been examined by the proposed approach on the triangular chart. Above all, the proposed diagram with more accuracy and less scattering than the previous charts is able to identify deltaic soil types. Of course, a large database with much more data delineation can improve the chart zonation and characterization.

References

Begemann, H.K.S., 1965. The friction jacket cone as an aid in determining the soil profile. In: Proceedings of the 6th International Conference on Soil Mechanics and Foundation Engineering, ICSMFE, Montreal, September 8–15, vol. 2, pp. 17–20.

Cai, G.J., Liu, S.Y., Puppala, A.J., Tong, L.Y., 2015. Identification of soil strata based on general regression neural network model from CPTU data. Marine Georesources & Geotechnology 33 (3), 229–238. https://doi.org/10.1080/1064119x.2013.843046.

Campanella, R.G., Gillespie, D., Robertson, P.K., 1982. Pore pressures during cone penetration testing. In: Proceedings of the 2nd European Symposium on Penetration Testing, ESOPT-2, Amsterdam, May 24–27, vol. 2, pp. 507–512.

Campanella, R.G., Robertson, P.K., 1988. Current status of the piezocone test. In: Proceedings of First International Symposium on Penetration Testing, ISOPT-1, Orlando, March 22–24, vol. 1, pp. 93–116.

Campanella, R.G., Robertson, P.K., Gillespie, D., Greig, J., 1985. Recent developments in in-situ testing of soils. In: Proceedings of 11th International Conference on Soil Mechanics and Foundation Engineering, vol. 2. ICSMFE, San Francisco, pp. 849–854.

Canadian Foundation Engineering Manual, CFEM, fourth ed., 2006. Canadian Geotechnical Society, BiTech Publishers, Vancouver. 488 pp.

Coduto, D.P., 2002. Foundation Design Principles and Practices, second ed. Prentice Hall, Inc, Upper Saddle River, NJ.

Douglas, B.J., Olsen, R.S., 1981. Soil classification using electric cone penetrometer. In: American Society of Civil Engineers, ASCE, Proceedings of Conference on Cone Penetration Testing and Experience, St. Louis, October 26–30, pp. 209–227.

Erwig, H., 1988. The Fugro guide for estimating soil type from CPT data. In: Proceedings of Seminar on Penetration Testing in the UK. Thomas Telford, London, pp. 261–263.

Eslami, A., 1996. Bearing Capacity of Piles from Cone Penetrometer Test Data. Ph.D. Thesis. University of Ottawa, Department of Civil Engineering, p. 516.

Eslami, A., Alimirzaei, M., Aflaki, E., Molaabasi, H., 2016. Deltaic soil behavior classification using CPTu records—proposed approach and applied to fifty-four case histories. Marine Georesources & Geotechnology 35 (1), 62–79.

Eslami, A., Fellenius, B.H., 1995. Toe bearing capacity of piles from cone penetration test (CPT) data. Report 3:95. In: Proceedings of the International Symposium on Cone Penetration Testing, CPT 95, Linköping, Sweden, October 4–5, vol. 2. Swedish Geotechnical Institute, SGI, pp. 453–460.

Eslami, A., Fellenius, B.H., 1996. Pile shaft capacity determined by piezocone (CPTu) data. In: Proceedings of 49th Canadian Geotechnical Conference, September 23-25, St. John's, Newfoundland, vol. 2, pp. 859–867.

Eslami, A., Fellenius, B.H., 1997. Pile capacity by direct CPT and CPTu methods applied to 102 case histories. Canadian Geotechnical Journal 34 (6), 880–898.

Eslami, A., Fellenius, B.H., 2004. CPT and CPTu data for soil profile interpretation: review of methods and proposed new approach. Iranian Journal of Science and Technology 28, 69–86.

Farhadi, H., Eslami, A., 2012. Evaluating the performance of deep soil improvement in problematic soils using CPT and CPTu. In: 9th International Congress on Civil Engineering, Isfahan, Iran.

Finno, R.J., 1989. Subsurface conditions and pile installation data. American Society of Civil Engineers. In: Proceedings of Symposium on Predicted and Observed Behavior of Piles, Evanston, June 30. ASCE Geotechnical Special Publication, pp. 1–74. GSP23.

Holtz, R.D., Kovacs, W.D., Sheahan, T.C., 2011. An Introduction to Geotechnical Engineering, second ed. Prentice Hall, Inc.

Jefferies, M.G., Davies, M.P., 1991. Soil classification using the cone penetration test. Discussion Canadian Geotechnical Journal 28 (1), 173–176.

Jefferies, M.G., Davies, M.P., 1993. Use of CPTu to estimate equivalent SPT N60. American Society for Testing and Materials, Geotechnical Testing Journal 16 (4), 458–468. December.

Jones, G.A., Rust, E., 1982. Piezometer penetration testing, CPTU. In: Proceedings of the 2nd European Symposium on Penetration Testing, ESOPT-2, Amsterdam, May 24–27, vol. 2, pp. 607–614.

Ku, C.S., Juang, C.H., Ou, C.Y., 2010. Reliability of CPT Ic as an index for mechanical behavior classification of soils. Géotechnique 60 (11), 861–875. https://doi.org/10.1680/geot.09.p.097.

Larsson, R., Mulabdic, M., 1991. Piezocone Tests in Clay. Swedish Geotechnical Institute, SGI. Report No. 42, 240 pp.

Li, J., Cassidy, M.J., Huang, J., Zhang, L., Kelly, R., 2015. Probabilistic identification of soil stratification. Géotechnique 66 (1), 16–26.

Mayne, P.W., 2008. Piezocone profiling of clays for maritime site investigations. In: 11th Baltic Sea Geotechnical Conference Geotechnics in Maritime Engineering, Gdansk, Poland, September 15–18.

Mayne, P.W., Kulhawy, F., Kay, J.N., 1990. Observations on the development of pore water pressure during piezocone penetration in clays. Canadian Geotechnical Journal 27 (4), 418–428.

Olsen, R.S., Farr, V., 1986. Site characterization using the cone penetration test. In: American Society of Civil Engineers, Proceedings of In-Situ 86, ASCE SPT 6, Blacksburg, June 23–25, pp. 854–868.

Olsen, R.S., Malone, P.G., 1988. Soil classification and site characterization using the cone penetrometer test. In: Proceedings of First International Symposium on Cone Penetration Testing, ISOPT-1, Orlando, March 22–24, vol. 2, pp. 887–893.

Olsen, R.S., Mitchell, J.K., 1995. CPT stress normalization and prediction of soil classification. In: Proceedings of International Symposium on Cone Penetration Testing, pp. 257–262. CPT95, Linköping, Sweden, SGI Report 3:95, 2.

Robertson, P.K., 1990. Soil classification using the cone penetration test. Canadian Geotechnical Journal 27 (1), 151–158.

Robertson, P.K., 2010. Soil behavior type from CPT. In: 2nd International Symposium on Cone Penetration Testing, Huntington Beach, CA, USA, May 2010.

Robertson, P.K., Campanella, R.G., Gillespie, D., Grieg, J., 1986. Use of piezometer cone data. In: Clemence, S. (Ed.), Proceedings of American Society of Civil Engineers, ASCE, In-Situ 86 Specialty Conference, pp. 1263–1280. Blacksburg, June 23-25, Geotechnical Special Publication GSP No. 6.

Sanglerat, G., Nhim, T.V., Sejourne, M., Andina, R., 1974. Direct soil classification by static penetrometer with special friction sleeve. In: Proceedings of the First European Symposium on Penetration Testing, ESOPT-1, June 5–7, Stockholm, 2.2, pp. 337–344.

Schmertmann, J.H., 1978. Guidelines for Cone Test, Performance, and Design. Report FHWA-TS-78209. Federal Highway Administration, Washington, 145 pp.

Searle, I.W., 1979. The interpretation of Begemann friction jacket cone results to give soil types and design parameters. In: Proceedings of 7th European Conference on Soil Mechanics and Foundation Engineering, ECSMFE, Brighton, vol. 2, pp. 265–270.

Vos, J.D., 1982. The practical use of CPT in soil profiling. In: Proceedings of the Second European Symposium on Penetration Testing, ESOPT-2, Amsterdam, May 24–27, vol. 2, pp. 933–939.

CPT in foundation engineering; scale effect and bearing capacity

6

6.1 Scale effect in correlation between pile, foundation, and CPT

Due to similarities between cone penetration test (CPT) and pile, cone penetrometer data can be employed for determination of pile capacity by applying some corrections related to geometry differences and limited displacement of pile. Several research studies show that cone point resistance and sleeve friction are not equal to pile base and shaft capacities, respectively, owing to differences between the cone and pile including geometry, critical depth, strain level, penetration rate, and failure mechanism of soil (Takesue et al., 1998; Fleming et al., 2009; Eslami et al., 2013).

In some direct methods that consider scale effects, the focus has been on geometry parameters, including pile diameter and embedment depth. However, it seems that factors such as penetration rate, strain level, nonhomogeneity, ultimate pile capacity, and friction fatigue are other important factors that must be considered in relating f_s and q_c to pile unit shaft capacity (r_s).

Fig. 6.1 illustrates the differences between pile and CPT in terms of material, penetration rate, and dimensions. As a result of the limited displacement of piles, ultimate soil strength cannot be mobilized and failure of soil does not occur (Fleming et al., 2009). In contrast, CPT can cause failure in the soil. In other words, penetration mechanisms of pile and CPT are different and CPT causes large displacements in soil,

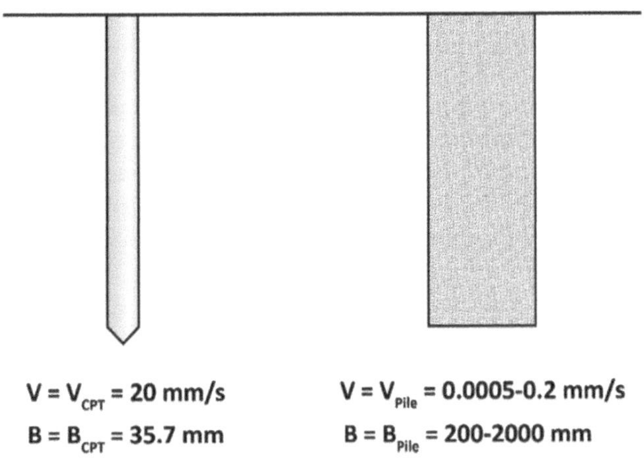

Figure 6.1 Schematic view of pile and cone penetration test differences in material, penetration rate, and dimensions.

Piezocone and Cone Penetration Test (CPTu and CPT) Applications in Foundation Engineering
https://doi.org/10.1016/B978-0-08-102766-0.00006-7

whereas the pile penetrates in a smaller scale into the soil. Moreover, the differences in materials can be attributed to the differences in rigidity and roughness.

6.2 CPT and pile scale effect for toe capacity

Various CPT-based design methods correlate the pile unit toe capacity, r_t, directly with the cone resistance, q_c. On the account of pile and CPT resemblance in terms of geometry and vertical penetration, numerous researchers such as De Beer (1963) and Eslami and Fellenius (1997) have proposed considering the unit toe capacity equal to cone resistance. On the other hand, however, some researchers including Meyerhof (1983), Gui and Bolton (1998), and Jardine et al. (2005) claim that the unit toe resistance is not necessarily equal to cone resistance. Some factors regarding the difference between pile and CPT are as follows:

1. Embedment depth
2. Influence zone
3. Data production processing and averaging
4. Diameter
5. Nonhomogeneous condition
6. Penetration rate and failure mechanism
7. Ultimate capacity interpretation

6.2.1 Embedment depth

Several studies such as De Beer (1963), Meyerhof (1983), and Gui and Bolton (1998) indicate that when penetrometer or a pile penetrates into a new layer, the toe capacity is not fully mobilized. The full mobilization occurs when the pile or penetrometer advances sufficiently in the target layer. The embedment depth is a function of pile diameter and is expressed as a multiple of pile diameter.

Via theoretical and analytical analysis based on conventional bearing capacity theory, Nottingham (1975) compared failure surface zone length for different foundations embedment as shown in Fig. 6.2. With increase in the embedment depth from shallow to deep, the failure mechanism transforms from shallow to deep as well. Accordingly, the failure zone extends, and, subsequently, the ultimate capacity increases.

As a pile has a greater diameter than a CPT instrument, a deeper embedment from the ground surface, or into a hard layer, is required to mobilize the "full" strength of that layer.

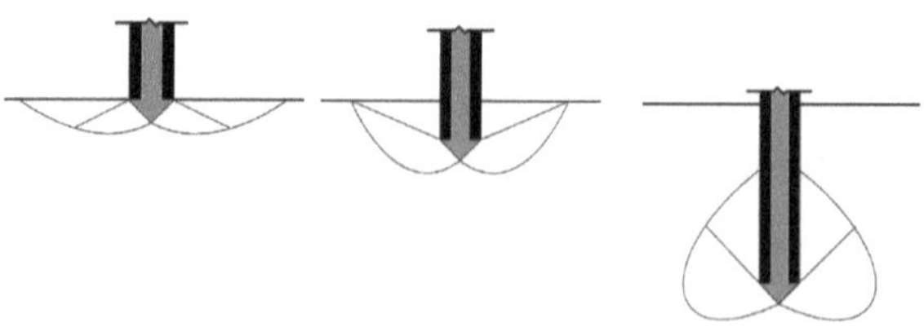

Figure 6.2 Schematic view of transformation of shear failure from shallow to deep.

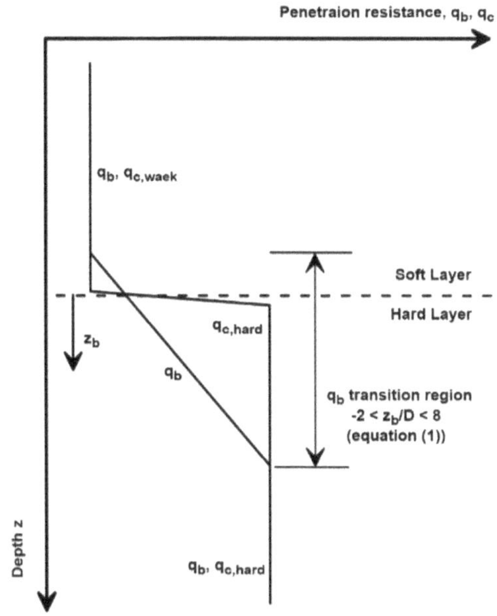

Figure 6.3 Partial embedment reduction factor on base resistance (White and Bolton, 2005).

Prior to sufficient penetration, r_t (q_b) will be less than q_c, as the previous layer will still be "felt" by the pile tip. This mechanism is illustrated in Fig. 6.3. Also, as the L/D ratio of a CPT exceeds that of a pile, the ratio of shaft to base area is higher, and hence, it is for the ratio of q_s/q_b. Analysis of the interaction between the shaft and base offers a mechanism by which the surcharge on the soil surrounding the base of a CPT is higher than around the base of a pile, leading to a corresponding decrease in q_b/q_c (White and Bolton, 2005).

6.2.2 Influence zone

To estimate the unit toe resistance of a pile (large-scale), from the average cone point resistance (a small-scale model pile), it is necessary to identify the failure zone around the pile toe. That is, the problem is to define the extent of the influence zone for averaging the cone resistance. However, no specific evidence exists to support whether the failure is local punching or general shear failure.

Experimental studies by Meyerhof (1956, 1976) indicate that a penetration distance into a soil layer of about 10 times of pile diameter is required to fully mobilize the ultimate unit toe resistance of a bearing layer. A similar depth was found by De Beer (1963) to be needed for developing resistance for penetrometers in sand. Experimental and numerical work by Altaee et al. (1993) on two concrete piles in medium dense sand shows that the zone influencing the toe resistance extends from 5b below to 5b above the pile toe.

Ghionna et al. (1993) show that both displacement and nondisplacement piles reach a value of ultimate unit penetration resistance that is approximately equal to the cone point unit resistance value. A small difference was found to be due to the slight difference in apex angles between the soil cone that forms beneath a pile and apex angle of the cone penetrometer. However, the ultimate resistance is reached for nondisplacement piles at a higher ratio of penetrating than for displacement piles.

According to Salgado (1995), performing tests on dry sand in calibration chambers, the penetration resistance for a penetrometer with a certain apex angle, is independent of the penetrometer diameter. They also found that the difference between the penetration resistance for a flat-ended and for the standard cone penetrometer (with the apex angle of 60 degree) is about 15% for dense sands at moderately low confining stress and is much smaller for loose sands under high confining stress.

In homogeneous soil, upon experimental and analytical study by Eslami (1996), a logarithmic spiral can be assumed to define the failure zone, as presented in Fig. 6.4. The failure pattern is general shear failure type, in which the rupture surfaces reach to the body of pile or penetrometer, similar to those proposed by Meyerhof (1951) and De Beer (1963).

The deepest point of the rupture surface below the pile toe is determined by maximizing the projection of the radius of the logarithmic spiral on the vertical axis.

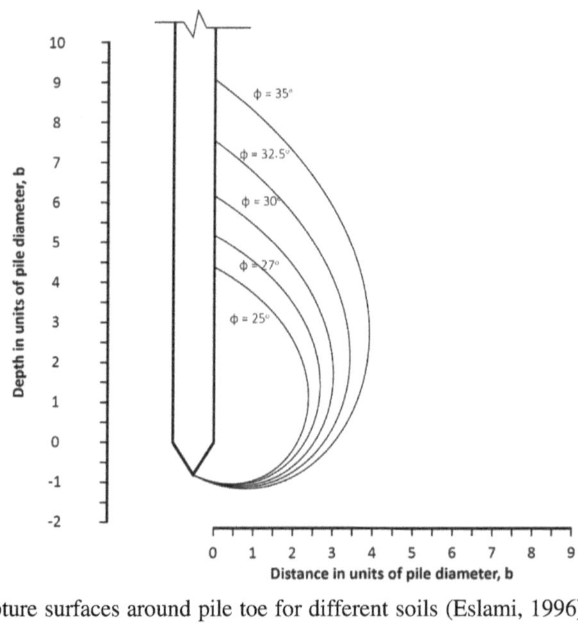

Figure 6.4 Rupture surfaces around pile toe for different soils (Eslami, 1996).

6.2.3 Nonhomogeneous condition

The cone resistance needs to be averaged to a representative value within the failure (influence) zone, and the logarithmic spiral is but one way to determine the zone. However, as mentioned, such rupture surfaces apply only to homogeneous soil, and the height of the zone is not critical for averaging cone resistance. To illustrate, the variation of the geometric average can be studied as a function of the extent of the failure zone above and below the pile toe. For the CPT result shown in Fig. 6.5,

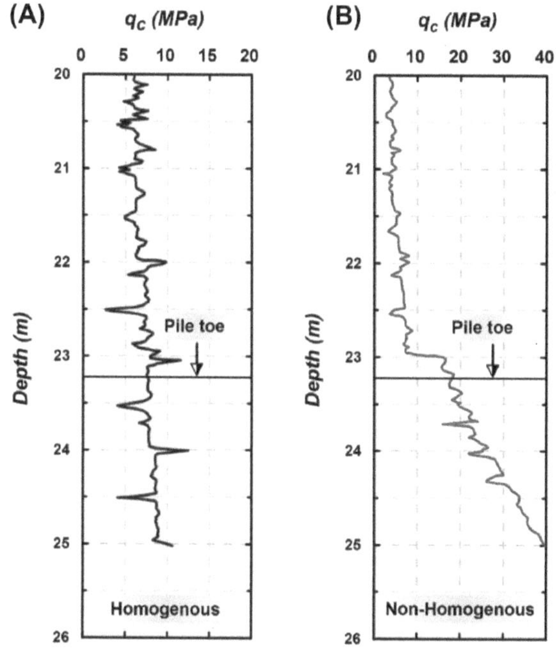

Geometric Avreage of Cone Resistance For Diffrent zones (b=350mm), MPa

Zone	Homogenous	Non-Homogenous
2b-2b	7.66	16.13
2b-4b	7.43	11.79
2b-6b	7.09	9.06
2b-8b	6.93	8.06
4b-2b	7.89	19.92
4b-4b	7.65	14.89
4b-6b	7.38	11.97
4b-8b	7.15	10.11

Figure 6.5 Comparison of pile unit toe resistance for different zones: (A) Homogeneous and (B) Nonhomogeneous (Eslami, 1996).

different zones for averaging cone resistance are considered. Fig. 6.5A presents CPT data from homogeneous soil, and Fig. 6.5B shows the records for a two-layered soil with a distinct increase of cone resistance in the lower layer (nonhomogeneous soil). A 350-mm-diameter pile (10 times the standard cone diameter) is assumed to have been installed a small distance into the lower layer and is used to illustrate the geometric average of q_c values in different height failure zones. For homogeneous soil, the geometric q_c values are insensitive to the height of the zone, see Fig. 6.5A.

In contrast, for a pile in nonhomogeneous soil when the pile toe is located in the vicinity of the boundary between two layers, significant difference exists between the geometric averages of cone resistance determined in zones of different heights.

As a result, in homogeneous soil, the extent of the influence zone for determining the average cone resistance to apply to pile unit toe resistance is not critical. In case of nonhomogeneous soil, detailed theoretical analysis does not apply. To calculate the cone resistance average, the distances to be 4b (b is pile diameter) below to 8b above pile toe which is used by current practice, when a pile is installed through a weak soil into a dense soil. When a pile is installed through a dense soil into a weak soil, decreasing the upper limit height from 8b to 2b appears appropriate.

The extent of 4b for influence zone below pile toe is due to the fact that the dominant extension of rupture surface below pile toe is horizontal in contrast to the zone above pile toe where it is vertical. Although, in theory the expansion of a rupture surface in a homogeneous soil below pile toe does not exceed about 1.5b (see Fig. 6.4), it is necessary to take into account the cone resistance in a zone somewhat deeper than 1.5b for two main reasons. First, because of the horizontal extension of the rupture surface which ranges from 2b through 5b, the failure path is longer than 1.5b. Second, to include the effect of weakening of the soil due to tension developing in the boundary to the weak soil punching effect.

6.2.4 Data processing and averaging

It is important to note that the variation of soil characteristics must be taken into account by averaging over a sufficient distance above and below the pile toe. The variation of soil characteristics is reflected in the variation of the cone resistance. For most natural soil deposits, a range of cone resistance exists from soft or loose layers into hard or dense layers and vice versa. In particular, natural sand deposits produce q_c profiles with many sharp peaks and troughs.

Along a specific well-defined rupture surface, when the unit force exerted by the pile toe approaches the ultimate unit toe resistance, no single friction angle and/or single fraction of mobilized q_c value applies to the whole rupture surface. Points at the rupture surface, located at different depths, may have different friction angle and/or mobilized q_c value due to the soil variation and confining stress. To obtain cone data for pile design especially for unit toe resistance, a representative cone resistance value (friction angle in an indirect method) must be sought. It is important to note that the larger the pile diameter, the greater the extent of rupture surface below and above the pile toe. Therefore, a procedure must be established to find the representative q_c value to relate to the unit toe resistance of a pile, and the value must be a function of the pile diameter.

To determine the pile unit toe resistance, all current direct CPT methods apply a correlation factor to the cone resistance after filtering the q_c values, "purging" the peak and trough values from the data. The mean of the resulting smoothened curve, the q_c average, is then determined over a certain zone, which differs in extent for the methods used in current practice. The correlation factor to apply to the q_c average differs between the methods.

Filtering the cone data is necessary because were a true mean produced from the data, the high and low values would have a disproportionate influence on the average. The filtering approach was developed when the CPT data were obtained in hard copy diagrams only, and it brings about a considerable judgment leeway in finding the average cone resistance. However, the subjective filtering is now not necessary because current tests produce digitized results in the form of tables of data, easily accessible for determining the average by direct computer manipulation, making working with hard copy diagrams redundant.

Two types of averaging the cone resistance can be considered: arithmetic average and geometric average. The arithmetic average is defined as:

$$q_{ca} = \frac{q_{c1} + q_{c2} + \cdots q_{cn}}{n} \tag{6.1}$$

where

q_{ca} = arithmetic average of values which range from q_{c1} through q_{cn} and
n = number of considered values.

Having the CPT data in the computer, the average q_c, according to Eq. (6.1), can be obtained automatically. However, without first excluding peaks and troughs, this average is only useful in very homogenous soils where the values are uniform.

In contrast, the natural variability of many sand deposits produces q_c profiles with many sharp peaks and troughs. Filtering is therefore necessary in most cases, and if done manually, it offsets the advantage of the computer. A filter effect can be achieved directly, however, by instead calculating the geometric average of the q_c values, which is defined as:

$$q_{cg} = \sqrt[n]{q_{c1} \cdot q_{c2} \cdot \cdots \cdot q_{cn}} \tag{6.2}$$

where

q_{cg} = geometric average of values which range from q_{c1} to q_{cn}.

The bias in the arithmetic mean arises from the influence of the absolute magnitude instead of ratios of variations (Kennedy and Neville., 1986). For example, the arithmetic average of the numbers 0.5 and 2.0 is 1.25, and the geometric average is 1.00. If the numbers are 0.33 and 3.00 instead, the arithmetic average becomes 1.65

while the geometric average is still 1.00. For a set of data made up of the numbers 0.33, 0.50, 2.00, and 3.00, the arithmetic and geometric averages are 1.46 and 1.00, respectively.

Assume that a set of values is as follows: 5, 5, 2, 5, 25, 5, 6, 1, 6, 6, 30, and 6, where the dominant values lie between 5 and 6. The arithmetic and geometric averages are 8.50 and 5.71, respectively. The result shows that the geometric average is closer to the dominant values, as opposed to the arithmetic average which is not representative for the dominant range. Thus, by taking the geometric average of q_c values in a zone at the vicinity of pile toe, a filtered representative value is obtained that is unaffected by operator's judgment and, therefore, repeatable.

As an example, a pile case history (POLA2, pile toe located in sand presented by Eslami and Fellenius, 1997), is given in Fig. 6.6.

The arithmetic q_c average by the Schmertmann (1978) and the European methods (applying minimum path rule) is 15.0 MPa, the average by the French method (eliminating peaks and troughs) is 24.7 MPa, and by the Meyerhof (1976) method is 17.5 MPa. The geometric q_c average (over a zone which extends 4b below to 8b above pile toe) is 11.5 MPa. These values and the specified zones recommended by methods for taking the average are indicated in Fig. 6.6. It is important to note that all methods relate the obtained arithmetic average of q_c to the unit toe resistance of pile by using correlation factors.

6.2.5 Diameter

Given the larger diameter of a pile compared to CPT, a pile requires larger penetration into a layer in order to mobilize the full capacity. However, considering smaller diameter of CPT, the penetrometer resistance is fully mobilized once it penetrates into a soil layer, as shown in Figure 6.7.

Meyerhof (1983) and Jardine et al. (2005) suggested that the pile unit toe capacity significantly decreases as the pile diameter increases, in a way that a penetrometer (B = 36 mm) has considerably higher unit toe resistance than a pile installed in the same soil. In this regard, using a database of 29 closed end piles with the diameters ranging from 0.1 to 0.9 m, Chow (1997) stated that the pile to cone unit resistance ratio is invariably less than unity, this ratio decreases according to the following equation (Axelsson, 2000):

$$\frac{r_t}{q_c} = 1 - 0.5 \log\left(\frac{B}{B_{CPT}}\right) \tag{6.3}$$

Eslami and Fellenius (1997) proposed a factor, C_t, in their method to correlate the pile unit toe resistance with effective cone resistance. They suggested the C_t value of unity for most cases. Actually, a one by one relation is between q_c and q_b (r_t). While, this relation is true that other factors affecting scale are realized.

White and Bolton (2005) reassembled the Chow (1997) database to examine the relationship between CPT resistance and toe capacity. They concluded that as long

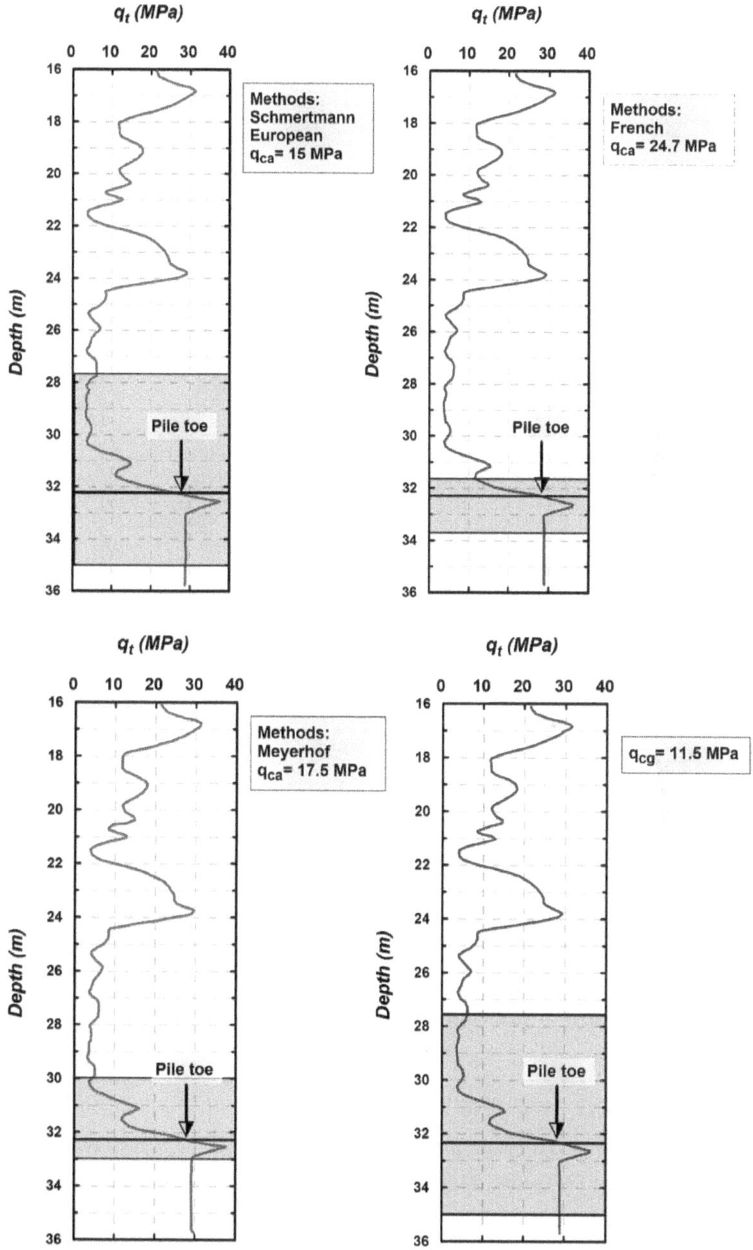

Figure 6.6 An example of comparison of average cone resistance for different cone penetration test methods (Eslami and Fellenius, 1997).

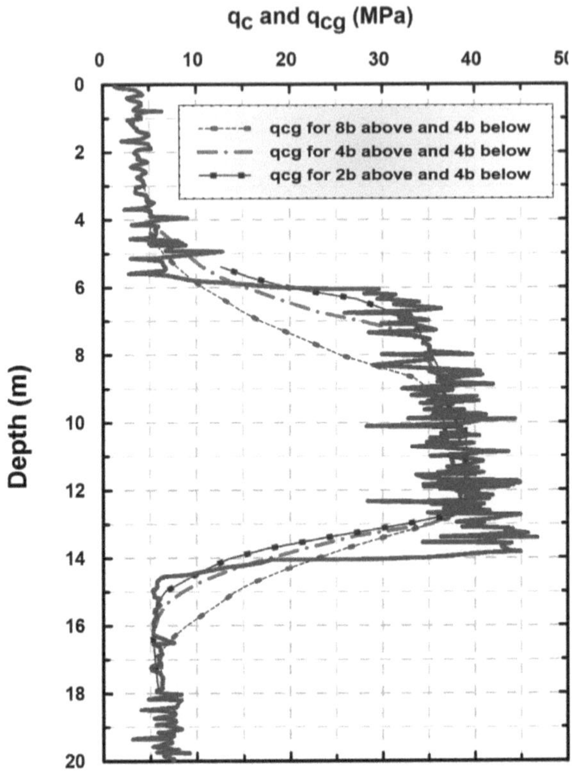

Figure 6.7 Comparison of cone resistance and pile mobilized capacity as penetrating into a soil layer (Eslami and Fellenius, 1997).

as the modification of embedment depth is applied and the loading test is performed according to the defined settlement criterion with residual stress modifications, the scale effect in terms of diameter would be negligible. In such conditions, the unit toe resistance is as high as 90% cone resistance.

6.2.6 Penetration rate and mechanism

Generally, in sands and clays, the CPT is considered to be in drained and undrained conditions, respectively (Campanella and Robertson, 1981). However, the lower penetration rate during static loading test provides sufficient time for the pore pressure to dissipate, and consequently, it appears to have fully drained conditions in sand and relatively drained condition in clays around the pile during a static load test.

Al-Mhaidib (1999), by performing static compression loading tests on a model pile with the diameter of 30 mm under various loading rate of 0.01, 0.05, 0.1, 0.5, and 1 mm/min, highlighted the effect of loading rate on the compressive capacity of piles, in a way that increasing the loading rate leads to an increase in pile capacity.

6.2.7 Ultimate capacity condition

In a static pile load test, the ultimate capacity is defined as when a rapid movement occurs under constant or slightly increased load. Because of large movements required for a pile to reach plunging mode failure, on most occasions a distinct plunging ultimate load is not obtained in the test (Fellenius, 2015). Therefore, several interpretation criteria have been proposed and are applied for determining the pile ultimate capacity.

It is necessary to apply a unique failure criterion in defining the ultimate capacity to make the load test results comparable. Six failure criteria have been summarized by Fellenius (2015). Among these approaches, the Davisson offset limit (Davisson, 1972), the Brinch Hansen 80% criterion (Hansen, 1963), and the Chin-Kondner extrapolation (Chin, 1978) are commonly used. The Davisson limit load usually reports loads in the lower part of the load–displacement diagram. On the other hand, the Chin-Kondner extrapolation assumes an asymptotic curve and the load is defined by extrapolation, and therefore, the results are always greater than the maximum load applied in the test. The Brinch Hansen 80% criterion normally agrees well with the intuitively perceived "plunging failure" of the pile (Fellenius, 2015). Two examples of interpretation of the pile capacity by these criteria are illustrated in Fig. 6.8, which is the load–displacement diagram of a closed-end driven pipe pile with the diameter of 300 mm and the embedment length of 16.2 m (Moshfeghi and Eslami, 2016). The pile capacity at displacement of 10% of the pile diameter is also shown on the diagrams.

Since the cone test is a high strain test, the plunging load should be considered as the ultimate capacity. Also, assessing the pile capacity based on the plunging load prevents

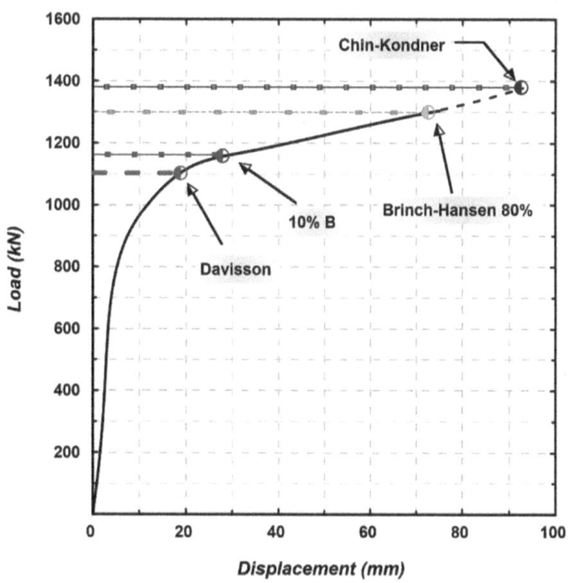

Figure 6.8 Interpretation of load–displacement diagram for Case 001-L&D31 (Moshfeghi and Eslami, 2016).

pile stiffness from clouding the measurements of ultimate pile capacity (White and Bolton, 2005). Therefore, the Brinch-Hansen 80% criterion is selected for interpretation of the load test results for justifying a reasonable comparison and evaluation.

Various CPT-based methods have introduced different load definitions for ultimate capacity. For instance, UniCone (Eslami and Fellenius, 1997) has defined the plunging load as the ultimate capacity, while some more recent methods such as Fugro-05 (Kolk et al., 2005), NGI-05 (Clausen et al., 2005), UWA-05 (Lehane et al., 2005), and ICP-05 (Jardine et al., 2005) have defined the ultimate capacity as the load at the toe or head displacement of 10% of the pile diameter. Some other CPT-based methods such as LCPC (Laboratoire Central des Ponts et Chaussees) have not mentioned a specific criterion for defining the ultimate capacity of piles.

In order to examine effects of the interpretation criterion which is applied to define the pile capacity, the results of the estimated capacity by selected methods are compared to two criteria: the Brinch Hansen 80% criterion and the load at the displacement of 10% of the pile diameter. The average and the standard deviation of the predicted to the measured capacity ratios for each of the methods were calculated and are presented in Fig. 6.9.

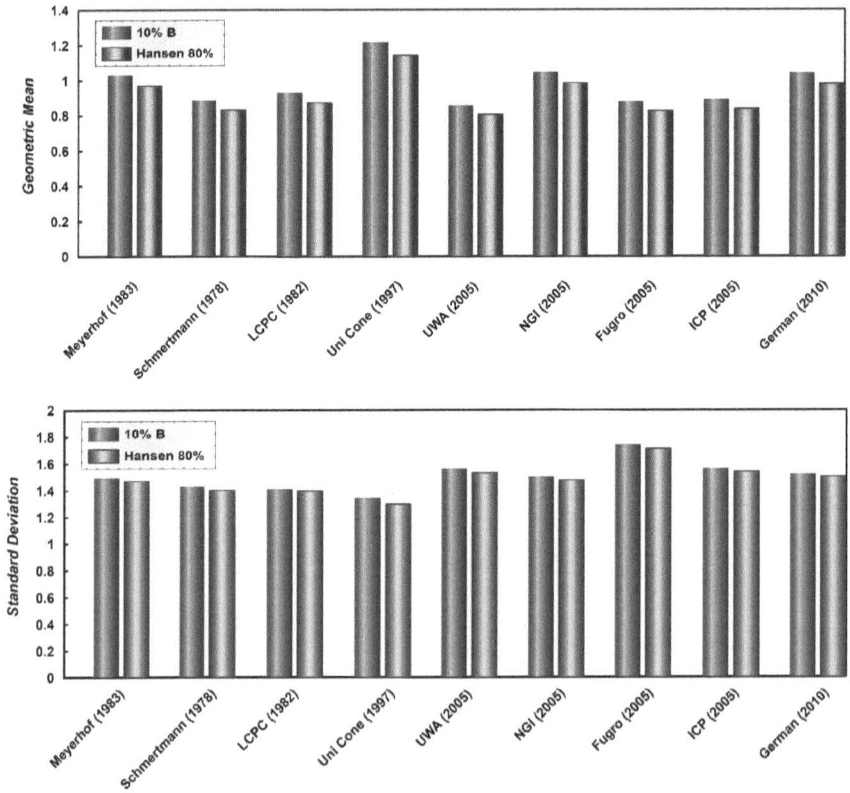

Figure 6.9 Performance of investigated methods against different load test interpretation criteria (Moshfeghi and Eslami, 2016).

The average results from both criteria are almost the same, but since the load at the displacement of 10% of the pile diameter is usually less than the load attained from the Brinch Hansen 80% criterion, the average ratios obtained from this criterion are about 6% less than that of the Brinch Hansen 80% criterion. The standard deviations are also quite close, but the Brinch Hansen 80% criterion essentially leads to less scatter. Thereby, since the Brinch Hansen 80% criterion showed more consistent evaluations, in this study, this criterion has been applied in the analyses.

6.3 Scale effect in estimation of shaft capacity

The studies on the effect of pile length on pile shaft capacity show that unit shaft capacity increases with depth and is proportional to the effective overburden stress. According to conventional interpretations, above a certain depth, called "critical depth," shaft capacity follows the effective stress principal and below the critical depth, shaft capacity would remain constant (Meyerhof, 1976). Also, De Beer (1963) found a reduction in unit shaft capacity with depth and attributed it to the arching that takes place close to the pile toe which results in decreased vertical stress at the pile toe level and leads to a reduced unit shaft capacity close to the pile toe. However, investigations demonstrate that pile shaft and toe capacity have an intermediate trend with depth rather than following a linear trend or remaining constant (Randolph et al., 1994).

The effect of diameter on shaft capacity has not been considered in any of the direct CPT methods except for the Schmertmann (1978) and Nottingham (1975) and UWA (Lehane et al., 2005) methods. It is expected that for two piles with the same soil profile and the same penetration rate, the pile with the larger diameter induces larger displacements in the surrounding soil and has a larger confinement pressure acting on the pile shaft, resulting in a larger pile shaft capacity. However, other parameters such as penetration rate may reverse this trend.

Brown and Hyde (2008) studied the impact of loading rate on the unit shaft capacity with two kinds of tests. Rapid load test, with load increasing from 1000 to 3000 kN, and load test with a constant rate of displacement (0.01 mm/s). They stated that with increasing the rate of loading, the unit shaft capacity increases. Moreover, Al-Mhaidib (1999) by conducting tensile load tests on piles with different load rates (0.01, 0.05, 0.1, 0.5, and 1 mm/s) remarked that increasing the rate of loading causes an increase in unit shaft capacity.

Sabatini et al. (2002) based on deviator stress and strain, shear modulus variations with strain rate, and considering the factor of safety conceived that common in situ tests are involved in large strain category for bearing capacity analysis, Fig. 6.10A and B. Accordingly, shear strength mobilization along pile or any penetrometer is a function of shear strain level as depicted in Fig. 6.10C.

Considering the abovementioned factors, a new approach was proposed by Eslami et al. (2013) that correlates pile unit shaft resistance to cone sleeve friction considering scale effects.

As a first step, a coefficient less than unity, named k, must be determined based on the pile penetration rate (V_{pile}), pile diameter D_{pile}, and strain level. To determine k, the

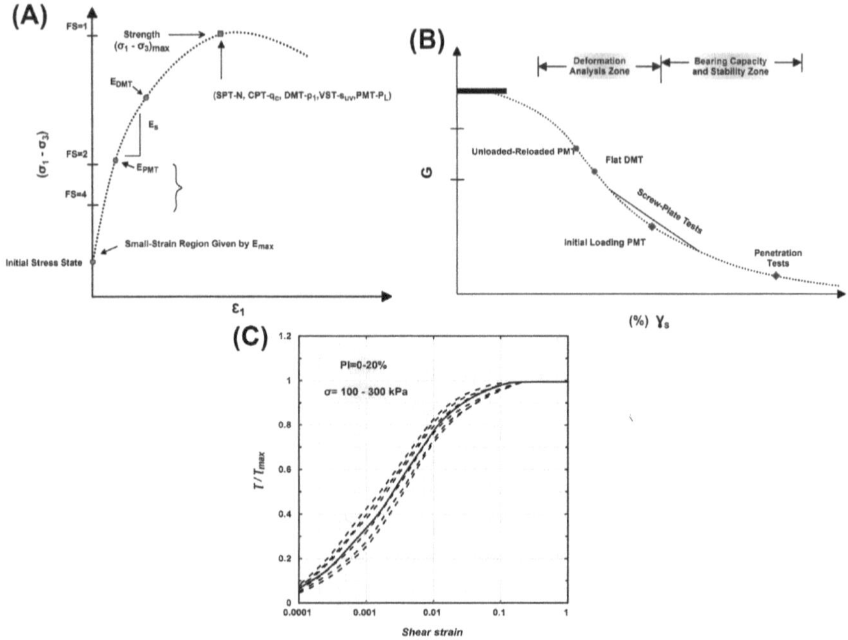

Figure 6.10 A) ideal soil stress-strain relation and soil modulus at various strains. B) variation of shear modulus with shear strain. C) Variation of shear stress with shear strain (Sabatani et al., 2002).

ratio of induced soil shear strain by pile, γ_{pile}, to induced shear strain in soil by CPT, γ_{CPT}, should be calculated. The final formula derived after calibration is as follows:

$$\frac{\gamma_{pile}}{\gamma_{CPT}} = \left(\frac{V_{pile}}{V_{CPT}}\right)^{0.61} \cdot \left(\frac{D_{pile}}{D_{CPT}}\right)^{0.5} \tag{6.4}$$

where, h is the embedment length, and D_{CPT} and V_{CPT} are CPT diameter and rate of penetration, respectively. It should be mentioned that in this formula, it is assumed that $h_{CPT} = h_{pile}$.

Considering $\gamma_{CPT} = 0.3$, $D_{CPT} = 35\ mm$, and $V_{CPT} = 20\ mm/s$, induced shear strain in soil by pile can be expressed as Eq. (6.5):

$$\gamma_{pile} = 0.0255(V_{pile})^{0.61}.(D_{pile})^{0.5} \tag{6.5}$$

where V_{pile} is the penetration rate of pile (mm/s) and D_{pile} is pile diameter (mm).

Now, assuming $\frac{\tau}{\tau_{max}} = k$ and using Ishibashi and Zhang's Chart (1993) for determining $\frac{\tau}{\tau_{max}}$ from shear strain, a new graph is proposed from which k can be obtained, as presented in Fig. 6.11. After calculating k, depending on pile diameter and penetration rate, pile unit shaft capacity, r_s, can be calculated from Eq. (6.6):

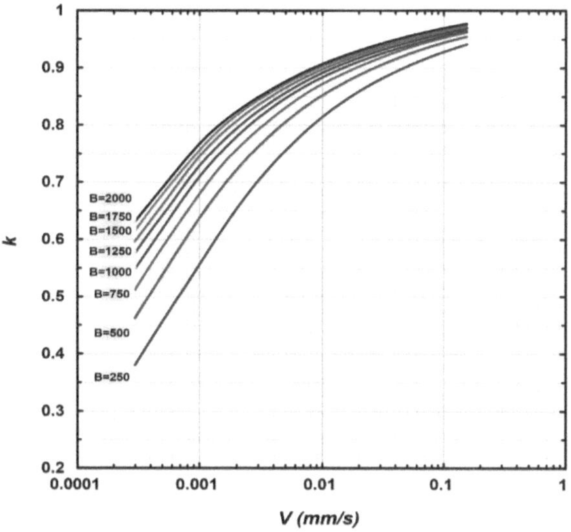

Figure 6.11 Determining k regarding pile diameter and pile penetration rate (Eslami et al., 2013).

$$r_s = k.f_s \tag{6.6}$$

As depicted in Fig. 6.12, for the investigated database of Eslami et al. (2013), the values of unit pile shaft resistance are invariably less that the cone sleeve friction for almost all the cases. A comparison of the load distribution along the pile is also delineated in Fig. 6.13.

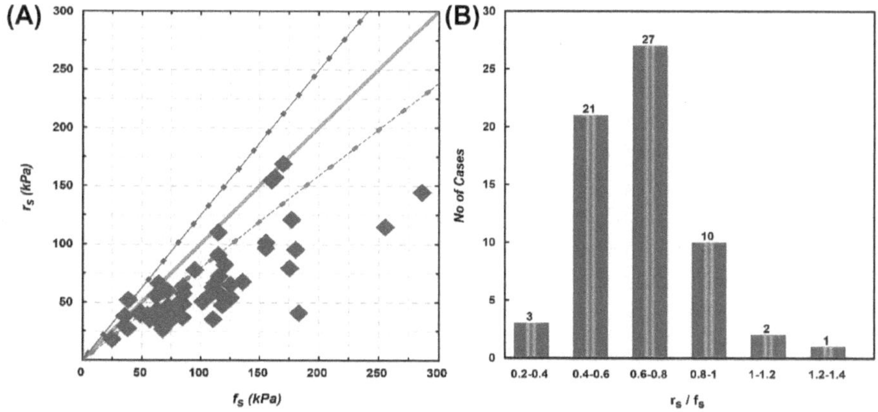

Figure 6.12 (A) Comparison of r_s and f_s, (B) distribution of r_s/f_s values for Eslami et al. (2013) database (Eslami et al., 2013).

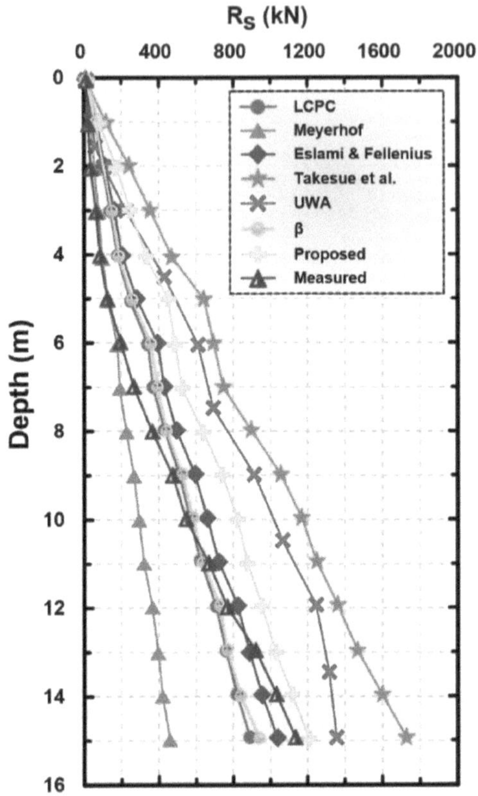

Figure 6.13 An example of shaft capacity distribution along the pile by cone penetration test–based methods and the measured values (Eslami et al., 2013).

6.4 Bearing capacity of shallow foundations

One of the main steps for safe and economic design of foundations is ultimate bearing capacity determination. The bearing capacity of foundations is defined as the maximum load that can be applied to subgrade soil from foundation without occurrence of shear or punching failure and keeping settlement in a limited range avoiding serviceability damages to super structure.

Four approaches currently are used to determine the bearing capacity of shallow and deep foundations; static analysis, in situ testing methods, full-scale loading tests, and using presumed values recommended by codes and handbooks. Among these approaches, the theoretical solution static analysis is more common and employed first, where other approaches are realized as supplementary of static analysis.

In contrast to the SPT (Standard Penetration Test), the CPT is simple, fast, relatively economical, and it supplies continuous records with depth. The results are interpretable on both empirical and analytical basis, and a variety of sensors can be incorporated with cone penetrometer. Evaluating bearing capacity of foundations from CPT data is one of the earliest applications of this sounding and includes two main approaches: direct and indirect methods. Direct CPT methods apply the measured values of cone bearing for bearing resistance with some modifications regarding scale effects, i.e., influence of foundation width to the cone diameter ratio. Indirect CPT methods employ friction angle and undrained shear strength values estimated from CPT data based on bearing capacity and/or cavity expansion theories.

The analogy of cone penetrometer and pile has been caused the concentration of research work in geotechnical practice, on application of CPT data for deep foundations. However, in practice and civil engineering projects, the majority of foundations are spread and shallow. Hence, in the following sections, for shallow foundations (footings), different CPT direct methods for determining bearing capacity are investigated. By utilizing a database including full scale loading test results of footings and CPT soundings performed close to footing locations, the capability of predictive methods has been compared and validated.

6.4.1 CPT indirect methods for bearing capacity of footings

Terzaghi (1943) presented an applicable formula, usually referred to as the "3N bearing capacity formula." For general shear failure mechanism beneath strip footings, the Terzaghi bearing capacity equation is as follows:

$$q_{ult} = CN_c + \overline{q}N_q + 0.5\gamma BN_\gamma \tag{6.7}$$

where

q_{ult} = ultimate unit resistance or bearing capacity of footing,
C = cohesion parameter,
$\overline{q}$ = surcharge around foundation, equals to γD_f,
γ = average effective unit weight of the soil below and around the foundation,
B = foundation width,
D_f = embedment depth of foundation,
N_c, N_q, N_γ = nondimensional bearing capacity factors as exponential functions of φ,
φ = soil internal friction angle.

Terzaghi also recommended a revised formula for different shape of foundations and also in case of occurring punching failure. Later on, Meyerhof (1956), Hansen (1963), and Vesić (1975) have applied modifications for shape, depth, inclination, ground, and base factors to Terzaghi original bearing capacity formula, which was conservative.

In common geotechnical engineering practice, following equation is used (CFEM, 2006):

$$q_{ult} = CN_c s_c i_c d_c + qN_q s_q i_q d_q + 0.5\gamma BN_\gamma s_\gamma i_\gamma d_\gamma \tag{6.8}$$

where

s_c, s_q, s_γ = nondimensional shape factors,

i_c, i_q, i_γ = nondimensional inclination factors,
d_c, d_q, d_γ = nondimensional depth factors.

The ultimate bearing capacity is divided in a safety factor usually equals to 3 and the result is called "safe bearing capacity," which is used for foundation design. However, to satisfy settlement criteria, it might be employed even greater values of safety factors, and the outcome is called "allowable bearing capacity."

Perhaps it should be emphasized that the forenamed equations pertain to the loading of footing under long-term service condition, that is, fully drained conditions. Rapid loading of foundations in normally consolidated clays and silts will create pore pressures that reduce soil strength. Practice has established rules for the analysis of such "undrained" conditions, normally assessing the stability in a slip circle analysis with input of undrained shear strength as follows:

$$q_{ult} = S_u N_c + \overline{q} \tag{6.9}$$

where

S_u = undrained shear strength,
N_c = cohesion coefficient ranges from 5 to 6 for shallow foundations.

Several correlations have been suggested for determining φ-angle or S_u from CPT data as presented in Chapter 4. In turn, the bearing capacity can be calculated from Eqs. 6.8 or 6.9. Other than empirical and semiempirical correlations between cone resistance and soil friction angle, the two theories, bearing capacity or cavity expansion, are used.

Numerous methods have been developed for evaluating φ-angle from q_c. Schmertmann (1978) proposed a relationship between the φ-angle and relative density (D_r) for different grain size characteristics. This approach requires the approximate nature to which D_r can be estimated from CPT data. Experience has shown that D_r−q_c correlations are sensitive to soil compressibility and in situ horizontal stresses. Robertson and Campanella (1983) compared measured cone resistance q_c to measured friction angle values which were obtained from drained triaxial compression tests performed at confining stresses equal to in situ horizontal stress σ'_{h0} in the calibration chamber. A set of simple relationship is shown in Fig. 6.14, where q_c increases linearly with effective overburden stress for constant φ-angle. The results were obtained for uncemented moderately incompressible silica sands. For highly compressible sands, the chart shown in Fig. 6.14 would tend to predict low friction angle.

Bearing capacity solutions are basically based on plane strain conditions, linear strength envelops incompressible materials. Two main available methods were developed by Janbu and Senneset (1974) and Durgunoglu and Mitchell (1975). The solution by Janbu and Senneset depends on the shape of failure zone, but the Durgunoglu and Mitchell (1975) method accounts for the effect of in situ horizontal stress and cone roughness. The cavity expansion method developed by vesić (1972) accounts for soil compressibility and volume change characteristics. Research by Mitchell and Keaveny (1986) has shown that the cavity expansion method appeared to model the

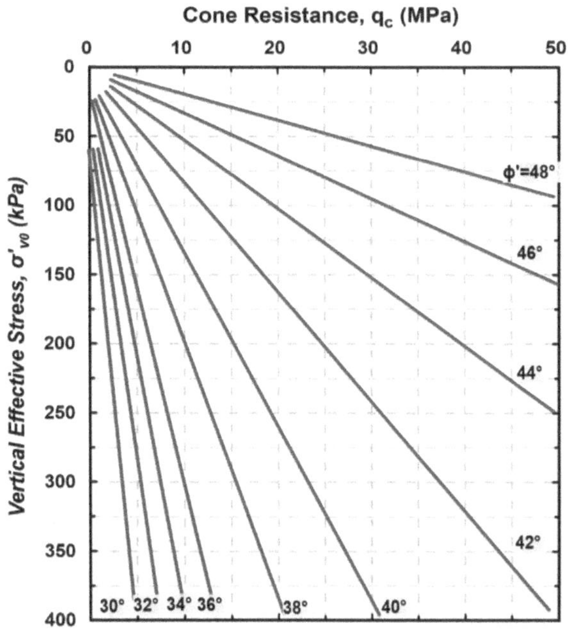

Figure 6.14 Correlation between peak friction angle φ and q_c for uncemented quartz sand (Durgunoglu and Mitchell, 1975).

measured response extremely well and could predict the response in compressible sands. However, the cavity expansion approach requires knowledge of soil stiffness and volumetric strains in the plastic region, which both are difficult to estimate or derive.

Also for interpretation of undrained shear strength of clays from CPT or CPTu results, a large amount of work has been reported based on theoretical solutions or empirical correlations. The relationship between cone resistance q_c and S_u from theory is as follows:

$$q_c = N_{Co} S_u + \sigma_0 \tag{6.10}$$

where

N_{Co} is the cone factor and σ_0 is the in situ total stress. Depending on which theory is used, σ_0 may be replaced by σ_{v0}, σ_{h0}, or σ_{mean}.

Since cone penetration performance is a complex phenomenon, the theoretical solutions make several simplifying assumptions regarding soil behavior, boundary conditions, and failure mechanism. The output of theoretical solutions requires to be verified from actual field or laboratory test data. Therefore, empirical correlations are preferred, the S_u from empirical approaches can be estimated from q_c, f_s, or u as follows (Robertson and Campanella, 1989)

$$S_u = \frac{q_c - \sigma_{v0}}{N_k} \tag{6.11}$$

$$S_u = \frac{\Delta_u}{N_u} \tag{6.12}$$

$$S_u = N_s f_s \tag{6.13}$$

where

N_k = cone factor ranging from 10 to 15 for N. C. clays and 15–20 for O. C. clays,
$\Delta_u = u - u_0$; u is mobilized penetration pore water pressure and u_0 is hydrostatic pressure,
N_u = a factor that ranges from 2 to 20 depending some soil parameters,
f_s = CPT sleeve friction,
N_s = a coefficient ranges from 0.8 to 1.

According to Lunne et al. (1997), the use of the piezocone to correct the cone resistance for pore pressure effects, and employing reference values of undrained shear strength have helped in improving the quality of correlations in general. Moreover, use of site-specific empirical correlations still seems to be the well-defined procedure for interpretation of S_u from CPT or CPTu data.

Above all, soil parameters estimated from the CPT data include errors due to disregard of horizontal stress, assumption of plane strain failure mechanism for axisymmetric nature of deep cone penetration, soil compressibility, rate of penetration, and strain softening.

6.4.2 *Empirical direct CPT methods for bearing capacity of footings*

Direct CPT methods which initiated theoretically or empirically, relate the ultimate bearing capacity of soils, q_{ult} to the cone point resistance, q_c with some modification factors. Among different methods, the followings are commonly used by geotechnical engineers.

Schmertmann (1978) proposed bearing capacity factors based on Terzaghi (1943) basic formula for noncohesive soils from CPT data as follows:

$$q_{ult} = \overline{q}N_q + 0.5\gamma B N_\gamma \tag{6.14}$$

$$N_q = N_\gamma = 1.25\overline{q}_c \tag{6.15}$$

$$\overline{q}_c = \sqrt{q_{c1} \times q_{c2}} \tag{6.16}$$

where

q_{c1} = arithmetic average of q_c values in an interval between footing base and 0.5B beneath footing base and
q_{c2} = arithmetic average of q_c values in an interval between 0.5B and 1.5B beneath footing base.

Meyerhof (1976) suggested a direct method for estimating q_{ult} from cone resistance in sands and clays as follows:

$$q_{ult} = \overline{q}_c \left(\frac{B}{12.2}\right)\left(1 + \frac{D_f}{B}\right) \tag{6.17}$$

where

$\overline{q}_c$ = arithmetic average of q_c values in a zone including footing base and 1.5B beneath footing.

Factor of safety of at least 3 is recommended by Meyerhof to obtain the allowable bearing pressure.

Bowles (1996) proposed separated equations for ultimate bearing capacity of sands as follows:

$$q_{ult} = 28 - 0.0052\left(300 - \overline{q}_c\right)^{1.5}, \text{ for strip footings} \tag{6.18}$$

$$q_{ult} = 48 - 0.0052\left(300 - \overline{q}_c\right)^{1.5}, \text{ for square footings} \tag{6.19}$$

where

$\overline{q}_c$ = the arithmetic average of q_c values in an interval between footing base and 1.5B beneath, in terms of kg/cm^2.

CFEM (2006) recommended an equation for evaluation of allowable bearing capacity using:

$$q_a = 0.10\overline{q}_c \tag{6.20}$$

Also, based on CFEM, the safety factor of 3 has been suggested. Hence, the ultimate bearing capacity is:

$$q_{ult} = 0.30\overline{q}_c \tag{6.21}$$

Eslaamizaad and Robertson (1996), based on Meyerhof method, (Eq. 6.22), by using some case histories and CPT soundings close to foundation locations in cohesionless soils, proposed a relationship between q_{ult} and q_c as follows:

$$q_{ult} = k\overline{q}_c \tag{6.22}$$

where

k is a correlation factor, and is a function of B/D_f, shape of footing and sand density, as shown in Fig. 6.15.

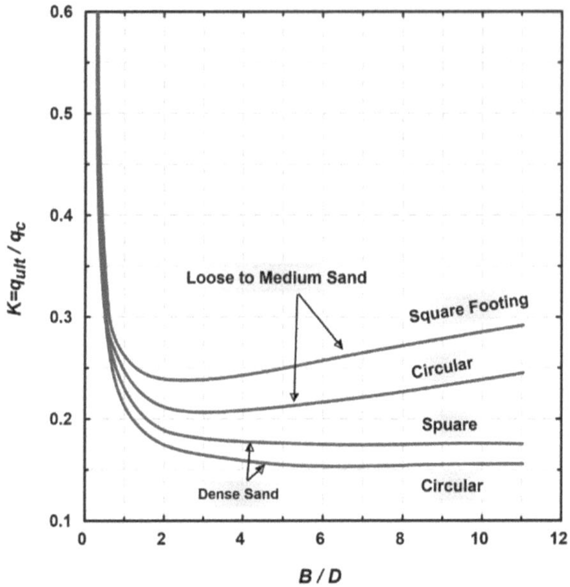

Figure 6.15 Correlation between ratio of bearing capacity to average cone resistance and width factor (Eslamizaad and Robertson, 1996).

Tand et al. (1994) employed a few full-scale load tests with CPT data and suggested the ultimate bearing capacity of shallow footings on lightly cemented medium dense sand by following equation:

$$q_{ult} = R_k q_c + \sigma_{v0} \quad (6.23)$$

where

R_k values ranges from 0.14 to 0.2, depending on the footing shape and depth, and σ_{v0} is the initial vertical stress at the footing base.

6.5 Analytical approach for shallow foundation bearing capacity: direct CPT method

The attempts to predict pile toe capacity from CPT data were generally successful, whether the unit toe resistance was taken directly from the q_c value at pile toe level or taken as the average of q_c values for some distance above and below the pile toe. However, there is still some skepticism regarding application of small-scale cone penetrometer to large-scale deep foundations, which referenced as scale effect. Considerable insight into these problems was provided by De Beer (1963) in a thorough study of the effect of pile or penetrometer size on the ultimate toe bearing capacity. By considering penetrometers of different size, De Beer showed that all penetrometers should ultimately attain the same point resistance and the depth at which this constant value occurs must be proportional to the penetrometer size.

The explanation offered by De Beer (1963) for this phenomenon was that the bearing capacity failure which occurs beneath the cone tip gradually changes from that for a shallow foundation to that of a deep one as shown in Fig. 6.2.

During the transition between the shallow- and deep-type base failure, the point resistance increases linearly with depth, but after the deep foundation condition is reached, point resistance only increases slightly with increased penetration. The depth in which this condition is met is termed the critical depth (D_c). The correct value of D_c is more difficult to determine. Meyerhof (1976), De Beer (1963), and Eslami and Fellenius (1997) have shown that the D_c is a function of the foundation size and soil density. Thus it is more convenient to express this depth in terms of the ratio of the foundation size to the critical depth, i.e., $(D/B)_c$. Both experimental and theoretical studies have denoted that $(D/B)_c$ for sands is a function of the sand density and varies from about 5 for loose sands to about 15 for dense sands.

Research by De Beer (1963), Meyerhof (1976), and Eslami and Fellenius (1997) indicated that a type of logarithmic spiral failure zone and attributed rupture surfaces to reach penetrometer body need a penetration depth of almost 10 times of penetrometer diameter to fully mobilize the ultimate unit toe resistance. In other words, penetration depth of at least 10 footing width or cone diameter is required to transform shallow to deep mechanism of rupture surface. In the former, the rupture surface extent to the ground level but for the latter, the rupture surface reaches to the penetrometer or footing body at deep soil level.

A direct CPT method by Eslami and Gholami (2006) was developed based on an analytical model for determining ultimate bearing capacity, q_{ult}, of shallow foundations from CPT cone resistance, q_c. Regarding the basic bearing capacity formula, the length (L) confining and depth of rupture surface, i.e., embedment depth (A), have a major role in mobilized foundation bearing capacity. These two key factors can be regarded in a rational matter for connection of shallow rupture surface to deep entirely mobilized failure zone. Therefore, the q_{ult} to q_c in direct approach can be correlated by a general rule as follows:

$$q_{ult} = \overline{\alpha} \times \overline{q}_{c,g} = (\alpha_1 + \alpha_2)/2 \times \overline{q}_{c,g} \quad (6.24)$$

$$\overline{\alpha} = (\alpha_1 + \alpha_2)/2 \quad (6.25)$$

$$\alpha_1 = \frac{L_\theta}{L_o} \quad (6.26)$$

$$\alpha_2 = \frac{A_\theta}{A_o} \quad (6.27)$$

where

$\overline{q}_{c,g}$ = geometric average of q_c values from footing base to 2B beneath footing depth,

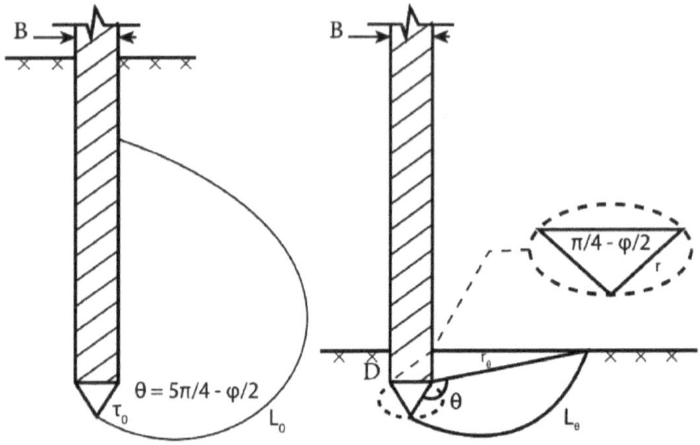

Figure 6.16 Comparison of rupture surface length for shallow and deep conditions (Eslami and Gholami, 2006).

α_1 = modification transforming length ratio,
α_2 = modification transforming area embedment (depth) ratio,
$\overline{\alpha}$ = average of length and area modification factor.
L_o, L_θ, A_o, and A_θ have been shown in Figs 6.16 and 6.17.

In the following, the details of determining forenamed parameters and ratios will be discussed.

In practice, natural soil deposits produce cone point resistance profile with high and low values. The cone resistance variations reflect the alteration of soil strength and stiffness. Therefore, when determining foundation bearing capacity, which is a

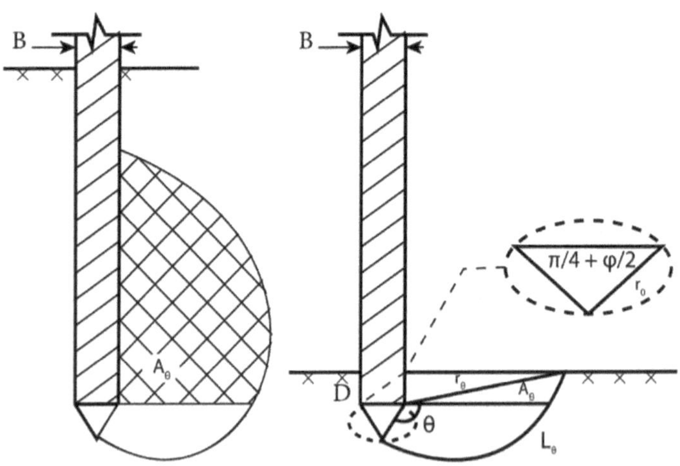

Figure 6.17 Comparison of embedment depth (surcharge) for shallow to deep conditions (Eslami and Gholami, 2006).

function of the subsoil conditions in a zone above and below the foundation base, an average must be determined that is representative for the influence zone. The extent of influence zone regarding scale effect should be a function of the foundation width, B. Based on the modified bearing capacity equations by Meyerhof (1976), Hansen (1963), and Vesić (1972), the extent of rupture surface beneath the foundation base is about B/2 tan (45+φ/2), which approximately recommends 1.5−2B in geotechnical practice. Consequently, the CPT q_c-values should be averaged along influence zone extending from 2B below foundation base to ground level.

According to the argument by Eslami and Fellenius (1997), filtering and smoothening of peaks and troughs for q_c-values can be achieved by calculating the geometrical average, $q_{c.g}$ (geometric mean), within influence zone. By geomean of q_c data, a filtered representative value is obtained which is unaffected by bias and therefore repeatable. The geomean of q_c values must be calculated in a zone at the vicinity of foundation base.

According to Fig. 6.16 and with assumption of log spiral general shear failure, the radius of rupture surface, r, by realizing geometric and trigonometric relationships can be calculated as:

$$r = r_0 e^{\theta \tan \varphi} \tag{6.28}$$

$$r = \frac{B}{2} \tan\left(\frac{\pi}{4} + \frac{\varphi}{2}\right) \times \frac{1}{\cos\left(\frac{\pi}{4} - \frac{\varphi}{2}\right)} e^{\theta \tan \varphi} \tag{6.29}$$

where r is the radius of the logarithmic spiral; r_0 is the radius of the log-spiral for $\theta = 0$, θ is the angle between a radius and r_0, as shown in Fig. 6.16; and φ is the angle between radius and normal, at any point on the spiral, assumed equal to the friction angle of the soil, as pointed out by Meyerhof (1976).

Fig. 6.16 presents pattern of rupture surfaces regarded in the proposed method. The depth of embedment is derived as:

$$D = y = r_0 e^{\theta \tan \varphi} \times \sin\left(\theta - \frac{\pi}{2} - \left(\frac{\pi}{4} - \frac{\varphi}{2}\right)\right) \tag{6.30}$$

By integrating the curve length and using substitution factor “m” as:

$$m = \frac{B}{2} \tan\left(\frac{\pi}{4} + \frac{\varphi}{2}\right) \times \frac{1}{\cos\left(\frac{\pi}{4} - \frac{\varphi}{2}\right)} \tag{6.31}$$

The relative depth, the ratio of foundation depth to foundation width, can be obtained by following equation:

$$\frac{D_f}{B} = \frac{\tan\left(\frac{\pi}{4} + \frac{\varphi}{2}\right) e^{\theta \tan \varphi}}{2 \cos\left(\frac{\pi}{4} - \frac{\varphi}{2}\right)} \sin\left(\theta - \frac{3\pi}{4} + \frac{\varphi}{2}\right) \tag{6.32}$$

Therefore, the length of rupture surface becomes:

$$L_\theta = \frac{m\sqrt{1+tan^2\varphi}}{tan\varphi} \times \left(e^{\theta tan\varphi} - 1\right) \tag{6.33}$$

As a result, the ratio of rupture surface length converting from shallow to deep conditions which was defined as α_1 is:

$$\alpha_1 = \frac{L_0}{L_{\theta=\left(\pi+\frac{\pi}{4}-\frac{\varphi}{2}\right)}} = \frac{e^{\theta tan\varphi} - 1}{e^{\left(\frac{5\pi}{4}-\frac{\varphi}{2}\right)} - 1} \tag{6.34}$$

In addition to length of rupture surface effect, the influence of surcharge around penetrometer (depth) can be considered as area (A) of surcharge in unit length and calculated as:

$$A_\theta = m^2/4 \tan\varphi \left(\frac{L_0}{L_{\theta=(\pi+\frac{\pi}{4}-\frac{\varphi}{2}}} = \frac{e^{\theta tan\varphi} - 1}{e^{\left(\frac{5\pi}{4}-\frac{\varphi}{2}\right)} - 1} \right) \tag{6.35}$$

$$A_\theta = m^2/4 \tan\varphi\left(e^{2\theta \tan\varphi} - 1\right) \tag{6.36}$$

Similarly, the ratio of rupture surface area converting from shallow to deep condition of rupture, which was defined as α_2, is:

$$\alpha_2 = \frac{A_o}{A_{\theta=(\pi+\pi/4-\varphi/2)}} = \frac{e^{2\theta \tan\varphi} - e^{2(3\pi/4-\varphi/2)}}{e^{2(5\pi/4-\varphi/2)} - e^{2(3\pi/4-\varphi/2)}} \tag{6.37}$$

Fig. 6.18A and B illustrates values of α_1 and α_2 versus φ and D/B parameters.

Based on Eq. (6.25), the variation of $\overline{\alpha}$ as average of α_1 and α_2 is illustrated in Fig. 6.18C as a function of equivalent φ and D/B values, in which the equivalent φ depends on effective stress and measured cone resistance, q_c, at depth.

Fig. 6.19 illustrates a simplified correlation for estimating equivalent φ angle from q_c and effective stress level which has been calculated from following equation based on bearing capacity theory on deep-seated failure zone.

$$\varphi = \frac{\log\left(\frac{\overline{q}_c}{\gamma' z}\right) + 0.5095}{0.0915} \tag{6.38}$$

This is obtained from Eq. (6.7) by q_c replaced q_{ult} and using standard cone penetrometer geometry.

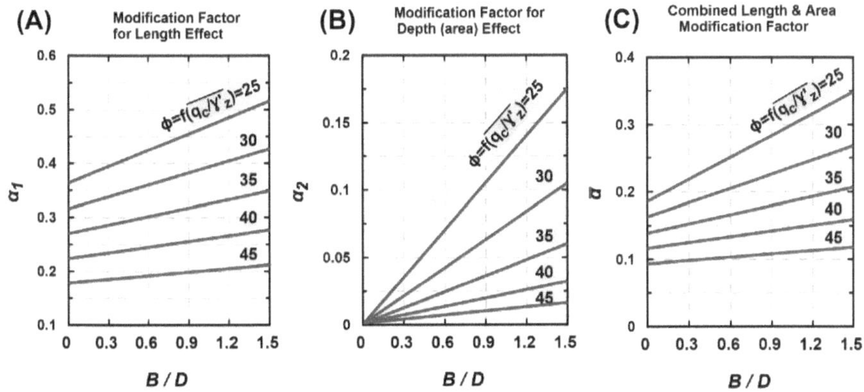

Figure 6.18 Bearing capacity correlation factor for relating q_{ult} to q_{cg} (Eslami and Gholami, 2006). (A) Modification factor for length effect; (B) modification factor for depth (area) effect; (C) combined length and area modification factor.

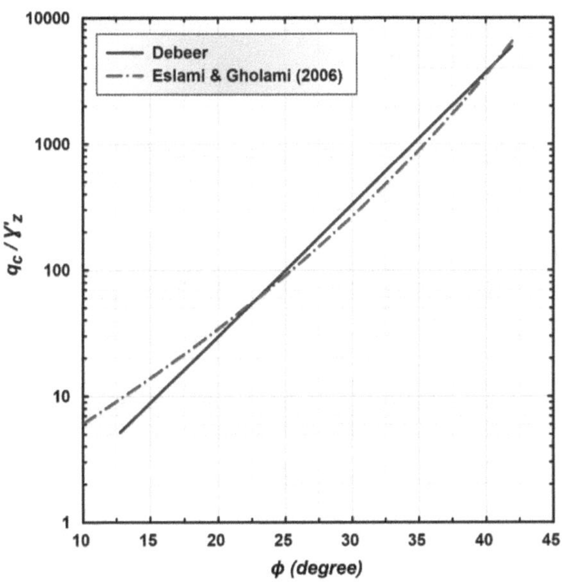

Figure 6.19 Correlation between friction angle and ratio of q_c to effective stress (after Eslami and Gholami, 2006).

Finally, summary of the Eslami and Gholami (2006) method in a step-by-step procedure is as follows:

1. The zone located between foundation base to 2B beneath can be divided in sublayers. The values of $\overline{q}_{cg}$ and $\left(\frac{\overline{q}_c}{\gamma' z}\right)_{cg}$ in this interval is calculated.

2. The average φ angle is obtained by using $\left(\frac{\overline{q_c}}{\gamma z}\right)_{cg}$ from Eq. (6.30) or Fig. 6.18.
3. Based on D/B and φ values, from Fig. 6.18, $\overline{\alpha}$ can be obtained.
4. Finally, the ultimate bearing capacity is calculated as $q_{ult} = \overline{\alpha} \times \overline{q}_{cg}$.

Example 6.1: According to the given graph of CPT, find bearing capacity of the square shallow foundation if B = 3 m, D = 0.76 m, GWT = 4.9 m, and $\gamma = 15.5\frac{kN}{m^3}$.

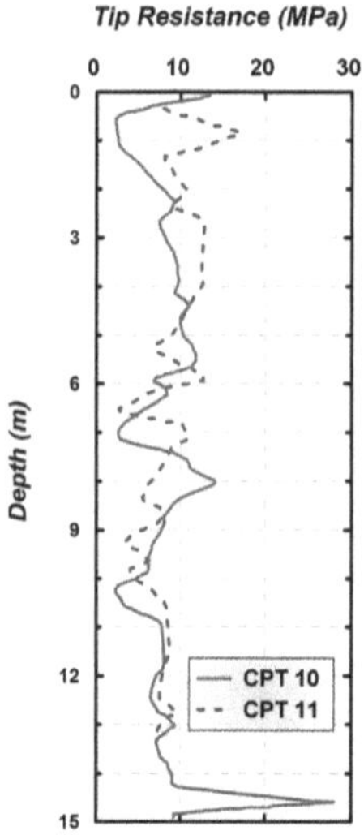

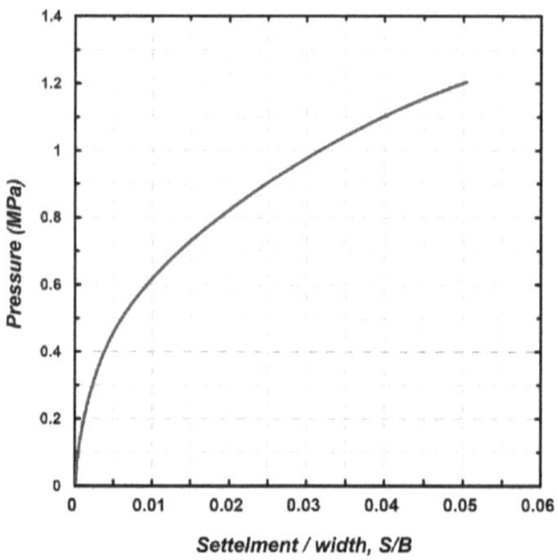

Eslami and Gholami (2006) method:

Step 1: Calculating $\overline{q_{cg}}$ and $\overline{q_c/\gamma' z}$ in the zone of influence from the ground level down to 2B under the foundation:

Number	Z (m)	$q_{c\text{-}1}$ (kPa)	$q_{c\text{-}2}$ (kPa)	$q_{c\text{-}avg}$ (kPa)	Y'z (kPa)	$q_{c\text{-}avg}$/Y'z
1	0.5	2258	9105	5681.5	7.75	733.1
2	1.0	2494	15,834	9164	15.5	591.2
3	1.5	5445	8751	7098	23.25	305.3
4	2.0	7806	10,522	9164	31	295.6
5	2.5	8161	10,876	9518.5	38.75	245.6
6	3.0	8161	12,528	10,344.5	46.5	222.5
7	3.5	9577	12,647	11,112	54.25	204.8
8	4.0	9577	12,056	10,816.5	62	174.5
9	4.5	10,167	10,640	10,403.5	69.75	149.2
10	5.0	10,167	9223	9695	27.5	352.5
11	5.5	11,820	8042	9931	30.25	328.3

Number	Z (m)	q_{c-1} (kPa)	q_{c-2} (kPa)	q_{c-avg} (kPa)	ϒ'z (kPa)	q_{c-avg}/ϒ'z
12	6.0	6862	12,056	9459	33	286.6
13	6.5	6744	4265	5504.5	35.75	154.0
14	7.0	2730	10,522	6626	385	17.2

$$\overline{q_{cg}} = \sqrt[12]{q_{c1} \times q_{c2} \times q_{c3} \times \ldots \times q_{c12}} = 8683.1\text{Kpa} = 86.83\frac{\text{kg}}{\text{cm}^2}$$

$$\overline{q_c/\gamma'z} = \sqrt[12]{\left(q_{c1}/\gamma' z_1\right) \times \left(q_{c2}/\gamma' z_2\right) \times \left(q_{c3}/\gamma' z_3\right) \times \ldots \times \left(q_{c12}/\gamma' z_{12}\right)}$$

$$= 227.6$$

Step 2: Calculating ϕ:

$$\phi = \left[log\left(\overline{q_c/\gamma' z}\right) + 0.5095\right] / 0.0915 = 31.33$$

Step 3: Calculating $\overline{\alpha}$ based on D = 0.76 m and B = 3 m, using Fig. 6.18:

$$\overline{\alpha} = 0.15$$

Step 4: Calculating q_{ult}:

$$q_{ult} = \overline{\alpha} \times \overline{q_{cg}} = 0.15 \times 86.83 = 1.302MPa$$

CFEM (2006):

Step 1: Calculating $\overline{q_c}$ from the bottom level of foundation down to 1B under the bottom level of foundation:

Number	z (m)	q_{c-1} (kPa)	q_{c-2} (kPa)	q_{c-avg} (kPa)
1	1.0	2494	15,834	9164
2	1.5	5445	8751	7098
3	2.0	7806	10,522	9164
4	2.5	8161	10,876	9518.5
5	3.0	8161	12,528	10,344.5
6	3.5	9577	12,647	11,112
7	4.0	9577	12,056	10,816.5

Step 2: Calculating q_{ult}:

$$q_{ult} = F.S. \times \alpha \times \overline{q_c} = 3 \times 0.10 \times 9.603 = 2.881MPa$$

6.6 Comparison of direct CPT methods via databases

A database has been compiled from six sites including 28 full-scale footings and/or plate load tests accompanying CPT soundings performed close to foundation locations. Following, the brief site specifications of cases are summarized:

Site No. I, located in Texas, USA, was reported by Briaud and Gibbens (1994). Five square spread footings with footing width B ranges from 1 to 3m and footing embedment D_f, from 0.7 to 0.9 m were constructed on uniform sand (SP). The measured q_{ult} was about 15 kg/cm^2 and the q_c values varies from 40 to 110 kg/cm^2.
Site No. II, reported by Amar (1979), four square 1m surface footings were loaded on silty (ML) soil. The measured q_{ult} ranges from 3 to 3.75 kg/cm^2 and the q_c values range from 17 to 28 kg/cm^2.
Site No. III is located in Texas, USA (Tand et al., 1994). Four circular footings with diameters of 1.75−2 m were located in depth of 2.16−2.35m. The subsoil type is silty sand (SM & SP). The q_{ult} values vary from 9.4 to 13.5 kg/cm^2, and the q_c values range from 10 to 180 kg/cm^2.
Site No. IV: Three load tests were performed on 0.6-m wide surface plates in Tabriz, northwest of Iran (Saniee, 1993). The site is formed of silty and clayey sands in which the measured q_{ult} values range from 12.6 to 12.8 kg/cm^2 and the q_c values vary from 30 to 100 kg/cm^2.
Site No. V: Consoli et al. (1998) reported a five-plate load test has been done on 0.3−0.6m diameter plates. The subsoil of footing is a mixture of clay and sand. The q_{ult} obtained from PLT was about 1.7 kg/cm^2, and the q_c values were in the range of 5−20 kg/cm^2.
Site No. VI: According to Tand et al. (1986), seven 0.6m plate load tests were performed on silty clay and stiff clay deposits at 1.5m depth. The measured bearings were in the range of 3.1−7 kg/cm^2, and the q_c values varied from 10 to 60 kg/cm^2.

Table 6.1 presents the general specifications of the case histories. An example of load−settlement diagram and q_c profile is shown in Fig. 6.20.

For ultimate bearing capacity, q_{ult}, when the trend of failure has not achieved in testing process, a bearing capacity corresponding to 10% of relative settlement was chosen as recommended by current methods. Utilizing 28 footing case histories, the predicted bearing capacity compared to the measured from test results. For validation of predictive method outputs, the numerical and graphical approaches have been used to evaluate the accuracy of results.

The numerical approach employs cumulative probability procedure to quantify the comparison of estimated and measured footing capacity for all of the case records in a linear plot. Comparison of methods will provide more insight when the values are plotted versus a cumulative average called "cumulative probability." For the current set of data, the ratio of calculated to measured footing capacity,

Table 6.1 Case records summary (Gholami, 2001).

Site No.	Reference	Location	Soil type	Footing shape	Footing size (m)B & D_f		Relative settlement (%)	q_u (kg/cm^2)
I	Briaud and Gibbens (1994)	USA	Sand, silty sand	Square	1	0.71	10	15
I	Briaud and Gibbens (1994)	USA	Sand, silty sand	Square	1.5	0.76	10	15
I	Briaud and Gibbens (1994)	USA	Sand, silty sand	Square	2.5	0.76	10	15
I	Briaud and Gibbens (1994)	USA	Sand, silty sand	Square	3	0.76	10	15
I	Briaud and Gibbens (1994)	USA	Sand, silty sand	Square	3	0.89	10	13
II	Amar (1979)	France	Silt	Square	1	0	6	3.75
II	Amar (1979)	France	Silt	Square	1	0	6	3.7
II	Amar (1979)	France	Silt	Square	1	0	6	3
II	Amar (1979)	France	Silt	Square	1	0	6	3.25
III	Tand et al. (1994)	USA	Sand, silty sand	Circular	1.75	2.35	10	11.5
III	Tand et al. (1994)	USA	Sand, silty sand	Circular	1.75	2.35	10	11.5
III	Tand et al. (1994)	USA	Sand, silty sand	Circular	2	2.2	10	7.5
III	Tand et al. (1994)	USA	Sand, silty sand	Circular	1.88	2.16	10	9.5
IV	Saniee (1993)	Iran	Silty sand, sand	Square	0.6	0	8	12.6
IV	Saniee (1993)	Iran	Silty sand, sand	Square	0.6	0	8	12.6
IV	Saniee (1993)	Iran	Silty sand, sand	Square	0.6	0	8	12.8
V	Consoli et al. (1998)	Brazil	Silty clay	Circular	0.3	0	1	1.7
V	Consoli et al. (1998)	Brazil	Silty clay	Circular	0.3	0	1	1.7
V	Consoli et al. (1998)	Brazil	Silty clay	Circular	0.45	0	1	1.7
V	Consoli et al. (1998)	Brazil	Silty clay	Circular	0.6	0	1	1.7

Continued

Table 6.1 Case records summary (Gholami, 2001).—cont'd

Site No.	Reference	Location	Soil type	Footing shape	Footing size (m)B & D_f		Relative settlement (%)	q_u (kg/cm^2)
V	Consoli et al. (1998)	Brazil	Silty clay	Circular	0.6	0	1	1.7
VI	Tand et al. (1986)	USA	Stiff clay	Circular	0.6	1.5	12	6
VI	Tand et al. (1986)	USA	Stiff clay	Circular	0.6	1.5	12	6
VI	Tand et al. (1986)	USA	Stiff clay	Circular	0.6	1.5	12	6
VI	Tand et al. (1986)	USA	Silty clay	Circular	0.6	1.5	11	5.2
VI	Tand et al. (1986)	USA	Silty clay	Circular	0.6	1.5	15	3.1
VI	Tand et al. (1986)	USA	Silty clay	Circular	0.6	1.5	15	3.1
VI	Tand et al. (1986)	USA	Stiff clay	Circular	0.6	1.5	10	6.9

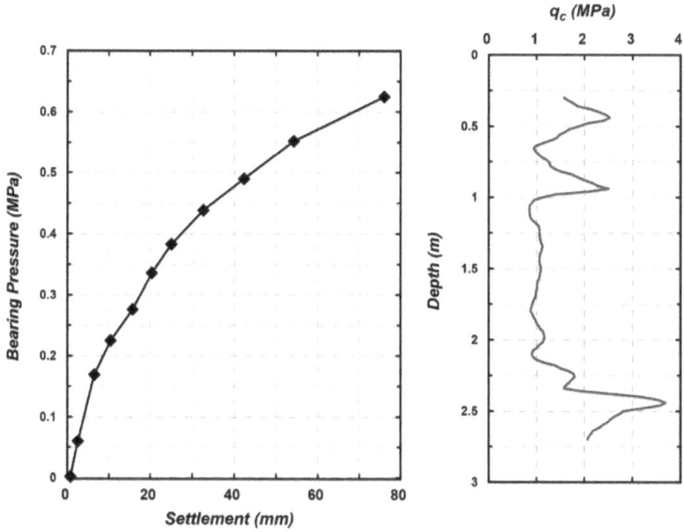

Figure 6.20 Example of full-scale footing load test results and q_c profile for site No. VI (Gholami, 2001).

$q_{ult,Cal.}/q_{ult,mes.}$, is arranged in ascending order (numbered, 1, 2, ..., i, ..., n) and a cumulative probability, P is determined for each capacity values as:

$$P = \frac{i}{n+1} \tag{6.39}$$

where

$q_{ult,Cal.}$ = calculated ultimate bearing capacity,
$q_{ult,Mes.}$ = measured bearing capacity,
P = cumulative probability,
n = number of total cases,
i = case number.

To assess the bias and dispersion associated with a particular predictive method, the following points are useful (Long and Shimel, 1989):

- The value of $q_{ult,Cal}/q_{ult,Mes.}$ at a probability of $P = 50\%$, is the measure of the tendency to overestimate or underestimate the bearing capacity. The more closer to a ratio of unity, the better the agreement.
- Log–normal distributed data will plot on a straight line.
- The slope of the line through the data points is a measure of the dispersion or "standard deviation." The flatter the line, the better the general agreement.

The results of comparison indicates that the $q_{ult,Cal}/q_{ult,Mes.}$ at a probability of 50% is closer to unity for proposed method than others as shown in Fig. 6.21. It means the new method predicts foundation capacity with less overestimation or underestimation

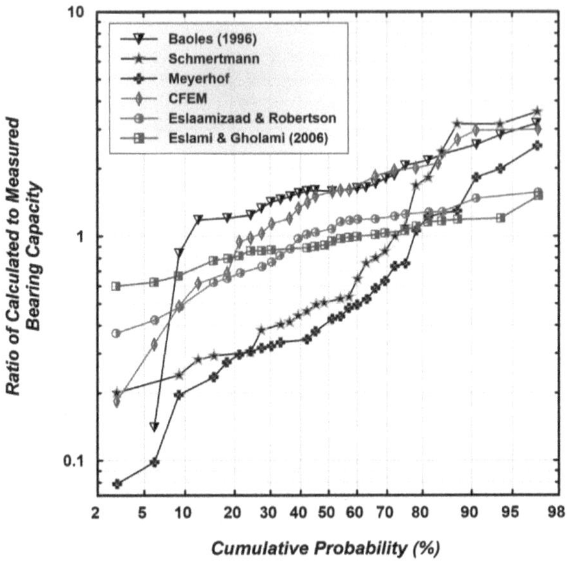

Figure 6.21 Comparison of predictive methods by cumulative probability approach.
After Eslami, A., Gholami, M., 2006. Analytical model for the ultimate bearing capacity of foundations from cone resistance. International Journal of Science and Technology, Scientica IRANICA 13(3), 223–233.

than other methods. Besides, the slope of line through points for current methods exhibits higher dispersion than the proposed method. Therefore, the results are closer to log–normal distribution for the proposed method.

References

AL-Mhaidib, A.I., 1999. Loading rate effect on piles in clay from laboratory tests. Journal of King Saud University 13, 39–55.

Altaee, A., Evgin, E., Fellenius, B.H., 1993. Axial load transfer for piles in sand. III: load transfer for piles in sand and the critical depth. Canadian Geotechnical Journal 30 (3), 455–463.

Amar, S., 1979. Experimental study in full scale of shallow foundation behavior. In: Proceeding of 6th European Conference on Soil Mechanics and Foundation Engineering, pp. 199–204.

Axelsson, G., 2000. Long-term Set-Up of Driven Piles in Sand. Ph.D. Thesis. Royal Institute of Technology, Stockholm. in partial fulfillment of the requirements for the degree of Doctor of Philosophy).

Bowles, J.E., 1996. Foundation Analysis and Design, fifth ed. the McGraw-Hill Companies, p. 266.

Briaud, J.L., Gibbens, R.M., 1994. Behavior of five large spread footings on sand. Journal of Geotechnical and Geoenvironmental Engineering, ASCE 125 (9), 787–796.

Brown, M.J., Hyde, A.F.L., 2008. Rate effects from pile shaft resistance measurements. Canadian Geotechnical Journal 45, 425–431.

Campanella, R.G., Robertson, P.K., 1981. Applied Cone Research. Symposium on Cone Penetration Testing and Experience. Geotechnical Engineering Division, ASCE, pp. 343–362.
Canadian Foundation Engineering Manual, CFEM, 2006, fourth ed. Canadian Geotechnical Society, BiTech Publishers, Vancouver. 488 pp.
Chin, F.K., 1978. Diagnosis of pile condition. In: Proceedings of the Institution of Civil Engineers-Geotechnical Engineering, vol. 9, pp. 85–104. London.
Chow, F.C., 1997. Investigations into the Behaviour of Displacement Piles for Offshore Foundations. Ph.D. thesis. Imperial College, London.
Clausen, C., Aas, P., Karlsrud, K., 2005. Bearing capacity of driven piles in sand, the NGI approach. In: Proceedings of International Symposium on Frontiers in Offshore Geotechnics, Perth, pp. 574–580.
Consoli, N.C., Schnaid, F., Milititisky, J., 1998. Interpretation of load tests of residual soils. Journal of Geotechnical and geoenvironmental Engineering, September 124 (9), 857–867.
Davisson, M.T., 1972. High capacity piles. In: Proceedings, Soil Mechanics Series on Innovation in Foundation Construction. ASCE, Illinois Section, Chicago, pp. 81–112.
De Beer, E.E., 1963. The scale effect in the transposition of the results of deep sounding tests on the ultimate bearing capacity of piles and caisson foundations. Géotechnique 13 (1), 39–75.
Durgunoglu, H.T., Mitchell, J.K., 1975. Static penetration resistance of soils. I-II. In: Proceedings of the ASCE Specialty Conference on in Situ Measurement of Soil Properties, Raleigh, North Carolina, vol. 1, pp. 151–189.
Eslaamizaad, S., Robertson, P.K., 1996. Cone penetration test to evaluate bearing capacity of foundations in sands. In: Proceeding of 49th Canadian Geotechnical Society, pp. 429–438.
Eslami, A., 1996. Bearing Capacity of Piles from Cone Penetrometer Test Data. Ph.D. Thesis. University of Ottawa, Department of Civil Engineering, 516pp.
Eslami, A., Fellenius, B.H., 1997. Pile capacity by direct CPT and CPTu methods applied to 102 case histories. Canadian Geotechnical Journal 34 (6), 880–898.
Eslami, A., Gholami, M., 2006. Analytical model for the ultimate bearing capacity of foundations from cone resistance. International Journal of Science and Technology, Scientica IRANICA 13 (3), 223–233.
Eslami, A., Lotfi, S., Eslami, M., 2013. Pile shaft capacity from cone penetration test (CPT) records; considering scale effects. In: Proceedings of the 38th Annual Conference on Deep Foundations. Deep Foundation Institute (DFI), Phoenix AZ, USA.
Fleming, K., Weltman, A., Randolph, M., Elson, K., 2009. Piling Engineering, third ed. CRC press, New York.
Fellenius, B.H., 2015. Basics of Foundation Design, Electronic Edition. 432 pp. www.Fellenius.net.
Ghionna, V.N., Jamiolkowsi, M., Lancellotta, R., Pedroni, S., 1993. Base capacity of bored piles in sands from in-situ tests. In: Van Impe (Ed.), Proceeding of 2nd International Geotechnic Seminar on Deep Foundations on Bored and Auger Piles. Balkema, Rotterdam, The Netherlands, pp. 67–74.
Gholami, M., 2001. Bearing Capacity of Shallow Foundations from CPT Data. MSc. Thesis. University of Guilan.
Gui, M.W., Bolton, M.D., 1998. Geometry and scale effects in CPT and pile design. In: Proceedings of the 1st International Conference on Site Characterization, Atlanta, pp. 1063–1068.
Hansen, B.J., 1963. Discussion on hyperbolic stress-strain response, cohesive soils. American Society of Civil Engineers, ASCE Journal of Soil Mechanics and Foundation Engineering 89 (SM4), 241–242.

Ishibashi, I., Zhang, X., 1993. Unified dynamic shear module and damping ratios of sand and clay. Soils and Foundation 33 (1), 182–191.

Janbu, N., Senneset, K., 1974. Effective stress interpretation of in-situ static penetration tests. In: Proceedings of the European Symposium on Penetration Testing, 2.2. ESOPT, Stockholm, pp. 181–193.

Jardine, J., Chow, F., Overy, R., Standing, J., 2005. ICP Design Method for Driven Piles in Sands and Clays. Thomas Telford Publishing Ltd., London, 105pp.

Kennedy, J.B., Neville, A.M., 1986. Basic Statistical Methods for Engineering and Scientists. Harper and Row, New York, 519pp.

Kolk, H., Baaijens, A., Senders, M., 2005. Design criteria for pipe piles in silica sands. In: Proceedings of International Symposium on Frontiers in Offshore Geomechanics (ISFOG 2005), Perth. Taylor and Francis Group, London, pp. 711–716.

Lehane, B.M., Scheider, J.A., Xu, X., 2005. CPT Based Design of Driven Piles in Sand for Offshore Structures. Internal Rep. No. GEO: 05345. University of Western Australia, Perth, Australia.

Long, J.H., Shimel, I.S., 1989. Drilled shafts, a data-base approach. American Society of Civil Engineers. In: Proceedings of Foundation Engineering Congress: Current Principles and Practices. ASCE Geotechnical Special Publication. June 25 - 29, GSP 22.

Lunne, T., Robertson, P.K., Powell, J.J., 1997. Cone Penetration Test in Geotechnical Practice. Blackie Academic & Professional.

Meyerhof, G.G., 1951. The ultimate bearing capacity of foundations. Géotechnique 2 (4), 301–332.

Meyerhof, G.G., 1956. Penetration tests and bearing capacity of cohesionless soils. Journal of Soil Mechanics and Foundation Engineering Division, ASCE 82 (SM1), 1–12.

Meyerhof, G.G., 1976. Bearing capacity and settlement of ASCE. Journal of Geotechnical Engineering 102 (GT3), 195–228.

Meyerhof, G.G., 1983. Scale effects of pile capacity. Journal of Geotechnical Engineering, ASCE 108 (GT3), 195–228.

Mitchell, J.K., Keaveny, J.M., 1986. Determining sand strength by cone penetrometer. In: Proceedings of the ASCE Specialty Conference in Situ 86: Use of in Situ Tests in Geotechnical Engineering, Blacksburg, pp. 823–839.

Moshfeghi, S., Eslami, A., 2016. Study on pile ultimate capacity criteria and CPT-based direct methods. International Journal of Geotechnical Engineering 12 (1), 28–39.

Nottingham, L.C., 1975. Use of Quasi-Static Friction Cone Penetrometer Data to Predict Load Capacity of Displacement Piles. Ph. D. Thesis. Dept. of Civil Engineering, University of Florida, 553pp.

Randolph, M.F., Dolwin, J., Beck, R., 1994. Design of driven piles in sand. Géotechnique 44 (3), 427–448.

Robertson, P.K., Campanella, R.G., 1983. Interpretation of cone penetrometer test: Part I: Sand. Canadian Geotechnical Journal 20 (4), 718–733.

Robertson, P.K., Campanella, R.G., 1989. Guidelines for Geotechnical Design Using the Cone Penetrometer Test and CPT with Pore Pressure Measurement. Soil Mechanics Series, No. 105, fourth ed. UBC, Dept. of Civil Engineering and Hogentogler and Company, Inc. 193 pp.

Sabatini, P.J., Bachus, R.C., Mayne, P.W., Schneider, J.A., Zettler, T.E., 2002. Geotechnical Engineering Circular No. 5: Evaluation of Soil and Rock Properties (No. FHWA-IF-02-034).

Salgado, R., 1995. Design of piles in sands based on CPT results. In: Proceeding of 10th Pan-Am. Conference of Soil Mechanics and Foundations. Engrg., Guada-Lajara, vol. 3, pp. 1261–1274.

Saniee, H., 1993. New method for determining bearing capacity of shallow foundation on cohesionless soils. In: Proceedings of the Second International Seminar on Soils Mechanics and Foundation Engineering of IRAN, Tehran, pp. 133–148.

Schmertmann, J.H., 1978. Guidelines for Cone Test, Performance and Design. Report FHWA-TS-78209. Federal Highway Administration, Washington, 145pp.

Takesue, K., Sasao, H., Matsumoto, T., 1998. Correlation between Ultimate Pile Skin Friction and CPT Data, Geotechnichal Site Characterization (2). Balkema, Rotterdam, pp. 1177–1182.

Tand, K., Funegard, E., Warden, 1986. Bearing Capacity of footing on clay CPT method. In: Proceeding of American Society of Civil Engineers ASCL, In-Situ 86 Specialty Conference, pp. 1017–1033.

Tand, K., Funegard, E., Warden, 1994. Footing load test on sand. Vertical and Horizontal Deformations of Foundations and Embankments 8 (40), 164–178. ASCE Geotechnical Special Publication, 1.

Terzaghi, K., 1943. Theoretical Soil Mechanics. John Wiley and sons, New York, p. 510p.

Vesić, A.S., 1972. Expansion of cavities in infinite soil masses. American Society of Civil Engineers, ASCE Journal of the Soil Mechanics and Foundations Division 96 (SM3), 265–290.

Vesić, A.S., 1975. Bearing capacity of shallow foundations. In: Winterkorn, H.F., Fang, H.-Y. (Eds.), Chapter 3 in Foundation Engineering Handbook. VanNostrand Company, pp. 121–147.

White, D.J., Bolton, M.D., 2005. Comparing CPT and pile base resistance in sand. Proceedings of the Institution of Civil Engineers-Geotechnical Engineering 158 (1), 3–14.

CPT and CPTu applications for deep foundation—bearing capacity

7

7.1 Introduction

Piles are structural members made of steel, concrete, and/or wood installed to transfer superstructure loads to competent subsurface soil layers. A pile subjected to axial load will carry the load partly by shear resistance generated along the shaft and the rest by pile toe or base resistance. Some conditions where application of deep foundations is inevitable may include, but are not limited to, the following:

1. Upper soil strata are problematic, i.e., compressible, expansive, collapsible, soft, or weak
2. Conventional shallow foundations cannot transmit heavy vertical loads, inclined, horizontal, or uplift forces
3. As a supplement of shallow mat foundations and forming hybrid systems to enhance bearing capacity and settlement aspects
4. Stability control of tall buildings, long span bridges, and high-rise structures
5. Special environmental conditions, extreme loading such as earthquake, floods, or scouring
6. Possible future excavation or other construction being carried out adjacent to the structure
7. Ground modification required, i.e., in loose, cohesionless deposits which are densified through a combination of pile penetration displacement and driving vibration
8. Sensitive serviceability purposes as the soils beneath industrial foundations need to be stiffened to control vibrations, shocks, displacements, etc.
9. Slope stability control against ground movement
10. As a crucial embedded element for onshore and offshore structure systems due to existing soft deposits and extraordinary hydrodynamic loading

Piles can be classified by a number of factors; Fig. 7.1 shows some of the more common pile types including driven, bored, and drilled displacement types. The choice of pile installation method as well as pile material type is influenced by subsurface conditions, location, and topography of the site and structural and geometrical characteristics of the project. Moreover, practical and economical aspects play an essential role in choosing the type and relative size of pile and installation method for a given application. As a substructure system, piles or deep foundations are employed to support different load combinations as single, group, and piled raft foundations. Regardless of which type of substructure system a pile is performing in, it can be subjected to axial loads (compression or tension) or lateral loading or more possibly a combination of these two simultaneously as illustrated in Fig. 7.2.

Piezocone and Cone Penetration Test (CPTu and CPT) Applications in Foundation Engineering
https://doi.org/10.1016/B978-0-08-102766-0.00007-9

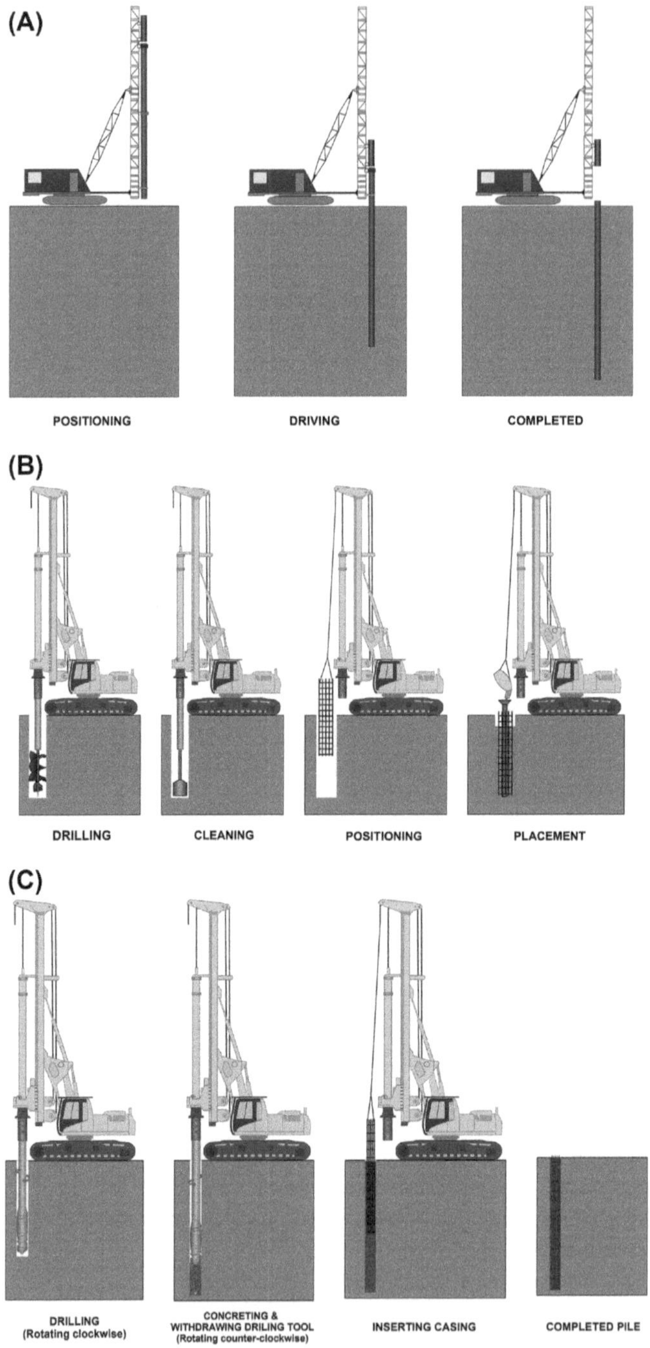

Figure 7.1 Different pile types: (A) driven, (B) bored, and (C) drilled displacement.

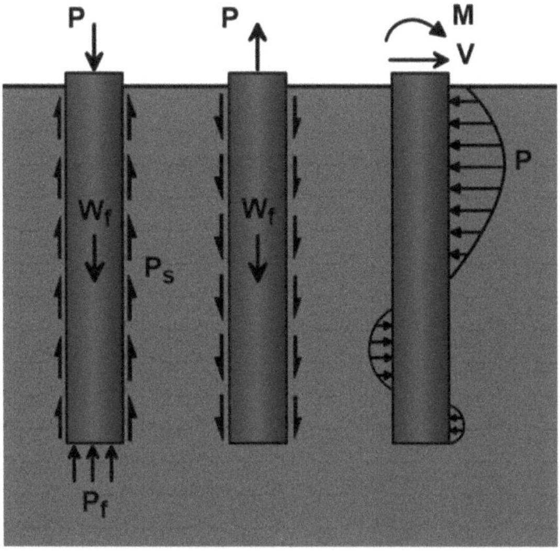

Figure 7.2 Piles' different loading conditions.

7.2 Pile design

Significant application of deep foundations features the necessity of precise investigations into design process. The designer of pile foundation must possess a variety of skills, experience, and considerable knowledge of foundation engineering. No single set of simple rules and procedures can cover the variety of conditions that affect deep foundation performance.

Generally, the design of any foundation system, shallow, semideep, and deep, consists of geotechnical, structural, and constructional aspects as mentioned in Chapter 2. Structural aspects are governed by internal interactions and material strength against compressive, tensile, and shear forces as well as bending moments. The pile material must have the ability to safely resist all loads imposed during installation and service conditions. If the applied load is properly determined from the applicable codes, structural failure is seldom a concern, unless the pile is very long or encounters severe corrosive conditions. Constructional aspects pertain to the performance of a foundation system in each specific site. The construction practice and installation methods are attributed to the project location, available equipment, pile geometry, i.e., length and diameter, underground conditions, and environmental concerns.

On the other hand, geotechnical capacity involves external interaction of foundation systems with surrounding and underlain soil in terms of settlement and bearing capacity. Many factors affect the settlement of a pile or pile group, and it is difficult to make accurate estimates of settlement of deep foundations. Generally, the actual load-movement response of a pile is a function of the relative contribution of shaft and toe resistances, soil conditions, pile material stiffness, and method of pile installation. Hence, a number of empirical and theoretical approaches have been developed to reasonably estimate the settlement of piles. It is important to note that the aspects of capacity, both geotechnical and structural, and settlement are interactive and cannot be separated in design (Fellenius, 1989).

7.3 CPT records for pile installation

From performance-based point of view, installation of deep foundations depends on external conditions such as material, geometry, and boring or driving techniques or equipment, environmental issues, and adjacent structures. However, the more critical factor is the type of geomaterial being classified as soft, medium, or hard subsurface soils. As discussed in Chapter 5, one of the very first applications of CPT data is subsurface profiling and soil behavior classification. Therefore, having CPT records available can help specify design aspects such as pile geometry, construction type, and the portion of bearing components including shaft and toe resistance. This concept is illustrated in CPT profiles as shown in Fig. 7.3 via an example.

Example 7.1. According to Fig. 7.3A, down to 7.5 m, the soil type is sand (high q_c and low R_f) and can be considered as a bearing layer; while, below 7.5 m, the soil type is clay with relatively low tip resistance and high friction ratio as indicated in q_c profile. Thus, as a first choice, it is more appropriate to locate the pile toe somewhere within the first 7.5 m to achieve the highest toe capacity (geometry dictation). However, if the placement of the pile toe in this stratum does not provide sufficient capacity, the pile must penetrate inside the clay layer. In this case, given the low resistance at the pile toe location (i.e., inside clay layer), the dominant bearing capacity of the pile would be from the shaft capacity, i.e., longer pile is required.

On the other hand, Fig. 7.3B depicts a clayey soil layer with limited resistance within first 15 m depth underlain by a sandy layer with higher capacity down to 28 m. In this case, to have optimal toe capacity, it is desirable to locate the pile toe in the sand layer. However, if for any reason (e.g., less required capacity, insufficient equipment, etc.) the pile toe has to be located at a shallower depth, the pile construction type must be opted in a way to also accompany improving the soil conditions. In such situations, bored piles are not recommended. Instead, driven or drilled displacement piles are more favored, considering the capability of enhancing soil conditions by vibration or displacements imposed on the surrounding soil during installation process in such methods.

7.4 Bearing capacity of deep foundations

The bearing capacity, also termed ultimate resistance or ultimate load, of a pile consists of fully developed shaft resistance (shear resistance generated along the shaft) and toe resistance (resistance generated at the base of the pile) in response to axial load and can be computed as shown in Eq. (7.1).

$$R_u = R_t + R_s = r_t A_t + r_s A_s \tag{7.1}$$

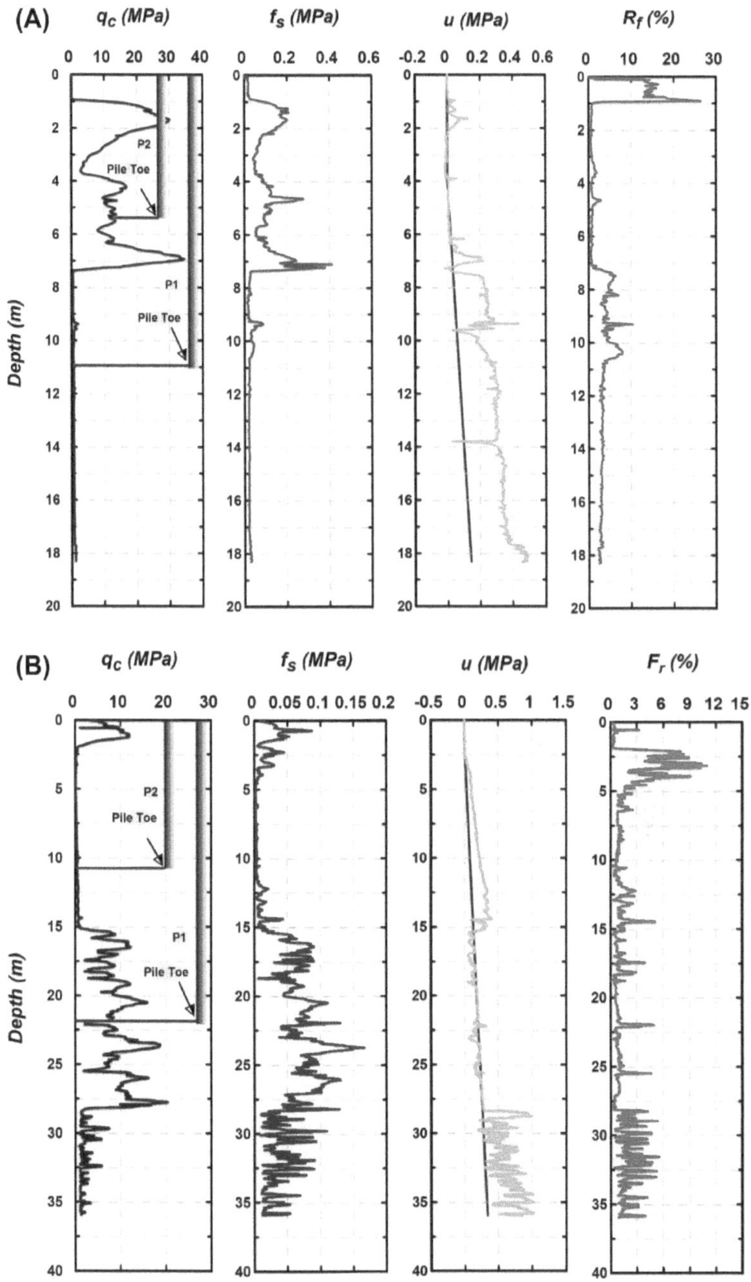

Figure 7.3 CPT profiles and pile alternatives: (A) NWU site and (B) UBC site. (A) CPT data from Finno, R.J. 1989. Subsurface conditions and pile installation data. American Society of Civil Engineers. In: Proceedings of Symposium on Predicted and Observed Behavior of Piles, Evanston, June 30, ASCE Geotechnical Special Publication, GSP23, pp 1–74; (B) CPT data from Eslami (1996).

where

R_u = pile-bearing capacity
R_t = toe resistance
R_s = shaft resistance
r_t = unit toe resistance
r_s = unit shaft resistance
A_t = pile toe cross-sectional area
A_s = pile unit circumferential area

The relative magnitude of the shaft and the toe resistances will depend on the geometry of the pile, interaction with soil, and the type of geomaterial profile. In cohesive soils, the shaft resistance generally dominates, whereas in cohesionless soils, toe resistance generates the main portion of the pile-bearing capacity.

Various approaches of determining the bearing capacity of piles fall within one of the five categories:

1. Static analysis: considering pile failure mechanism and by applying soil parameters in effective stress or total stress analysis (ESA or TSA)
2. Methods using penetrating in situ tests as a complementary approach to static analyses—direct or indirect methods
3. Full-scale pile static load test
4. Dynamic analysis (based on wave equation analysis) and dynamic loading test [by means of the pile-driving analyzer (PDA) or rapid load test and analysis of the measurements]
5. Numerical methods, modeling based on various soil constitutive models

Statics analysis defines a failure mechanism for piles and is founded on the bearing capacity or cavity expansion theory. However, these theories do not account for the effects of soil compressibility, volume change associated with shear, and the dependency of the friction angle on mean pressure, as well as soil dilatancy. Hence, without relevant experience or lack of adequate geotechnical engineering judgment, estimated capacities may be erroneous. Furthermore, the difficulty of determining a reliable and representative in situ value of soil parameters from laboratory tests casts doubts about the applicability of the conventional static analysis approaches in pile foundation analysis and design (Eslami and Fellenius, 1997; Randolph, 2000).

As a complementary tool to static analysis for pile capacity determination, **in situ test methods** can be realized. In recent years, in situ testing techniques have shown an increase in use for geotechnical design. This is due to rapid development of in situ testing instruments, an improved evaluation of the real behavior of soils, and subsequent realization of some limitations and inadequacies of conventional laboratory testing (Wroth, 1988).

Among many in situ test techniques for site investigation, the standard penetration test (SPT) and the cone penetration test (CPT) are more common for pile design. The SPT is still the most commonly used in situ test during geotechnical investigation. However, many issues and limitations exist for the use of SPT with respect to interpretation and repeatability. In contrast, the CPT is simple, fast, relatively economical, and provides continuous records with depth. The results are interpretable on both empirical and analytical basis. Therefore, the CPT is becoming the preferred type of penetration test for pile analysis.

Because of similarities between CPT and pile, and advantages of CPT as a useful geotechnical investigation tool, the estimation of pile capacity from CPT data has been one of the earliest applications of the cone penetrometer. It is important to note that during site investigation activities, particularly in difficult or problematic underground conditions, the CPT sounding provides continuous records in depth, a large source of data, which can be employed for bearing-capacity estimation.

By analogy of cone penetrometer as a model pile, the measured cone resistance and sleeve friction can be employed for estimation of unit toe and shaft resistances, respectively. It is possible to develop almost a continuous profile by CPT measurement with depth. Consequently, the problems of providing representative undisturbed samples, along with determining correct values of required soil parameters by conventional laboratory testing, will be eliminated. Furthermore, there is no need for estimating intermediate parameters such as K_s (shaft coefficient) and N_q (toe coefficient) for pile capacity estimation.

Static and dynamic loading tests, through interpretation and analysis or numerical simulations, involve significant advantages and of course a few limitations. At the initial geotechnical design stage, the piling configuration, including type, length, diameter, group pattern, and installation, is determined and assessed by static analysis via analytical approaches or in situ testing methods.

It should be noted that, despite the limitations of static analysis approaches, they should be considered integral and inseparable steps for piled foundation design. In other words, they cannot be fully substituted by neither dynamic approaches nor static loading tests. Even though in projects that performing pile load testing is economically justifiable, an initial estimation of pile capacity for determining approximate pile characteristics is done by static analysis approaches.

7.5 Pile axial—bearing capacity from indirect CPT—based approaches

In recent years, implementation of in situ tests in geotechnical designs has increased significantly, given the fast improvement of in situ testing equipment and elimination of difficulties regarding soil sampling as well as insufficiency of conventional laboratory tests. There are two approaches in estimating the axial-bearing capacity of piles by means of CPT:

1. Direct approach: the pile unit shaft and toe resistances are directly estimated from CPT records.
2. Indirect approach: CPT records are initially used to estimate the soil strength parameters such as cohesion (c) and internal friction angle (φ) or any other pertinent soil parameters such as density (γ), Young's modulus (E), subsequently, the pile unit shaft, and toe resistances are calculated using static analysis methods.

The direct methods are more common than the indirect approaches in geotechnical practice. This section briefly introduces the indirect approaches first, and the following section specifically focuses on direct approaches for estimating the pile-bearing capacity.

7.5.1 Pile unit toe resistance

The basic bearing capacity equation can be generalized for deep foundations as well (Eq. 7.2). For deep foundations, this equation can be assessed in two approaches: ESA for noncohesive soils and TSA for cohesive soils, as presented in Eqs. (7.3) and (7.4), respectively.

$$r_t = cN_C + \overline{q}N_q + 0.5\gamma BN_\gamma \tag{7.2}$$

$$r_t = c \cdot N_C \tag{7.3}$$

$$r_t = \overline{q} \cdot N_q \tag{7.4}$$

where c is soil cohesion or undrained shear strength, and N_c and N_q and N_γ are bearing capacity factors, a function of the soil friction angle, and q is surcharge stress. In indirect CPT approaches, cohesion and/or friction angles are estimated using CPT-based correlations. Then, the unit toe resistance is calculated using the above equations.

7.5.2 Pile unit shaft resistance

For shaft resistance calculation, the basic soil shear strength is assessed in two conditions, drained and undrained, similar to the unit toe resistance:

$$\tau = c + \sigma_z' \cdot \tan \varphi \tag{7.5}$$

$$r_s = \alpha \cdot c \tag{7.6}$$

$$r_s = \beta \cdot \sigma_z' \tag{7.7}$$

where, α is adhesion coefficient, and β is cohesion coefficient which is a function of the friction angle.

Once again, the unit shaft resistance is calculated by the above equations, using the cohesion or friction angle values derived from CPT-based correlations.

Example 7.2. According to the following CPTu profiles, the bearing capacity of a driven pile with given characteristics in Fig. 7.4, would be as follows, using TSA approach:

(the pile load–displacement diagram is presented for comparison)

Calculating undrained shear strength using CPT-based correlations:

$$\text{For shaft resistance : } S_u = \frac{q_t - \sigma_{v_o}}{N_k} = \frac{2.15 \times 10^3 - 94}{15} = 137kPa$$

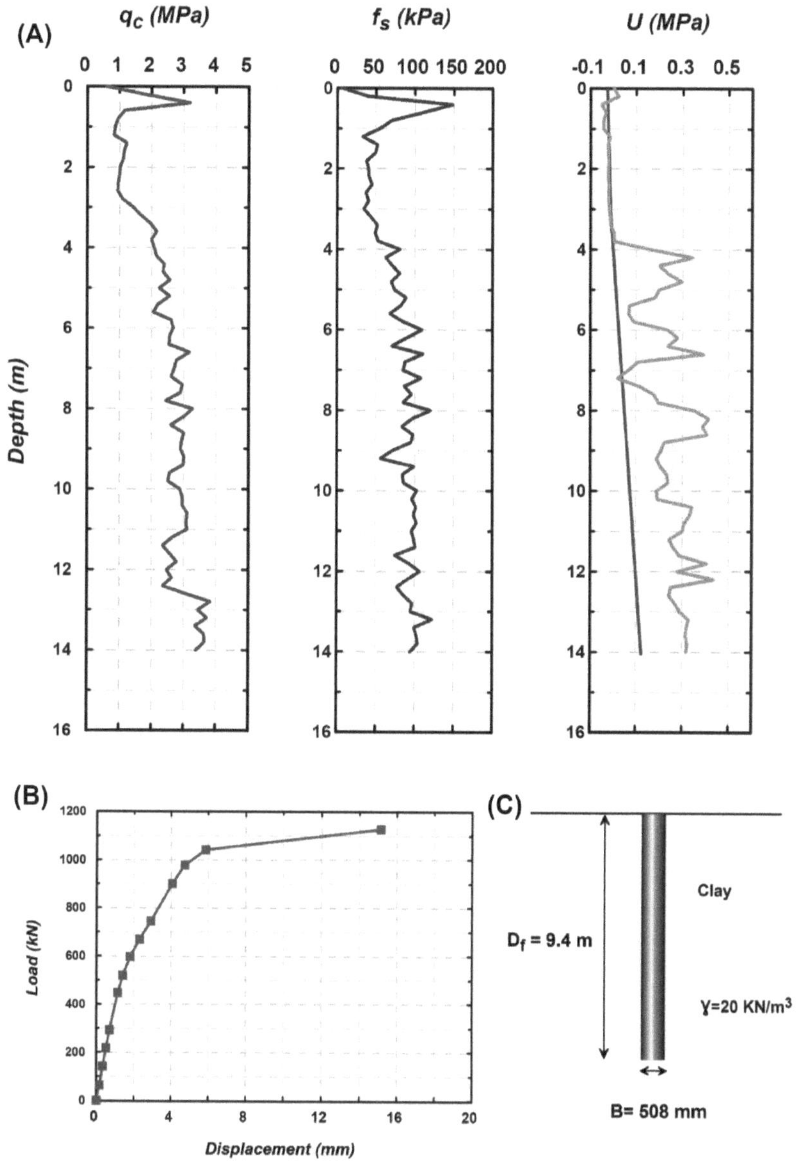

Figure 7.4 (A) CPT profiles, (B) load–displacement diagram, and (C) soil and pile characteristics.
Data from Niazi, F.S. 2014 Static Axial Pile Foundation Response Using Seismic Piezocone Data (Ph.D. thesis). Georgia Institute of Technology, School of Civil and Environmental Engineering, 926 pp.

$$\text{For toe resistance}: S_u = \frac{2.99 \times 10^3 - 188}{15} = 186kPa$$

Toe capacity:

$$r_t = N_c s_u = 9s_u$$

$$\text{Toe Area}: A = \frac{\pi D^2}{4} = 0.2m^2$$

$$r_t = 9s_u = 9 \times 186 = 1674\ kPa$$

$$R_t = r_t A = 1674 \times 0.2 = 334.8kN$$

Shaft capacity:
Circumferential area: $P = \pi D = 1.6\ m$

$$r_s = \alpha s_u \rightarrow \begin{cases} \psi \leq 1 \rightarrow \alpha = 0.5\psi^{-0.5} \\ \psi > 1 \rightarrow \alpha = 0.5\psi^{-0.25} \end{cases}$$

$$\psi = \frac{s_u}{\sigma'_{v_0}} = \frac{137}{60.8} = 2.25$$

$$\alpha = 0.5\psi^{-0.5} = 0.44$$

$$r_s = \alpha s_u = 0.44 \times 137 = 60.28\ kPa$$

$$R_s = r_s \pi DL = 60.28 \times 1.6 \times 9.4 = 906.6\ kN$$

$$R_u = R_t + R_s = 1241\ kN = 124.1\ ton$$

Pile load test: $R_u = 112.7\ ton$.

Example 7.3. According to the following CPTu profiles, the bearing capacity of a pile with given characteristics in Fig. 7.5 would be as follows, using ESA approach:
(the pile load–displacement diagram is presented for comparison)
Calculating friction angle using CPT-based correlations:
Kulhawy and Mayne (1990):

$$\varphi = 17.6 + 11 \log q_{c_1N}$$

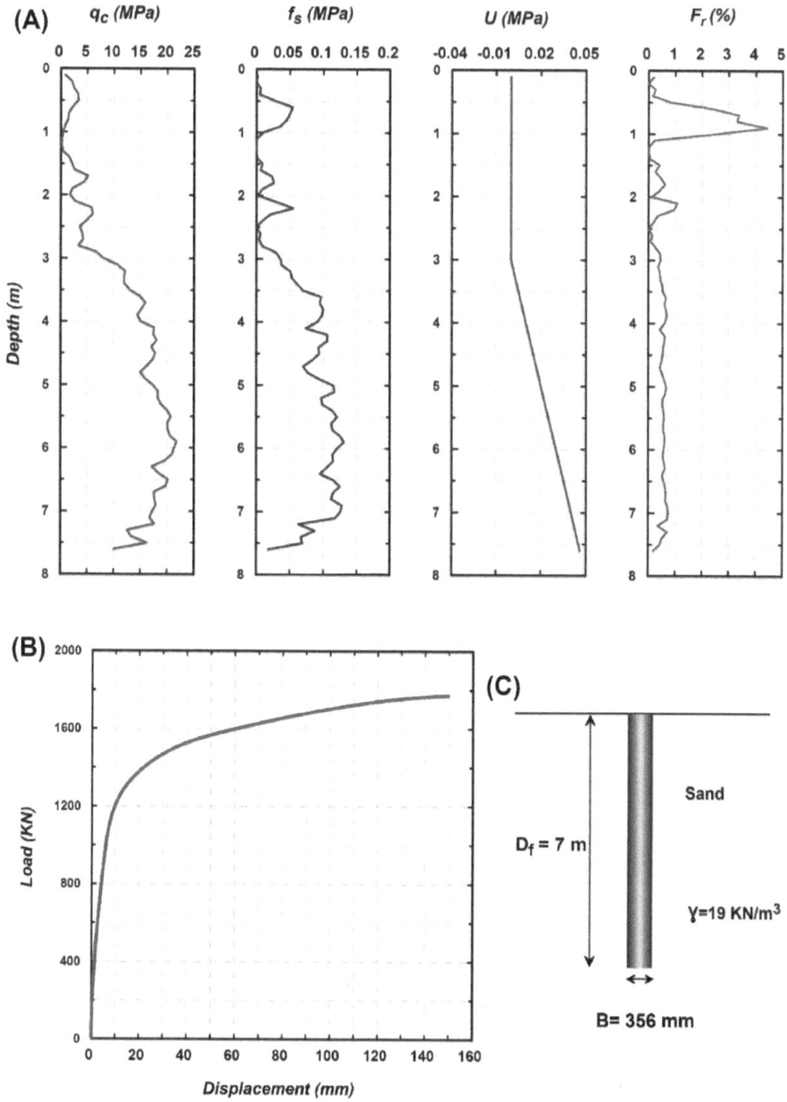

Figure 7.5 (A) CPT profiles, (B) load–displacement diagram, and (C) soil and pile characteristics.
Data from Paik, K., Salgado, R., Lee, J., Kim, B. 2003. Behavior of open-and closed-ended piles driven into sands. Journal of Geotechnical and Geoenvironmental Engineering 129(4), 296–306.

$$q_{c_1N} = \frac{\frac{q_t}{\sigma_{atm}}}{\left(\frac{\sigma'_{v0}}{\sigma'_{atm}}\right)^{0.5}} = \frac{\frac{11.11 \times 1000}{100}}{\left(\frac{59.21}{100}\right)^{0.5}} = \frac{111.1}{0.769} = 144.4$$

$$\varphi = 17.6 + 11 \log 144.4 = 41^\circ$$

Toe capacity:

$$r_t = N_t \sigma'_v$$

$$N_t = 50 - 120 \quad for\ \varphi = 32 - 42^\circ$$

$$\text{Assume}: \ N_t = 115 \rightarrow q_t = 115 \times 108.5 = 12477\ kPa$$

$$\text{Toe Area}: \ A = \frac{\pi D^2}{4} = \frac{\pi \times 0.356^2}{4} = 0.1\ m^2$$

$$R_t = r_t A = 12477 \times 0.1 = 1247.7 kN$$

Shaft capacity:

$$r_s = \frac{\sigma'_v}{2} \cdot k_s \cdot \tan \delta$$

$$k_s = 2k_0 = 2(1 - \sin 41) = 0.69$$

$$\beta = k_s \tan \delta = 0.6$$

$$\delta = (0.5 - 1)\varphi = 1\varphi \rightarrow \tan \delta = \tan \varphi = \tan 41^\circ = 0.87$$

$$\sigma'_v = 108.5\ KPa$$

$$r_s = \frac{108.5}{2} \times 0.6 = 32.55\ kPa$$

Circumferential area: $P = \pi D = 1.12\ m$

$$R_s = r_s \pi D L = 32.55 \times 1.12 \times 7 = 255.2\ kN$$

$$R_u = R_t + R_s = 1502\ kN = 150.2\ ton$$

Pile load test: $R_u = 174.3\ ton.$

7.6 Commonly used direct CPT–based methods for pile axial–bearing capacity

Given the penetrometer analogy with the pile, it can be considered as model pile. Fig. 7.6 illustrates how closely the CPT tip resistance and sleeve friction resemble the pile toe and shaft capacity components, i.e., r_t and r_s.

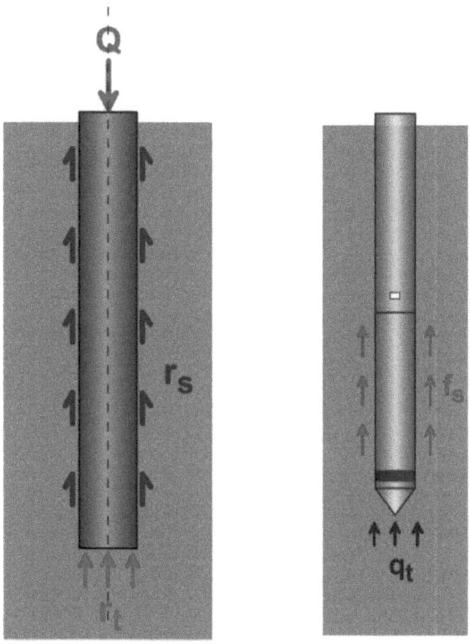

Figure 7.6 Analogy between pile and penetrometer.

Over 30 approaches have been proposed for estimating the pile axial–bearing capacity by CPT records within the last 30 years. A summary of such methods has been gathered by Niazi and Mayne (2013). Table 7.1 demonstrates a list of these methods. Below we discuss the more commonly used methods.

7.6.1 Nottingham (1975) and Schmertmann (1978).

The Schmertmann and Nottingham method is based on a summary of the work on model and full-scale piles presented by Nottingham (1975). The unit toe resistance of a pile, r_t, in sand and in clay, is taken as equal to the average of the cone resistance. The average q_c value is determined from the graphical representation of the CPT measurement in a zone ranging from 8B above (B is pile diameter) and from 0.7B through 4B below pile toe (the actual value depends on the trend of q_c values), see Fig. 7.7. This zone is defined by the failure pattern for the pile toe, which is assumed to follow a log-spiral shape, as originally introduced by Begemann (1963). An upper limit of 15 MPa is imposed for the unit toe resistance.

The unit shaft resistance of a pile, r_s, is determined from either sleeve friction or cone resistance:

$$r_s = Kf_s \tag{7.8}$$

where K = dimensionless coefficient, for clay defined as the reverse ratio in Fig. 7.8.

Depending on pile shape and material, cone type, and embedment ratio, in sand, the K coefficient ranges from 0.8 through 2.0, and in clay, it ranges from 0.2 through 1.25.

Table 7.1 List of common CPT- and CPTu-based methods for pile-bearing capacity estimation.

No.	Method/Reference	No.	Method/Reference
1	Begemann (1969)	16	UCD-05 (Gavin and Lehane, 2005)
2	Meyerhof (1983)	17	ICP-05 (Jardine et al., 2005)
3	Aoki and Velloso (1975)	18	UWA-05 (Lehane et al., 2005)
4	Nottingham (1975), Schmertmann (1978)	19	NGI-05 (Clausen et al., 2005)
5	Penpile (Clisby et al., 1978)	20	Cambridge-05 (White and Bolton, 2005)
6	Dutch (de Ruiter and Beringen, 1979)	21	Togliani (2008)
7	Philipponnat (1980)	22	German (Kempfert and Becker, 2010)
8	LCPC (Bustamante and Gianeselli, 1982)	23	UCD-11 (Igoe et al., 2010, 2011)
9	Cone-m (Tummay and Fakhroo, 1982)	24	V−K (Van Dijk and Kolk, 2011)
10	Price and Wardle (1982)	25	SEU (Cai et al., 2011, 2012)
11	Gwizdala (1984)	26	HKU (Yu and Yang, 2012)
12	Eslami and Fellenius (1997), Fellenius (2002)	27	UWA-13 (Lehane et al., 2013)
13	KTRI (Takesue et al., 1998)	28	Probabilistic approach by Chen and Zhang (2013)
14	TCD-03 (Gavin and Lehane, 2003)	29	Enhanced UniCone method (Niazi and Mayne, 2016)
15	Fugro-05 (Kolk et al., 2005)	30	Valikhah et al. (2018)

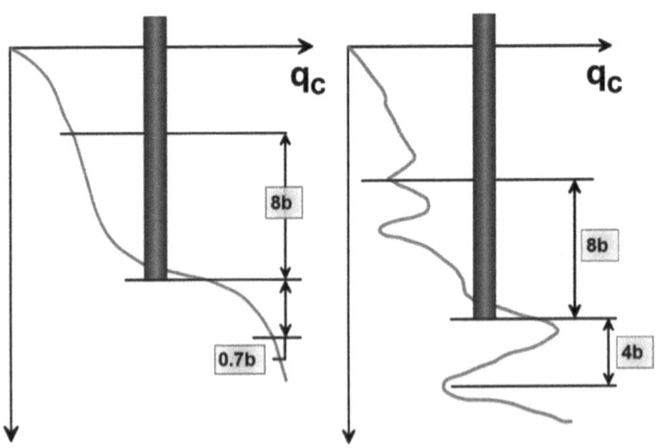

Figure 7.7 Application of CPT data for pile design (after Schmertmann, 1978).

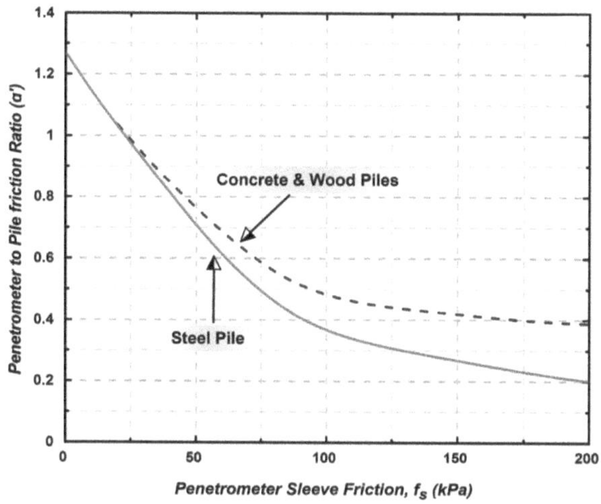

Figure 7.8 Design curves for pile shaft resistance in clay (Schmertmann, 1978).

Above an embedment depth of eight pile diameters, the unit shaft resistance is linearly interpolated from zero at the ground surface to the value of Eq. (7.8). In sand, but not in clay, the shaft resistance may be determined from the cone resistance instead:

$$r_s = \bar{c}q_c \tag{7.9}$$

where

$\bar{c}$ = a dimensionless coefficient
q_c = cone resistance (total; uncorrected for pore pressure)

The $\bar{c}$ coefficient is a function of the pile type and ranges from 0.8% through 1.8%. An upper limit of 120 kPa is imposed on the unit shaft resistance, r_s, regardless whether it is determined from Eq. (7.8) or from Eq. (7.9). For tension capacity, the shaft resistance is reduced to 70% of that determined by Eqs. (7.8) and Equation (7.9).

7.6.2 LCPC (Bustamante and Gianeselli, 1982)

The LCPC (Laboratoire Central des Ponts et Chausees) method (also called the French method) is based on experimental work of Bustamante and Gianeselli (1982) for the French Highway Department. Both the unit toe and unit shaft resistances are determined from the cone resistance, q_c. The sleeve friction, f_s, is not used. The unit toe resistance is determined as a range of 40% thxrough 55% of the q_c, averaged within a zone of 1.5B above and 1.5B below the pile toe, see Table 7.2 and Fig. 7.9.

The unit shaft resistance is determined from Eq. (7.3) with the c coefficient ranging from 0.5% through 3.0% as governed by rules for the magnitude of the cone resistance, type of soil, and type of pile. Upper limits of the unit shaft resistance are imposed, ranging from 15 kPa through 120 kPa depending on soil type, pile type, and pile installation method, see Table 7.2.

Table 7.2 Empirical coefficients for French method for pile capacity calculation using CPT.

			Driven piles				Bored piles			
			K_s [b]		Upper limit of r_s (kPa)		K_s [b]		Upper limit of r_s (kPa)	
Nature of soil	q_c (MPa)	K_c [a]	A	B	A	B	A	B	A	B
Soft clay and mud	<1	0.5	30	30	15	15	30	30	15	15
Moderately compact clay	1 to 5	0.45	40	80	35 (80)	35 (80)	40	80	35 (80)	35 (80)
Compact to stiff clay and compact silt	>5	0.55	60	120	35 (80)	35 (80)	60	120	35 (80)	35 (80)
Silt and loose sand	<5	0.5	60	120	35	35	60	150	35	35
Moderately compact sand and gravel	5 to 12	0.5	100	200	80 (120)	80	100	200	80	35 (80)
Compact to very compact sand and gravel	>12	0.4	150	200	120 (150)	120	150	300	120 (150)	80 (120)

Category:
A, Driven precast piles, Prestressed tubular piles, and Jacked concrete piles; *B*, Driven metal piles and Jacked metal piles.
Note: Bracket value for r_s: apply to careful execution and minimum disturbance of soil due to disturbance.
[a] $r_t = K_c\, q_{c,s}$ ($q_{c,s}$ = average q_c values).
[b] $r_s = q_c/K_s$
Modified after Fellenius, B.H., 2015, Basics of foundation design. Electronic Edition, April, 2015.

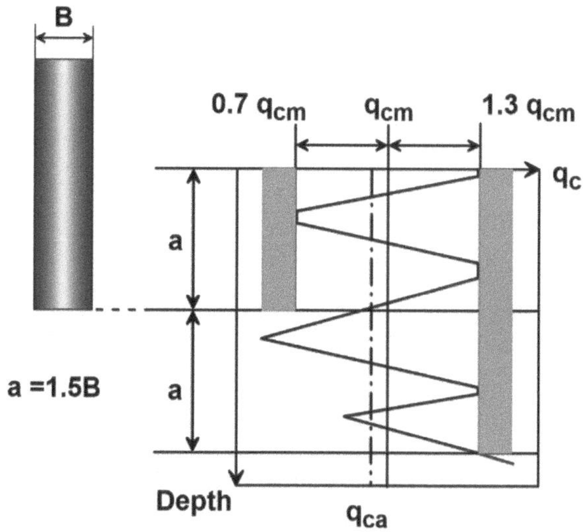

Figure 7.9 LCPC CPT method for determining equivalent cone resistance at pile toe.

7.6.3 Meyerhof (1976, 1983)

The Meyerhof method is based on theoretical and experimental studies of deep foundations in sand. For unit toe resistance, the influence of scale effect of piles and shallow penetration in dense sand strata have been considered. The unit toe resistance in sand is determined from cone resistance through an empirical relation as follows:

$$r_t = q_{c,a} C_1 C_1 \tag{7.10}$$

where

$q_{c,a}$ = arithmetic average of q_c values in a zone ranging from "1b" below through "4b" above pile toe

$C_1 = \left[(b + 0.5)/_{2B}\right]^n$; modification factor for scale effect when $b > 0.5$, otherwise $C_1 = 1$

$C_2 = D_b/_{10B}$; modification factor for penetration into dense strata when $D_b < 10b$, otherwise $C_2 = 1$

B = pile diameter (m)

n = an index; 1 for loose sand, 2 for medium dense sand, and 3 for dense sand

D_b = embedment of pile (m) in dense sand strata

The unit toe resistance of a bored pile is reduced to 30% of that determined from Eq. (7.10).

The unit shaft of pile resistance is determined from either Eq. (7.8) with K = 1 or Eq. (7.9) with C = 0.5%. For bored piles, reduction factors of 70% and 50%, respectively, are applied to the calculated values.

7.6.4 Tummay and Fakhroo (1982)

The Tummay and Fakhroo method is based on an experimental study on a number of different piles in clay soils in Louisiana, USA. The unit toe resistance is determined as the same procedure recommended by Schmertmann (1978). The unit shaft resistance is determined according to Eq. (7.8) with the K coefficient determined from the following relation, in which the coefficient is no longer dimensionless.

$$K = 0.5 + 9.5e^{-90f_s} \tag{7.11}$$

where

f_s = sleeve friction in MPa

For a sleeve friction ranging from 10 kPa through 50 kPa, Eq. (7.8) results in a K coefficient ranging from about 4.5 through 0.6.

7.6.5 Price and Wardle (1982)

The Price and Wardle (1982) method employs both cone resistance and sleeve friction for determining pile capacity.

$$r_t = k_b \cdot q_{ca} \leq 15MPa \tag{7.12}$$

$$r_s = k_s \cdot f_s \leq 120kPa \tag{7.13}$$

k_b = 0.35 for driven piles and 0.35 for jacked piles
k_s = 0.53 for driven piles, 0.62 for jacked piles, and 0.49 for bored piles.

7.6.6 Penpile

The Penpile method, developed by Clisby et al. (1978) for Mississippi Department of Transportation, uses both cone resistance and sleeve friction for estimating the pile capacity as follows:

$$r_t = 0.25q_{ca}\ldots\text{(in clay)} \tag{7.14}$$

$$r_t = 0.125q_{ca}\ldots\text{(in sand)} \tag{7.15}$$

$$r_s = \frac{f_{sa}}{1.5 + 14.47f_{sa}} \tag{7.16}$$

f_{sa} = average cone sleeve friction in MPa.

7.6.7 Eslami and Fellenius (1997)

The Eslami and Fellenius method (Eslami, 1996, Eslami and Fellenius, 1995, 1996) is based on CPTu data but can also make use of CPT data. The method is developed from a large database of CPT and CPTu tests coupled with results from tests for axial pile capacity.

In the Eslami and Fellenius CPTu method (in contrast to other methods), the data are unfiltered. Instead, a potentially disproportionate influence of odd "peaks and troughs" is reduced by means of employing the geometric average of the cone stress as opposed to the arithmetic average used by the CPT methods. Furthermore, the cone stress is transferred to the "effective" cone stress, q_E, by subtracting the measured pore pressure, u_2, from the measured total cone stress (corrected for pore pressure acting against the shoulder). The pile unit toe resistance is the geometric average of the "effective" cone stress over an influence zone that depends on the soil layering.

When a pile is installed through weak soil into dense soil, the average is determined over an influence zone extending from 4B below the pile toe through a height of 8B above the pile toe. On the other hand, when a pile is installed through a dense soil into a weaker soil, the average is determined over an influence zone extending from 4B below the pile toe through a height of 2B above the pile toe. The relation is given in the following equation:

$$r_t = C_t q_{Eg} \tag{7.17}$$

where C_t is toe correlation coefficient, equal to unity in most cases, and q_{Eg} = geometric average of the cone point resistance over the influence zone after correction for pore pressure on shoulder and adjustment to "effective" stress.

Also the pile unit shaft resistance is correlated to the average "effective" cone point stress with a modification according to soil type per the approach detailed below. The C_s correlation coefficient is determined from the soil profiling chart (Fig. 7.10), which uses both cone stress and sleeve friction. However, because the sleeve friction is a

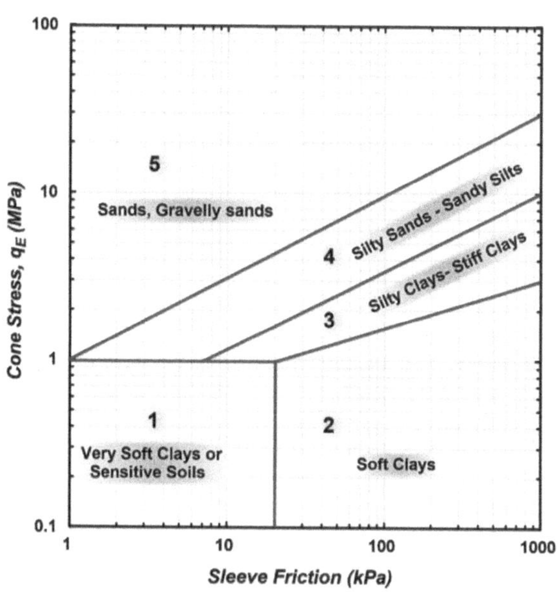

Figure 7.10 The Eslami and Fellenius (1997) chart for soil classification.

Table 7.3 Shaft coefficient correlation, C_s (Eslami and Fellenius, 1997).

Soil type	Cs
Soft sensitive soils	8.0%
Clay	5.0%
Stiff clay and mixture of clay and silt	2.5%
Mixture of silt and sand	1.0%
Sand	0.4%

more variable measurement than the cone point resistance, the sleeve friction value is not applied directly.

$$r_s = C_s q_E \tag{7.18}$$

where

r_s = pile unit shaft resistance
C_s = shaft correlation coefficient, which is a function of soil type determined from the soil profiling chart
q_E = cone point resistance after correction for pore pressure on the cone shoulder and adjustment to "effective" stress (Table 7.3)

7.6.7.1 UniCone

The UniCone software, developed by UniSoft, is a program for analysis and design of piles using CPT and CPTu records (Fellenius, 2002). This program determines the bearing capacity of piles by six CPT- and CPTu-base methods including Eslami and Fellenius (1997), European or Dutch (De Ruiter and Beringen, 1979), French or LCPC (Bustamante and Gianeselli, 1982), Mehyerhof (1983), Schmertmann (1978), and Tummay and Fakhroo (1982). An example of the program output is shown in Fig. 7.11.

In this software, a modification factor, C_t, is applied for the Eslami and Fellenius (1997) method. The toe correlation coefficient, C_t, also called toe adjustment factor, is a function of the pile size (toe diameter). The larger the pile diameter, the larger the movement required to mobilize the toe resistance.

Therefore, the "useable" pile toe capacity diminishes with increasing pile toe diameter. For pile diameters larger than about 0.4 m, the adjustment factor should be determined by the relation given in Eq. (7.19).

$$C_t = \frac{1}{3B} (B \text{ } in \text{ } meter), C_t = \frac{12}{B} (B \text{ } in \text{ } inches) \tag{7.19}$$

where

B = pile diameter in units of either meter (or inches)

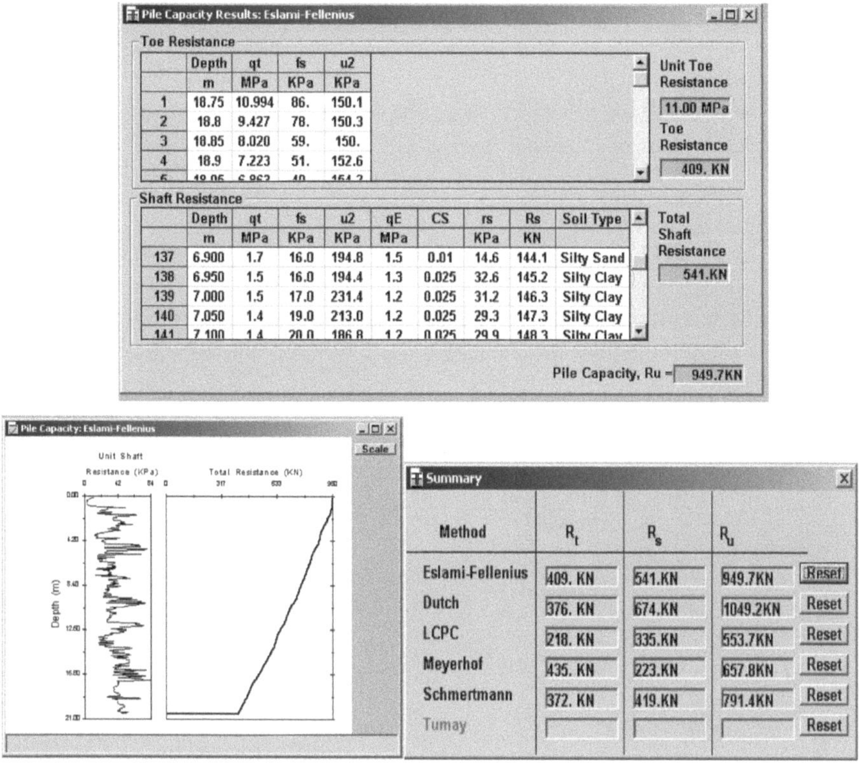

Figure 7.11 An example of the UniCone output.

7.6.8 Takesue et al. (1998)

In this method, the unit shaft capacity of piles is estimated exclusively by means of CPT skin friction (f_s) as well as pore pressure (u). The data used in development of this method consist of driven bored piles installed in clay, sand, and mixed soils.

$$\Delta u < 300kPa \Rightarrow r_s/f_s = \frac{\Delta u}{1250} + 0.76 \tag{7.20}$$

$$300 < \Delta u < 1200kPa \Rightarrow r_s/f_s = \frac{\Delta u}{200} - 0.5 \tag{7.21}$$

7.6.9 NGI (Clausen et al., 2005)

The NGI method, proposed by Clausen et al. (2005), in fact, uses CPT measurements for estimating the soil relative density. The relative density is then employed in the calculation of the unit shaft and toe resistance. The pile capacity is defined as the load mobilized at the pile head displacement of 10%B. the cone resistance at pile toe is used in the formulas and no averaging procedure is done. Eq. (7.22) is the unit toe resistance for a close-end pile according to the NGI approach.

$$\frac{q_{b0.1}}{q_{c,tip}} = \frac{0.8}{1 + D_r^2} \tag{7.22}$$

where $q_{b0.1}$ is pile unit toe resistance, $q_{c,tip}$ is cone resistance at pile toe elevation, and D_r is soil relative density calculated by Eq. (7.23).

$$D_r = 0.4\left[\ln\frac{q_{c,tip}}{22\left(\sigma'_{v0}p_{ref}\right)^{0.5}}\right] \tag{7.23}$$

The toe capacity for open-end piles is assessed in two conditions, plugged and unplugged; the minimum value obtained from these two conditions is considered as the pile toe capacity.

$$q_{b0.1} = \min\left(q_{b0.1,plugged}\ , q_{b0.1,\ unplugged}\right) \tag{7.24}$$

$$\frac{q_{b0.1}}{q_{c,tip}} = \frac{0.7}{1 + 3D_r^2}\ldots(\text{plugged}) \tag{7.25}$$

$$q_{b0.1} = q_{b,ann}A_r + q_{b,plug}(1 - A_r)\ldots(\text{unplugged}) \tag{7.26}$$

$$q_{b,ann} = q_{c,tip} \tag{7.27}$$

$$q_{b,plug} = \frac{12\tau_{f,avg}D}{\pi B_i} \tag{7.28}$$

$q_{b,ann}$ is the unit resistance of pile annulus, $q_{b,plug}$ is the unit resistance of soil plug, $\tau_{f,avg}$ is average external sleeve friction, and B_i is pile inner diameter.

The unit shaft resistance is estimated by the following equation. The minimum value of 10% effective overburden stress is imposed as the lower boundary for the shaft resistance.

$$r_s = \left(\frac{z}{Lp_{ref}F_{D_r}F_{sig}F_{tip}F_{load}F_{mat}}\right) \geq \tau_{min} \tag{7.29}$$

$$F_{D_r} = 2.1(D_r - 0.1)^{1.7} \tag{7.30}$$

$$F_{sig} = \left(\frac{\sigma'_{v0}}{p_{ref}}\right)^{0.25} \tag{7.31}$$

$$\tau_{min} = 0.1\sigma'_{v0} \tag{7.32}$$

where z is given depth, L is pile embedment length, $F_{tip} = 1.0$ for open end–driven piles and 1.6 for closed end–driven piles, $F_{load} = 1.0$ for piles in tension, 1.3 for piles in compression, and $F_{mat} = 1.0$ for steel piles and 1.2 for concrete piles.

7.6.10 UWA (Lehane et al., 2005)

This approach is only applicable for driven piles in sand. The main difference of this approach with other methods is considering the effect of partial plugging (Fig. 7.12). In this regard, incremental filling ratio (IFR) is introduced. IFR is the variation in soil plug thickness per pile penetration into the soil.

Fig. 7.12 demonstrates a schematic view of soil plug formation in an open-end pile. In a fully plugged condition, IFR is zero, and, in an unplugged condition, IFR equals unity. For intermediate partial plugging conditions, IFR varies between 0 and 1.0.

$$IFR = \frac{H - H_0}{L - L_0} \tag{7.33}$$

The pile capacity is defined as the load at pile toe displacement of 10%B. The cone resistance used in the formula is the average of the cone resistance within 4B below and 8B above the pile toe, using the minimum path rule.

According to the UWA method, the ratio of the unit toe capacity of a closed-end pile to the average of cone resistance is a constant value equal to 0.6. The proposed formula for the unit toe capacity of open-end piles is as follows:

$$\frac{q_{b0.1}}{q_{c,avg}} = 0.15 + 0.45A_{rb,eff} \tag{7.34}$$

$$A_{rb,eff} = 1 - FFR\left(\frac{B_i^2}{B^2}\right) \tag{7.35}$$

In the above equation, *FFR* is final filling ratio which is the average of IFR over the last 3B of pile penetration.

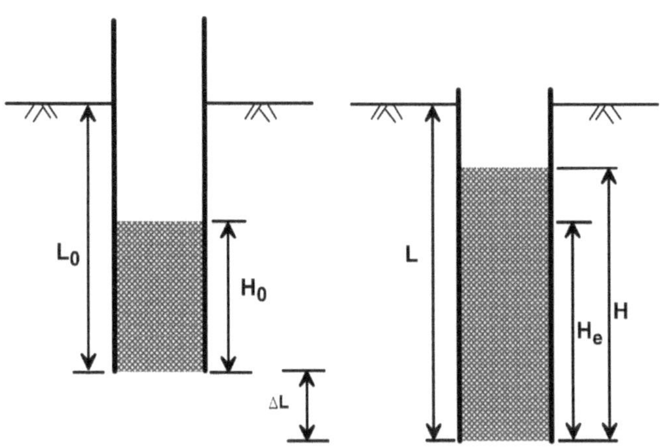

Figure 7.12 A schematic view of soil plug formation in an open-end pile.

For open-end piles, it is required to use the effective diameter for defining the averaging zone.

$$B_{eff} = B\sqrt{A_{rb,eff}} \tag{7.36}$$

As mentioned previously, the unit toe resistance of a pile depends on FFR. Because measuring this parameter is not common in practice, the following equation is proposed for estimating FFR:

$$FFR \approx min\left[1, \left(\frac{B_i(m)}{1.5(m)}\right)^{0.2}\right] \tag{7.37}$$

The unit shaft resistance of a pile can be calculated using Eq. (7.44).

$$r_s = \frac{f_t}{f_c}\left[0.03q_c A_{rs,eff}^{0.3}\left[\max\left(\frac{h}{B}, 2\right)\right]^{-0.5} + \Delta\sigma'_{rd}\right]\tan\delta_f \tag{7.38}$$

$$A_{rs,eff} = 1 - IFR\left(\frac{B_i}{B}\right)^2 \tag{7.39}$$

where $\frac{f_t}{f_c} = 1.0$ in compression and 0.75 in tension, $h =$ distance from pile toe, $q_c =$ cone resistance at a given depth, $\Delta\sigma'_{rd} =$ effective radial stress caused by dilation during loading defined by Eq. (7.20), and $\delta_f =$ friction angle between soil and pile.

The average IFR over the last 20B of pile penetration can be estimated by Eq. (7.40).

$$IFR_{mean} \approx min\left[1, \left(\frac{B_i(m)}{1.5(m)}\right)^{0.2}\right] \tag{7.40}$$

The radial effective stress due to dilation effect during loading is defined by the following equation:

$$\Delta\sigma'_{rd} = \frac{4G}{B}\Delta r \tag{7.41}$$

$$G \approx 185 q_c q_{c1N}^{-0.7} \tag{7.42}$$

$$q_{c1N} = \frac{(q_c/p_{ref})}{(\sigma'_{v0}/p_{ref})^{0.5}} \tag{7.43}$$

7.6.11 ICP (Jardine et al., 2005)

The ICP[1] approach (Jardine et al., 2005) provides separate equations for sand and clay. In sands, this method is applicable for driven piles for which the SML (slow maintained load) test is conducted around 10 days after installation. Friction fatigue and the pile tip condition are considered in terms of h/r* and A_r, respectively. Changes in radial stress, the interface friction angle, and the direction of loading are taken into account in this method.

The pile capacity is defined as the capacity mobilized at the pile head settlement equal to 10% pile diameter. The averaging zone is 1.5 pile diameter above and below the pile tip for the cone resistance in the unit toe resistance formula. The unit toe resistance for close-end piles is calculated by the following equation:

$$\frac{q_{b0.1}}{q_{c,avg}} = max\left[1 - 0.5 \log\left(\frac{B}{B_{CPT}}\right), 0.3\right] \tag{7.44}$$

For open-end piles, two conditions, plugged and unplugged, are considered. To calculate the toe capacity of an open-end piles, it is necessary to, first, define whether the pile is plugged or unplugged, then use the corresponding formula for toe capacity calculation. If one of the following equations are satisfied, the pile is plugged. The other way round is unplugged.

$$\begin{aligned} &B_i < 0.02(D_r - 30) \quad \text{or} \\ &B_i < 0.083\left(\frac{q_{c,avg}}{p_{ref}}\right)B_{CPT} \end{aligned} \tag{7.45}$$

where D_r is relative density in percent.

In case of unplugged condition, the pile unit toe capacity is calculated by the following equation:

$$\frac{q_{b0.1}}{q_{c,avg}} = max\left[0.5 - 0.25 \log\left(\frac{B}{B_{CPT}}\right), 0.15, A_r\right] \tag{7.46}$$

where A_r is effective area ratio defined as Eq. (7.47)

$$A_r = 1 - \left(\frac{B_i^2}{B^2}\right) \tag{7.47}$$

For plugged piles, the unit toe resistance is determined as follows:

[1] Imperial College Pile.

$$\frac{q_{b0.1}}{q_{c,avg}} = A_r \tag{7.48}$$

The mobilized unit shaft friction can be estimated by Eq. (7.19).

$$r_s = a\left[0.029bq_c\left(\frac{\sigma'_{v0}}{p_{ref}}\right)^{0.13}\left[\max\left(\frac{h}{R^*}, 8\right)\right]^{-0.38} + \Delta\sigma'_{rd}\right]\tan\delta_f \tag{7.49}$$

where a is 0.9 for open-end piles in tension and 1.0 for other conditions, and b is 0.8 for piles in tension and 1.0 for piles in compression. Other parameters in this equation are similar to Eq. (7.38). In addition, for open-end piles, effective pile radius (R^*) is used in calculations which is determined as follows:

$$R^* = \sqrt{R^2 - R_i^2} \tag{7.50}$$

7.6.12 Fugro (Kolk et al., 2005)

The Fugro method proposed by Kolk et al. (2005) is a design criterion for driven piles in silica sand and is mostly applicable for offshore piles. In this method, the pile–soil interface friction angle is assumed to be relatively constant, and the changes in radial stress are ignored. Friction fatigue is considered in term of the ratio of the height above the pile tip (h) to the modified radius (r*) and the end condition is defined by the area ratio (A_r).

The toe capacity is defined as the load at the displacement of 10%B. According to the Fugro method, it is assumed that open-end piles are fully plugged during the static loading test. This assumption materializes if the pile toe is sufficiently (about 5B) penetrated into the bearing layer.

The following equation is employed for calculation of toe capacity for both open- and closed-end piles.

$$\frac{q_{b0.1}}{q_{c,avg}} = 8.5\left(\frac{p_{ref}}{q_{c,avg}}\right)^{0.5} A_r^{0.25} \tag{7.51}$$

The dilation effect caused by loading is neglected in this method. Moreover, the friction angle between pile and soil is assumed to be constant and equal to 29°.

The proposed equations for unit shaft capacity are as follows.

$$r_{s,comp} = 0.08q_c\left(\frac{\sigma'_{v0}}{p_{ref}}\right)^{0.05}\left(\frac{h}{R^*}\right)^{-0.90} \quad for\ \frac{h}{R^*} \geq 4 \tag{7.52}$$

$$r_{s,comp} = 0.08q_c\left(\frac{\sigma'_{v0}}{p_{ref}}\right)^{0.05}(4)^{-0.90}\left(\frac{h}{4R^*}\right) \quad for\ h/R^* \leq 4 \tag{7.53}$$

$$r_{s,tension} = 0.045q_c \left(\frac{\sigma'_{v0}}{p_{ref}}\right)^{0.15} \left(\frac{h}{R^*}\right)^{-0.85} \quad for\ h/R^* \geq 4 \tag{7.54}$$

$$r_{s,tension} = 0.045q_c \left(\frac{\sigma'_{v0}}{p_{ref}}\right)^{0.15} (4)^{-0.85} \quad for\ h/R^* \leq 4 \tag{7.55}$$

7.6.13 Togliani (2008)

Togliani developed a direct CPTu method formulated based on his experience of designing and predicting the pile behavior. He proposed a simplified direct CPT–based method for displacement and nondisplacement, cylindrical, as well as tapered piles. In particular, this method is based on experiences combined with the LCPC method and certain previous recommendations for calculating the capacity of tapered piles to provide design equations for both r_s and r_t (Niazi and Mayne, 2013). The averaging zone of q_c for pile toe capacity is 8B above and 4B below the pile toe.

The toe resistance is calculated by applying a factor, k_3, to the average measured q_c value for a range between 8B above the pile toe to 4B below the pile toe. The factor k_3 depends on pile installation method and pile geometry.

$$r_t = k_3 q_t \tag{7.56}$$

$$k_3 = \lambda + [0.01(L_{pile}/d_{toe})] \tag{7.57}$$

where λ is considered to be 0.2 for driven piles and 0.1 for bored piles, and L_{pile}/d_{toe} is pile slenderness ratio.

The shaft resistance calculation applies a factor, k_1, which is based on the calculated CPTu friction ratio, R_f, and pile installation method.

$$r_s = k_1\ q_t^{0.5} \tag{7.58}$$

$$k_1 = \overline{\beta}\left\{1.2\left(0.8 + \frac{R_f}{8}\right)\right\} \quad for\ R_f < 1 \tag{7.59}$$

$$k_1 = \overline{\beta}\{1.1(0.4 + Ln(R_f))\} \quad for\ R_f \geq 2 \tag{7.60}$$

$$k_1 = \overline{\beta}\left\{1.2\left(0.8 + \frac{R_f}{8}\right) + 1.1(0.4 + Ln(R_f))\right\}/2 \quad for\ 1 \leq R_f < 2 \tag{7.61}$$

where $\overline{\beta}$ is considered as 1 for driven displacement piles, 0.6 for nondisplacement driven and CFA, and 0.5 for bored piles.

The value of r_s is modified for overconsolidation ratio (OCR), and for piles with a driven length below a total overburden pressure of 100 kPa, for "overlength (OL)" (Togliani, 2010) by the term: $[(OCR)^{0.13}][(\sigma'_v)^{0.052}]$.

For tapered piles,

$$Q_{shaft} = Q_{cylinder} + Q_{taper} \quad (7.62)$$

$$Q_{cylinder} = \pi.d_{mean}.L.f_{cylinder} \quad (7.63)$$

where $f_{cylinder}$ is estimated based on Eq. (7.58).

$$Q_{taper} = \pi/4\left[d_{top}^2 - d_{bottom}^2\right].f_{taper} \quad (7.64)$$

$$f_{taper} = q_c(k_2 d_{mean}/d_{base}); k_2 = 1.2 \text{ for } qc \leq 3 \text{ MPa; and } 1.0 \text{ for } qc > 3 \text{ MPa} \quad (7.65)$$

7.6.14 German

The German method, proposed by Kempfert and Becker (2010), is based on the results of loading tests on roughly 1000 case histories of bored, precast concrete, pipe, and screw cast in place piles and micropiles. According to this method, unit shaft and toe resistance of a pile in a noncohesive soil is calculated in terms of cone resistance. While in cohesive soils, the pile capacity is calculated using undrained shear strength (s_u) and is not directly correlated to the CPT measurements. This method is presented in form of r_s and r_t curves in terms of q_c or s_u.

As depicted in Fig. 7.13, the bearing capacity for each of the five pile-type categories is determined by the proposed intervals within two upper and lower boundaries.

7.6.15 Enhanced UniCone (Niazi and Mayne, 2016)

This method is a modified version of the Eslami and Fellenius (1997) method, using a database of 153 pile load tests introduced by Niazi and Mayne (2016). The modifications are offered to derive coefficients C_{se} for shaft component and C_{te} for toe

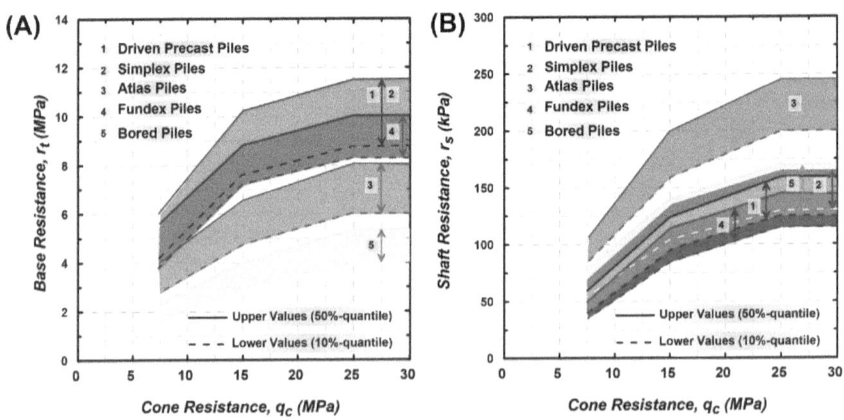

Figure 7.13 Upper and lower boundaries for pile capacity calculation according to the German method (Kempfert and Becker, 2010).

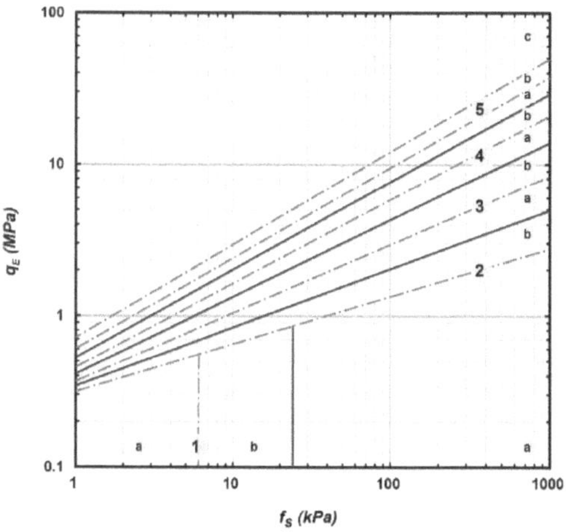

Zone No.	Soil Type	Average C_{se} (%)
1a	Very Soft Sensitive Clay	8.32
1b	Soft Clay and Silt	6.57
2a	Silty Marine and Varved Clays	4.69
2b	Stiff Weathered Clay, Clay Till	3.63
3a	Firm to Medium Soft Silty Clay	3.10
3b	Clayey Silt, Mudstone	2.47
4a	Sandy Silt, Medium Dense Silt	1.58
4b	Silt Sand, Very Dense Silt	1.09
5a	Uniform Loose Sand	0.82
5b	Medium Dense Sand	0.61
5c	Dense Sand, Gravel-Sand Mix	0.34

Figure 7.14 Modified UniCone CPTu chart for zone numbers, soil types, and pile side friction coefficient (C_{se}) (Niazi, 2014).

component of the axial capacity from CPTu data. In addition, a step-by-step methodology to use the method is presented in a flowchart by them.

To estimate Q_{tn}, F_r, and I_c, it is necessary to apply a trial and error procedure to find "n" as stated below (Zhang et al., 2002):

Step 1—Assume n = 1 and calculate Q_{tn}, F_r, and I_c
Step 2—If $I_c \leq 1.64$, n = 0.5
If $1.64 < I_c < 3.30$, $n = (I_c - 1.64)*0.3 + 0.5$
If $3.30 \leq I_c$, n = 1
Iterate until the change in n, $\Delta n < 0.01$.
If $\sigma'_{v0} > 300\ kPa$, let n = 1 for all soils.

Based on the information of soil classification types and subtypes from the database, new sets of boundaries and subboundaries were delineated on the diagram, thus proposing 11 soil zones, in contrast to the five originally presented by the UniCone method (see Fig. 7.14). It may also be noticed that based on this larger and later database, the new boundaries are slightly displaced from those of the original UniCone method (Fellenius, 2002). This process provided an improved CPT soil classification chart.

7.6.16 Summary of the CPT-based methods

Fifteen methods have been reviewed in this section, a summary of which is presented in Table 7.4.

Table 7.4 Summary of CPT- and CPTu-based methods.

No.	Method/references	Pile unit side resistance (r_s)	Pile unit end bearing (r_t)
1	Nottingham (1975) Schmertmann (1978)	$r_s = C_s q_c$ $r_s = K f_s$ $C_s = 0.8{-}1.8\%$ $K = 0.8{-}2(sand)$	$r_t = C_t q_{ca}$
2	Meyerhof (1976)	$r_s = k f_s$ $k = 1$ $r_s = \bar{c} q_c$ $\bar{c} = 0.5\%$	$r_t = q_{c.a} c_1 c_2$ $c_1 = \left(\frac{B+0.5}{2B}\right)^n$ $c_2 = \frac{D_b}{10B}$ D_b *bearing embedment depth* $n = 1$ (*loose*), 2 (*medium dense*), 3 (*dense*)
3	LCPC (Bustamante and Gianeselli, 1982)	$r_s = \frac{1}{k_s} q_c$ $k_s = 30{-}150$	$r_t = k_b q_{eq}$ $k_b = 0.4{-}0.55$
4	Tummay and Fakhroov (1982)	$r_s = K f_s$ $K = 0.5 + 9.5e^{-90 f_s}$	$r_t = C_t q_{ca}$
5	Price and Wardle (1982)	$r_s = k_s \cdot f_s \le 120\ kPa$ $k_s = 0.53$ for driven piles, 0.62 for jacked piles and 0.49 for bored piles.	$r_t = k_b \cdot q_{ca} \le 15\ MPa$ $k_b = 0.35$ for driven piles, 0.35 for jacked piles
6	Penpile (Clisby et al., 1978)	$r_s = \frac{f_{sa}}{1.5+14.47 f_{sa}}$ f_{sa} = average cone sleeve friction in MPa.	$r_t = 0.25 q_{ca}$(in clay) $r_t = 0.125 q_{ca}$(in sand)
7	Unicone (Eslami and Fellenius, 1997)	$r_s = c_{se} \times q_E$ $q_E = q_t - u_2$ $c_{se} = 0.3{-}8\%$	$r_t = c_{te} \times q_{Eg}$ $q_{cg} = (q_{c1} \times q_{c2} \times q_{c3} \times \ldots \times q_{cn})^{\frac{1}{n}}$ $c_{te} = 1$
8	Takesue et al. (1998)	$\Delta u < 300kPa \quad r_s/f_s = \frac{\Delta u}{1250} + 0.76$ $300 < \Delta u < 1200kPa \quad r_s/f_s = \frac{\Delta u}{200} - 0.5$	-

9	UWA method (Lehane et al., 2005)	$r_s = \frac{f_t}{f_c}\left[0.03 q_c A_{rs,eff}^{0.3}\left[\max\left(\frac{h}{D}, 2\right)\right]^{-0.5} + \Delta\sigma'_{rd}\right]\tan\delta_f$ $A_{rs,eff} = 1 - IFR\left(\frac{B_i}{B}\right)^2$ $IFR_{mean} \approx min\left[1, \left(\frac{B_i(m)}{1.5(m)}\right)^{0.2}\right]$	$\frac{r_{t0.1}}{q_{c,avg}} = 0.15 + 0.45 A_{rb,eff}$ $A_{rb,eff} = 1 - FFR\left(\frac{B_i^2}{B^2}\right)$ $FFR \approx min\left[1, \left(\frac{B_i(m)}{1.5(m)}\right)^{0.2}\right]$
10	Fugro method (Kolk et al., 2005)	Compression loading: $h/R^* \geq 4$: $r_s = 0.08 q_c \left(\frac{\sigma'_{v0}}{p_{ref}}\right)^{0.05}\left(\frac{h}{R^*}\right)^{-0.90}$ $h/R^* \leq 4$: $r_s = 0.08 q_c \left(\frac{\sigma'_{v0}}{p_{ref}}\right)^{0.05}(4)^{-0.90}\left(\frac{h}{4R^*}\right)$ Tension loading: $r_s = 0.045 q_c \left(\frac{\sigma'_{v0}}{p_{ref}}\right)^{0.15}\left(\max\left(\frac{h}{R^*}, 4\right)\right)^{-0.85}$	$r_{t0.1} = 8.5 q_{c,avg}\left(\frac{p_{ref}}{q_{c,avg}}\right)^{0.5} A_r^{0.25}$ $A_r = 1 - \left(\frac{B_i^2}{B^2}\right)$
11	ICP method (Jardine et al., 2005)	$r_s = a\left[0.029 b q_c \left(\frac{\sigma'_{v0}}{p_{ref}}\right)^{0.13}\left[\max\left(\frac{h}{R^*}, 8\right)\right]^{-0.38} + \Delta\sigma'_{rd}\right]\tan\delta_f$ $a = 0.9$ (OE piles in tension), 1.0 (all other cases) $b = 0.8$ (tension), 1.0 (compression) δ_f measured or estimated as fctn(d_{50})	$\frac{r_{t0.1}}{q_{c,avg}} = max\left[1 - 0.5\log\left(\frac{B}{B_{CPT}}\right), 0.3\right]$ The pile is fully plugged if: $B_i < 0.02(D_r - 30)$ *or* $B_i < 0.083\left(\frac{q_{c,avg}}{p_{ref}}\right)B_{CPT}$ *Fully Plugged:* $\frac{r_{t0.1}}{q_{c,avg}} = max\left[0.5 - 0.25\log\left(\frac{B}{B_{CPT}}\right), 0.15, A_r\right]$ *Coring:* $\frac{q_{b0.1}}{q_{c,avg}} = A_r$

Continued

Table 7.4 Summary of CPT- and CPTu-based methods.—cont'd

No.	Method/references	Pile unit side resistance (r_s)	Pile unit end bearing (r_t)
12	NGI method (Clausen et al., 2005)	$r_s = \left(\frac{z}{Dp_{ref}F_{D_r}F_{sig}F_{tip}F_{load}F_{mat}}\right) \geq 0.1\sigma'_{v0}$ $F_{D_r} = 2.1(D_r - 0.1)^{1.7} F_{sig} = \left(\frac{\sigma'_{v0}}{p_{pa}}\right)^{0.25}$ $F_{tip} = 1.0$ (*driven OE*), 1.6 (*driven CE*) $F_{load} = 1.0$ (*tension*), 1.3 (*compression*) $F_{mat} = 1.0$ (*steel*), 1.2 (*concrete*)	Closed-ended pile: $\frac{r_{t0.1}}{q_{c,tip}} = \frac{0.8}{1+D_r^2}$ Open-ended pile *Plugged* : $\frac{r_{t0.1}}{q_{c,tip}} = \frac{0.7}{1+3D_r^2}$ *Unplugged*: $r_{t0.1} = r_{t,ann}A_r + r_{t,plug}(1-A_r)$ $r_{t,ann} = q_{c,tip}$, $r_{t,plug} = \frac{12r_{s,avg}L}{\pi B_i}$ $r_{t0.1} = \min(r_{t0.1,plugged}, r_{t0.1,\ unplugged})$
13	Togliani (2008)	$r_s = k_1\, q_t^{0.5} k_1 = \beta\left\{1.2\left(0.8 + \frac{R_f}{8}\right)\right\}$ *for* $R_f < 1$ $k_1 = \beta\{1.1(0.4 + \mathrm{Ln}(R_f))\}$ *for* $R_f \geq 2$ $k_1 = \beta\left\{1.2\left(0.8 + \frac{R_f}{8}\right) + 1.1(0.4 + Ln(R_f))\right\} \Big/ 2$ *for* $1 \leq R_f < 2$ $\beta = f$(pile installation type)	$r_t = k_3 q_t$ $k_3 = \lambda + [0.01(L_{pile}/B_{toe})]$ λ is considered to be 0.2 for driven piles and 0.1 for bored piles; L_{pile}/B_{toe} is pile slenderness ratio
14	German method (Kempfert and Becker, 2010)	Defined based on the proposed chart of q_c-r_s	Defined based on the proposed chart of q_c-r_t
15	Enhanced UniCone (Niazi and Mayne, 2016)	According to the chart presented by Niazi and Mayne (2016)	According to the chart presented by Niazi and Mayne (2016)

Note: D = *pile embedment depth*, B = *pile diameter*, B_i = *pile inner diameter*, R = *pile radius*, R_i = *pile inner radius*, h = *height above pile tip*, $B_{CPT} = 0.036\ m$, $p_{ref} = 100\ kPa$, $R^* = \sqrt{(R^2 - R_i^2)}$, $A_r = 1 - \left(\frac{D_i^2}{D^2}\right)$, $\Delta\sigma'_{rd} = \frac{2G}{R^*}\Delta y$; $\Delta y \approx 2\, R_a \approx 0.02\ mm$, $G = 185 q_c q_{c1N}^{-0.7}$, or $G = q_c\left[0.0203 + 0.00125 q_{c1N} - 1.216e^{-6} q_{c1N}^2\right]^{-1}$, $q_{c1N} = \frac{(q_c/p_{ref})}{(\sigma'_{v0}/p_{ref})^{0.5}}$, and $D_r = 0.4\left[\ln \frac{q_{c,tip}}{22(\sigma'_{v0}p_{ref})^{0.5}}\right]$

7.7 Piles under uplift loading

Among 15 methods reviewed in this chapter, the Fugro method have proposed an exclusive correlation for piles uplift capacity and some methods have specified a reduction factor be imposed on shaft capacity correlations to account the pile capacity under tension loading such as UWA, NGI, ICP, and Nottingham and Schmertmann. Some others, however, have not recommended any considerations. Because the mobilized shaft resistance in tension is less than the bearing capacity in compression loading due to Poisson's effect, it can be suggested to bear this fact in mind when using these methods for piles in tension.

7.8 Comments on the methods

1. A few methods such as those developed in 1970s and 1980s do not consider the more accurate measurements achievable by CPTu, since, it was before the piezocone was generally available.
2. While the recommendations are specified to soil type (clay and sand) for a few methods, none of them, except for Eslami and Fellenius (1997) and enhanced UniCone (Niazi and Mayne, 2016), include a means for identifying the soil type from CPT data. Instead, the soil profile governing the coefficients relies on information from conventional boring and sampling, and laboratory testing, which may not be fully relevant to the CPT data.
3. All of the CPT-based methods include random smoothing and filtering of the CPT data, that is, elimination of peaks and troughs that exposes the results to considerable subjective operator influence.
4. The cone resistance (total resistance) has not been corrected for the pore pressure on the cone shoulder, and, therefore, the data behind the methods include errors—smaller in sand, larger in clay. This matter, i.e., penetration pore pressure, u_2, is realized by Eslami and Fellenius (1997).
5. In the French method, the extent of the zone above and below the pile toe in which the cone resistance is averaged appears to be too limited. As considered in the Schmertmann method, particularly if the soil strength decreases below the pile toe, the soil average must include the conditions over a depth larger than 1.5B distance below the pile toe.
6. The methods employ total stress values, whereas in long term, effective stress governs pile capacity.
7. Some of the methods are locally developed, that is, they are based on limited types of piles and soils, such as Schmertmann (1978) and Tummay and Fakhroo (1982).
8. The upper limit resistance imposed on the unit toe resistance in the Schmertmann, European, and Tummay methods is not reasonable in very dense sands where values of pile unit toe resistance, r_t, higher than 15 MPa frequently occur. All methods developed before 1990 except Meyerhof (1976) method impose an upper limit also to the unit shaft resistance, which cannot be justified by effective stress approach. Values higher than those recommended occur frequently.
9. Most of the direct methods involve engineering judgment in selecting the coefficient to apply to the average cone resistance to arrive at the unit toe resistance.
10. Some methods such as Eslami and Fellenius (1997), NGI, ICP, UWA, and Fugro specify a certain criterion for evaluating the pile capacity from static loading test results that can be

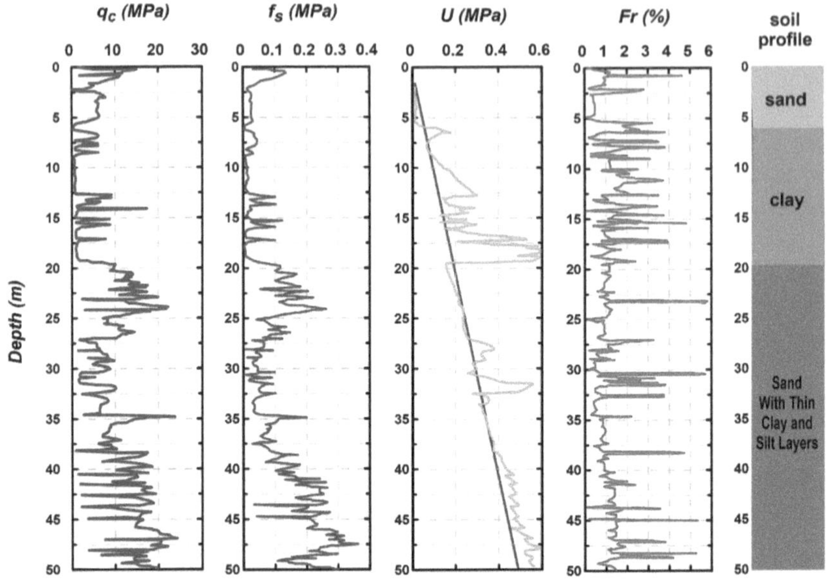

Figure 7.15 CPTu profile for Fraser Delta site, B.C., Canada.
Data from Eslami, A. 1996. Bearing Capacity of Piles from Cone Penetrometer Test Data. (Ph.D. thesis), University of Ottawa, Department of Civil Engineering, 516p.

used as reference to the pile capacity estimated from CPT data. While other methods have not introduced any criteria for pile ultimate capacity. Yet, the capacity of a pile is determined from the results of static loading tests and varies considerably with the method used to evaluate the test.

11. The NGI (2005), ICP (2005), Fugro (2005), and UWA (2005) methods are included in the commentary of the new 22nd edition of the API RP 2A Recommendations (2006) and are applicable for displacement piles in sand. They are more or less following a similar format. For instance, they all consider the effects of friction fatigue and toe condition in open-end piles. In addition, except for the Fugro method, the dilation effects during pile loading are accounted.
12. The distinction between the UWA (2005) method and other methods is claimed to be considering the effects of partial plugging in open-end piles.
13. Unlike the majority of CPT-based methods, German and NGI have not proposed averaging zone for CPT profiles in toe capacity calculation.
14. Methods such as Meyerhof (1983), Schmertmann (1978), Eslami and Fellenius (1997), Togliani (2008), and German involve routine procedures, excluding any complexities for pile capacity estimation, using CPTu and CPT records.

Example 7.4. The CPTu results of Fraser Delta, B. C., Canada, is as shown below. The digitized CPTu in 0.5m intervals is also presented in Fig 7.15. The bearing capacity of a pile with given specifications would be as follows using Eslami and Fellenius, Meyerhof, LCPC, and Schmertmann:

The pile is driven with the diameter of 324 mm and embedment length of 13.7, 16.1, and 31.1.

The results derived from static pile load test is also presented for comparison.

(A) *Unit shaft resistance for the pile with the length of 31.1 m. Geometric mean of cone resistance from surface to pile toe:*According to the CPTu profile, two layers are considered, one from 0 to 17m and the other from the depth of 17.0 m below.

$$R_s = C_s \cdot q_E$$

In soft clay (0−17.0 m): Cs = 0.08, $\overline{q}_E = 0.1895\ MPa$
In sand (17.0−31.0 m): Cs = 0.003, $\overline{q}_E = 9.017\ MPa$

$$r_{1s} = 0.185 \times 0.08 = 0.0152\ MPa$$

$$r_{2s} = 9.017 \times 0.003 = 0.027\ MPa$$

$$R_s = r_s \times A_s$$

$$R_{1s} = 0.0152 \times \pi \times 0.324 \times 17 = 0.263\ MN = 263\ kN$$

$$R_{2s} = 0.027 \times \pi \times 0.324 \times (31.1 - 17) = 0.387\ MN = 387\ kN$$

$$R_s = R_{1s} + R_{2s} = 263 + 387 = 650\ kN$$

(B) Unit toe resistance:

- the averaging zone is 8B above and 4B below the pile toe. Considering the pile embedment depth, the averaging zone depth is from 29.5 to 33.5 m.

$$r_t = C_t \times \overline{q}_{E(geo)}$$

$$\overline{q}_{E(geo)} = 1.587 MPa$$

$$r_t = 1 \times 1.587 = 1.587$$

$$R_t = A_t \times r_t = \frac{\pi \times 0.324^2}{4} \times 1.587 = 0.131 MN = 131\ kN$$

$$R = R_t + R_s = 1005 + 131 = 1136\ kN$$

The summary of the results for other methods and pile embedment depths is shown in Table 7.5.

Example 7.5. The bearing capacity of a closed end −driven steel pipe pile with the length and diameter of 22.5 m and 273 mm and the CPTu profile shown in Fig 7.16 by some of the presented methods is presented in Table 7.6.

Table 7.5 Pile capacity estimations using CPT-based direct methods.

Bearing Capacity (kN)					
Measured Capacity	Eslami and Fellenius	Meyerhof	LCPC	Schmertmann	Pile Number
290	244	–	217	141	1
630	630	–	633	431	2
1100	1136	1001	1053	825	3

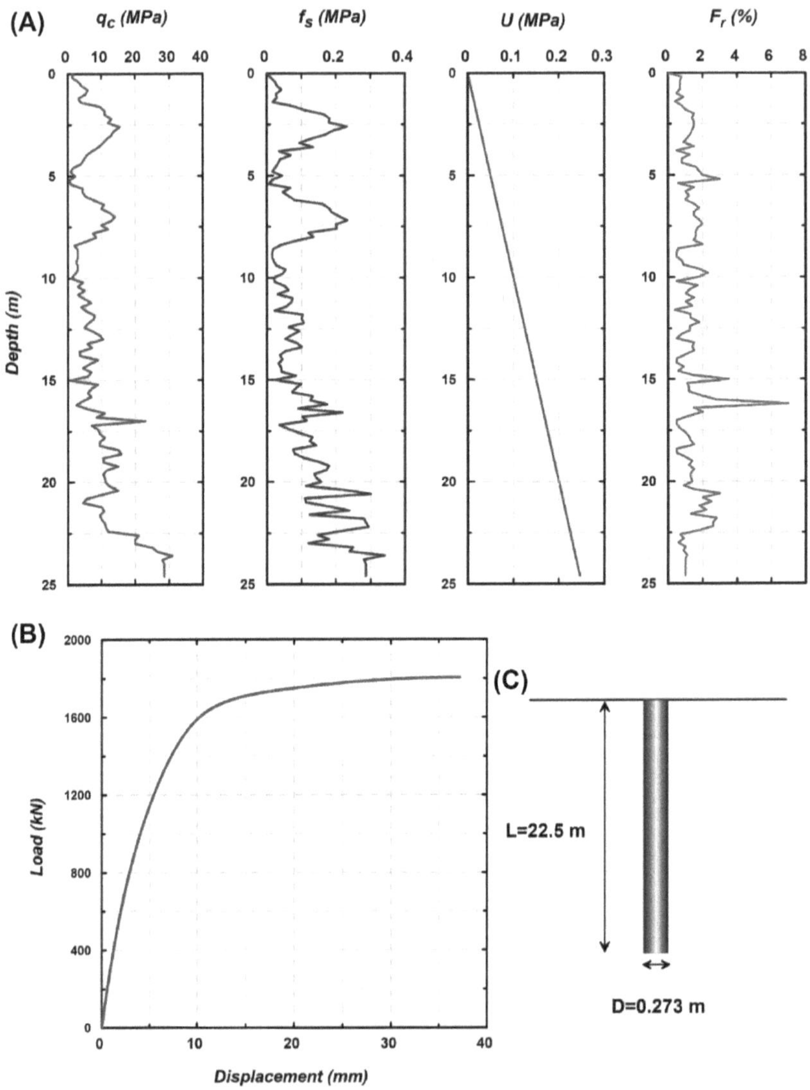

Figure 7.16 (A) CPT profiles, (B) load–displacement diagram, and (C) soil and pile characteristics.

Data from Eslami, A. 1996. Bearing Capacity of Piles from Cone Penetrometer Test Data. (Ph.D. thesis), University of Ottawa, Department of Civil Engineering, 516p.

Table 7.6 Pile capacity estimations using CPT-based direct methods.

Method	Bearing capacity (kN)
Brinch Hansen 80% criterion	1620
Eslami and Fellenius (1997)	1713
LCPC	1161
Meyerhof	1451
Schmertmann	2078
UWA	1379
NGI	1410
Fugro	1162
ICP	1438
German-Upper Bound	1836
German-Lower Bound	1291
German-Average	1564

Table 7.7 The characteristics of the pile for Example 7.6

Case ID	134-1-FITTJA C
Reference	Axelsson (1998)
Location	Sweden
Shape	Square
Material	Concrete
Installation	Driven
Embedment length, D (m)	19.0
Diameter, B (mm)	235
Cross-sectional area, A_t (m^2)	0.055
Perimeter (m)	0.940
GWL (m)	1.0

Example 7.6. The bearing capacity of a square concrete pile with the characteristics described in Table 7.7 with the CPTu profile illustrated in Fig 7.17 is calculated in this example and the results are presented in Tables 7.8 and 7.9.

Toe capacity

Calculating the average cone resistance using the minimum path rule.

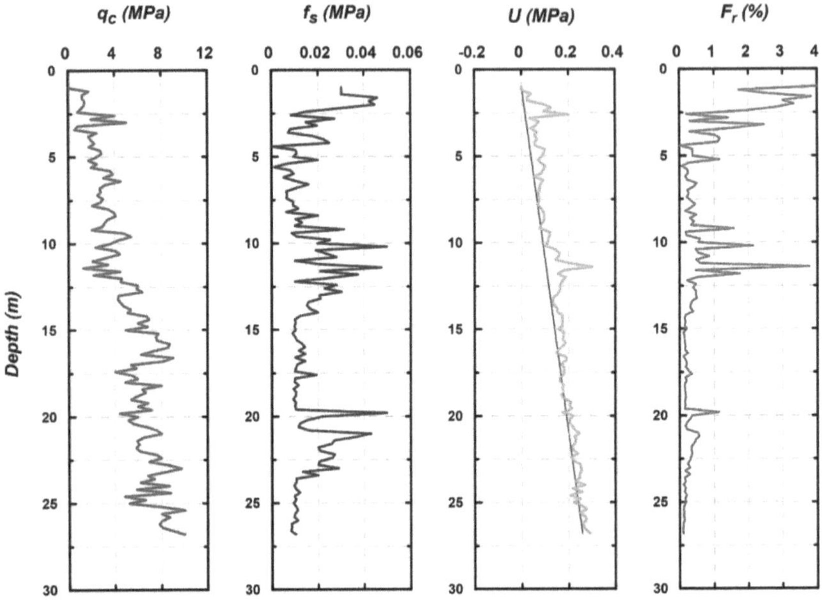

Figure 7.17 CPT profiles for Example 7.6.
Data from Eslami A, Sekhavatian A. (2014). Geotechnical engineering: principles, investigations, Amirkabir University of Technology Publication, In Persian.

$$q_{c,\ avg\ (0.7B\ below\ pile\ tip)} = 5.38\ MPa$$

$$q_{c,\ avg\ (4B\ below\ pile\ tip)} = 5.96\ MPa$$

$$\rightarrow q_{c,\ avg\ (0.7B\ bpt)} < q_{c,avg\ (4B\ bpt)}$$

$$q_{c,\ min\ (0.7B\ below\ pile\ tip)} = 5.38\ MPa$$

$$q_{c,\ avg\ (8B\ above\ pile\ tip)} = 4.96\ MPa$$

$$\rightarrow q_{c,\ avg} = \frac{\frac{5.38 + 5.38}{2} + 4.96}{2} = 5.17\ MPa$$

The pile is closed end $\rightarrow \dfrac{q_{b0.1}}{q_{c,avg}} = 0.6 \rightarrow q_{b0.1} = 0.6 \times 5.17 = 3.102\ MPa$

$$= 3102\ kPa$$

$$Q_b = 3102 \times 0.055 = 171\ kN$$

Table 7.8 Calculation of the pile capacity using the UWA method.

Z (m)	h/B	q_{c1N}	G (MPa)	$\Delta\sigma'_{rd}$ (kPa)	r_s (kPa)	ΔZ(m)	P(m)	ΔR_S (kN)	R_S (kN)
1.00	76.596	12,364.7	0.01	0.00	0.07	0	0.940	0.00	0.00
1.40	74.894	63.37	12.19	4.15	4.33	0.40	0.940	1.63	1.63
1.80	73.191	52.46	16.28	5.54	5.46	0.40	0.940	2.05	3.68
2.20	71.489	32.03	17.20	5.86	4.99	0.40	0.940	1.88	5.56
2.60	69.787	105.87	28.43	9.68	12.55	0.40	0.940	4.72	10.27
3.00	68.085	118.73	32.90	11.20	15.36	0.40	0.940	5.78	16.05
3.40	66.383	11.36	17.82	6.07	4.17	0.40	0.940	1.57	17.62
3.80	64.681	34.34	26.83	9.13	8.10	0.40	0.940	3.05	20.67
4.20	62.979	42.38	30.55	10.40	9.89	0.40	0.940	3.72	24.38
4.60	61.277	48.60	33.76	11.49	11.50	0.40	0.940	4.32	28.71
5.00	59.574	40.62	33.72	11.48	10.91	0.40	0.940	4.10	32.81
5.40	57.872	39.30	35.02	11.92	11.28	0.40	0.940	4.24	37.05
5.80	56.170	55.13	40.49	13.78	14.73	0.40	0.940	5.54	42.59
6.20	54.468	44.53	39.53	13.46	13.45	0.40	0.940	5.06	47.65
6.60	52.766	44.84	41.10	13.99	14.13	0.40	0.940	5.31	52.96
7.00	51.064	38.53	40.66	13.84	13.39	0.40	0.940	5.04	58.00
7.40	49.362	39.71	42.37	14.42	14.21	0.40	0.940	5.34	63.34
7.80	47.660	25.99	38.46	13.09	11.41	0.40	0.940	4.29	67.63
8.20	45.957	48.94	47.85	16.29	17.55	0.40	0.940	6.60	74.23

Continued

Table 7.8 Calculation of the pile capacity using the UWA method.—cont'd

Z (m)	h/B	q_{c1N}	G (MPa)	$\Delta\sigma'_{rd}$ (kPa)	r_s (kPa)	ΔZ(m)	P(m)	ΔR_S (kN)	R_S (kN)
8.60	44.255	40.74	46.53	15.84	16.15	0.40	0.940	6.07	80.30
9.00	42.553	35.10	45.65	15.54	15.22	0.40	0.940	5.72	86.03
9.40	40.851	55.20	53.58	18.24	21.22	0.40	0.940	7.98	94.01
9.80	39.149	48.00	52.59	17.90	19.98	0.40	0.940	7.51	101.52
10.20	37.447	25.60	44.53	15.16	13.84	0.40	0.940	5.20	106.72
10.60	35.745	47.81	54.87	18.68	21.33	0.40	0.940	8.02	114.74
11.00	34.043	21.84	44.27	15.07	13.39	0.40	0.940	5.04	119.78
11.40	32.340	13.00	38.64	13.15	10.30	0.40	0.940	3.87	123.65
11.80	30.638	21.81	45.99	15.65	14.22	0.40	0.940	5.35	129.00
12.20	28.936	44.98	58.19	19.81	23.42	0.40	0.940	8.81	137.80
12.60	27.234	57.69	63.80	21.72	28.95	0.40	0.940	10.88	148.69
13.00	25.532	41.89	58.95	20.07	23.90	0.40	0.940	8.99	157.68
13.40	23.830	42.44	60.16	20.48	25.01	0.40	0.940	9.40	167.08
13.80	22.128	50.12	64.25	21.87	29.25	0.40	0.940	11.00	178.07
14.20	20.426	61.82	69.49	23.66	35.60	0.40	0.940	13.39	191.46
14.60	18.723	54.07	67.76	23.07	33.60	0.40	0.940	12.63	204.09
15.00	17.021	43.92	64.59	21.99	30.11	0.40	0.940	11.32	215.41
15.40	15.319	66.09	74.05	25.21	43.14	0.40	0.940	16.22	231.63
15.80	13.617	75.55	78.14	26.60	50.75	0.40	0.940	19.08	250.72

16.20	11.915	62.93	74.97	25.52	46.59	0.40	0.940	17.52	268.23
16.60	10.213	75.98	80.36	27.36	58.24	0.40	0.940	21.90	290.13
17.00	8.511	44.78	69.45	23.64	41.07	0.40	0.940	15.44	305.57
17.40	6.809	33.07	64.20	21.86	35.42	0.40	0.940	13.32	318.89
17.80	5.106	48.65	72.96	24.84	54.27	0.40	0.940	20.41	339.30
18.20	3.404	64.35	80.28	27.33	81.99	0.40	0.940	30.83	370.13
18.60	1.702	52.89	76.57	26.07	87.08	0.40	0.940	32.74	402.87
19.00	0.000	42.29	72.41	24.65	72.27	0.40	0.940	27.18	**430.05**

Summary of the bearing capacity values from CPT-based methods: value-bearing capacity values from CPT-based methods: n or tension) or lateral loading.

Table 7.9 Summary of pile capacity estimations using CPT-based direct methods.

	UWA-05	NGI-05	ICP-05	German method		
				Upper bound	**Lower bound**	**Average value**
R_s (kN)	430	408	419	519	266	392
R_t (kN)	171	223	234	266	181	223
R_u (kN)	601	631	653	784	447	615

Shaft capacity

$$A_{rs,eff} = 1, B_{eff} = B = 0.265\ m, \Delta r = 0.02\ mm,\ \tan\delta_f = 0.52$$

7.9 Case study: Urmia Lake causeway (Eslami et al., 2011)

This project has been conducted with the aim of connecting two cities, Tabriz and Urmia, in the northwestern Iran. The narrowest part of the lake has been opted for construction and from two east and west coast sides 13.5 km were filled by earth works. Given the considerably thick layers of fine and sensitive soils and occurrence of sliding in the middle section as well as environmental issues, it was not pragmatic to continue filling the rest of the path. Consequently, a causeway was designed to connect two fills. In this regard, eight piezocone tests were carried out along the path for determining soil parameters and profiling. The geographical location of the site is depicted in Fig. 7.18.

In this case study, the applicability of five CPT- and CPTu-based methods of bearing capacity were assessed for 10 driven steel pipe piles, and the designs were consistent with site-specific conditions. Considering the difficulties regarding the performance of static loading test in the marine environment, dynamic load tests while driving the piles as well as pile integrity test were the most appropriate options, economically and practically, for analyzing the capacity of the piles. Thus, these tests were selected as the reference test in the analyses.

All of the 10 piles are open end–driven steel pipe piles with constant section along the pile length. The pile diameters vary between 685 and 913 mm with the embedment depth ranging from 41 to 85 m. The cone tests were performed using an electrical piezocone. An example of CPTu profiles recorded in the site along with soil profiling by the Eslami and Fellenius (1997) chart is shown in Fig. 7.19. For each of the piles, a corresponding CPT profile was carried out in the adjacency of the pile location.

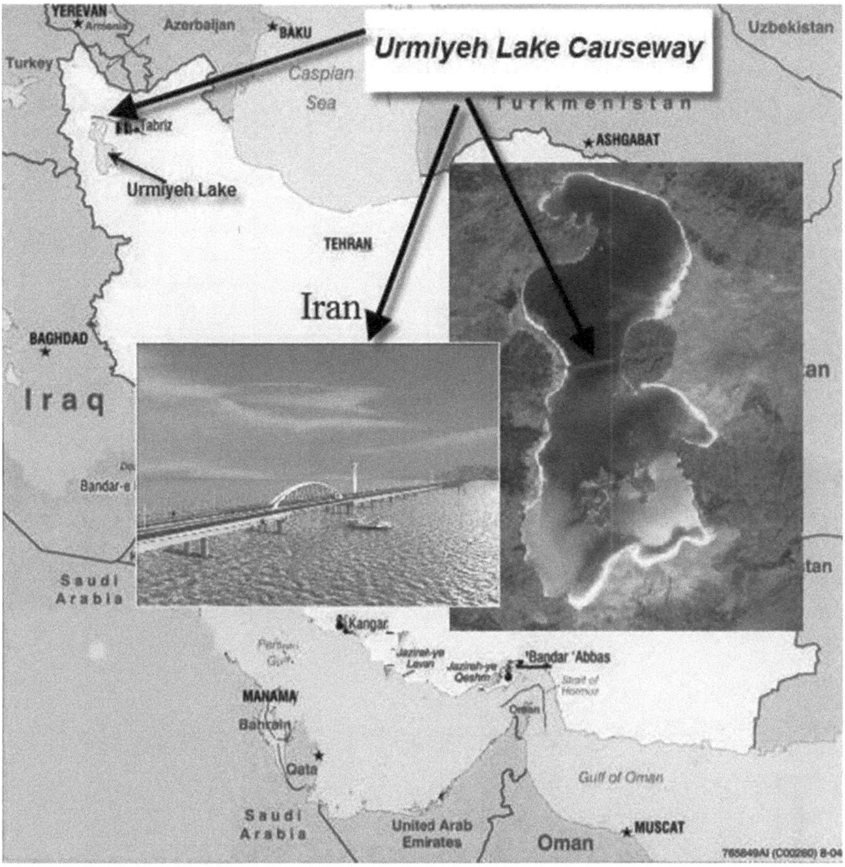

Figure 7.18 The geographical location of the causeway site (Eslami et al., 2011).

Moreover, the PDA (Pile Driving Analyzer) test has been performed on each pile. The piles-bearing capacity derived from dynamic load tests varied between 4700 and 16,500 kN. Fig. 7.20 illustrates the load–displacement diagrams from static loading tests for two piles.

The assessment of various CPT- and CPTu-based methods as well as static and dynamic load tests shows predicted to measured capacity ratio ranging from 0.92 to 1.08 which implies the satisfactory accuracy of the investigated methods. Albeit, this interval was obtained by imposing a correction/calibration factors, ranging from 0.5 to 1.3 for the Dutch, LCPC, and Schmertmann methods. Considering the notable values of pore pressures, the correction factors were imposed on the methods developed based on mechanical cone which neglect the effect of pore pressure on their predictions. Overall, the precise records of CPTu can be deemed to be an applicable tool for evaluation of the pile capacity in soft and loose deposits, particularly in marine

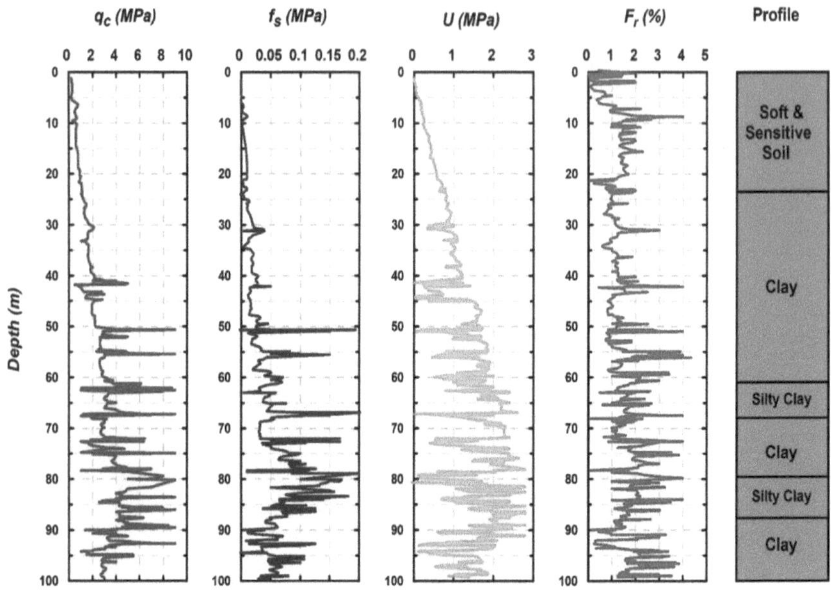

Figure 7.19 A typical CPTu profile of the Urmia site (Eslami et al., 2011).

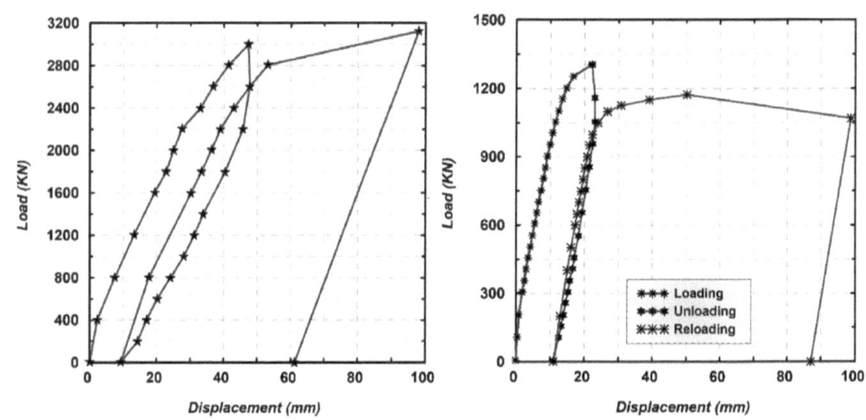

Figure 7.20 Results of static loading tests on two piles (Eslami et al., 2011).

environment. The soil type was characterized using the Robertson and Campanella (1989) as well as Eslami and Fellenius (1997) charts, according which the site soil consists 90% of sensitive soft clayey soils, clay and stiff clay, and 10% of silt and sandy silt as well as sand gravel. The soil classification diagrams are depicted in Fig. 7.21.

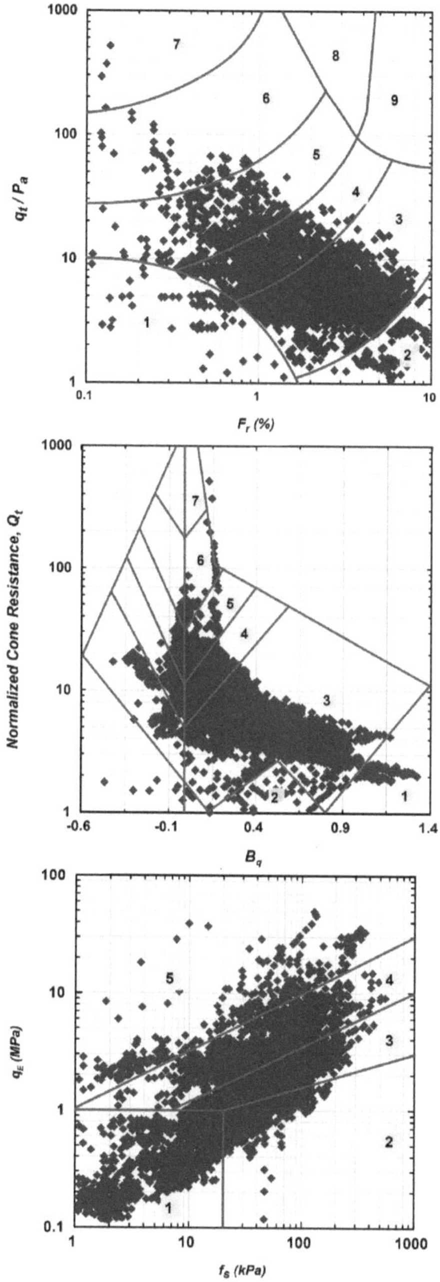

Figure 7.21 Soil behavior classification of the Urmia site (Eslami et al., 2011).

References

American Petroleum Institute (API), 2006. Recommended Practice for Planning, Designing and Constructing Fixed Offshore Platforms-Working Stress Design. API RP 2A-WSD, Washington, DC.

Aoki, N., Velloso, D.A., 1975. An approximate method to estimate the bearing capacity of piles. In: Proceedings of 5th Pan-American Conference of Soil Mechanics and Foundation Engineering, Buenos Aires, pp. 367–376.

Axelsson, G., 1998. Long-term increase in shaft capacity of driven piles in sand. In: Fourth International Conference on Case Histories in Geotechnical Engineering, St. Louis, Missouri, USA.

Basu, P., Prezzi, M., Basu, D., 2010. Drilled displacement piles–current practice and design. DFI Journal-The Journal of the Deep Foundations Institute 4 (1), 3–20.

Begemann, H.K.S.P., 1963. The use of the static penetrometer in Holland. New Zealand Engineering 18 (2), 41.

Begemann, H.K.S.P., 1969. The Dutch static penetration test with the adhesion jacket cone. Delft Soil Mechanics and Delft Hydraulics Laboratory XIII (1), 1.

Brown, D.A., Turner, J.P., Castelli, R.J., 2010. Drilled Shafts: Construction Procedures and LRFD Design Methods. US Department of Transportation, Federal Highway Administration (FHWA).

Bustamante, M., Gianeselli, L., 1982. Pile bearing capacity predictions by means of static penetrometer CPT. In: Proceedings of 2nd European Symposium on Penetration Testing (ESOPT-II), Amsterdam, pp. 493–500.

Cai, G., Liu, S., Puppala, A.J., 2011. Evaluation of pile bearing capacity from piezocone penetration test data in soft Jiangsu Quaternary clay deposits. Marine Georesources & Geotechnology 29, 177–201.

Cai, G., Liu, S., Puppala, A.J., 2012. Reliability assessment of CPTu-based pile capacity predictions in soft clay deposits. Engineering Geology 141, 84–91.

Campanella, R.G., Robertson, P.K., Davies, M.P., SY, A., 1989. Use of in-situ tests in pile design. In: Proceedings of 12th International Conference on Soil Mechanics and Foundation Engineering, vol. 1. ICSMFE, Rio de Janeiro, pp. 199–203.

Chen, J.J., Zhang, L., 2013. Effect of spatial correlation of cone tip resistance on the bearing capacity of piles, Journal of Geotechnical and Geoenvironmental Engineering, ASCE, 139, 494–500.

Clausen, C.J.F., Aas, P.M., Karlsrud, K., 2005. Bearing capacity of driven piles in sand, the NGI approach. In: Proceedings of International Symposium on Frontiers in Offshore Geomechanics (ISFOG 2005), Perth. Taylor & Francis, London, pp. 677–681.

Clisby, M.B., Scholtes, R.M., Corey, M.W., Cole, H.A., Teng, P., Webb, J.D., 1978. An Evaluation of Pile Bearing Capacities, vol. I. Mississippi State Highway Department. Final Report.

De Ruiter, J., Beringen, F.L., 1979. Pile foundations for large North Sea structures. Marine Geotechnology and Georesources 3 (3), 267–314.

Eslami, A., 1996. Bearing Capacity of Piles from Cone Penetrometer Test Data. Ph.D. Thesis. University of Ottawa, Department of Civil Engineering, p. 516p.

Eslami, A., Aflaki, E., Hosseini, B., 2011. Evaluating CPT and CPTu based pile bearing capacity estimation methods using Urmiyeh Lake Causeway piling records. Scientia Iranica 18 (5), 1009–1019.

Eslami, A., Fellenius, B.H., 1995. Capacity of Piles from CPT Data. Deep Foundation Institute, DFI, USA, pp. 9–12.
Eslami, A., Fellenius, B.H., 1996. Pile shaft capacity determined by Piezocone (CPTu) data. In: Proceeding of 49th Canadian Geotechnical Conference, St. Jonn's Newfoundland, vol. 2, pp. 859–867.
Eslami, A., Fellenius, B.H., 1997. Pile capacity by direct CPT and CPTu methods applied to 102 case histories. Canadian Geotechnical Journal 34 (6), 880–898.
Fellenius, B.H., 1989. Unified Design of Piles and Pile Groups. Transportation Research Board, Washington, pp. 75–82. TRB Record 1169.
Fellenius, B.H., 2002. Excerpt from Chapter 6 of the red book: direct methods for estimating pile capacity. In: Background to UniCone. http://www.fellenius.net.
Finno, R.J., 1989. Subsurface conditions and pile installation data. American Society of Civil Engineers. In: Proceedings of Symposium on Predicted and Observed Behavior of Piles, Evanston, June 30. ASCE Geotechnical Special Publication, GSP23, pp. 1–74.
Gavin, K.G., Lehane, B.M., 2003. The shaft capacity of pipe piles in sand. Canadian Geotechnical Journal 40, 36–45.
Gavin, K.G., Lehane, B.M., 2005. Estimating the end bearing resistance of pipe piles in sand using the final filling ratio. In: Proceedings of International Symposium on Frontiers in Offshore Geomechanics (ISFOG 2005), Perth. Taylor and Francis Group, London, pp. 717–723.
Gwizdala, K., 1984. Large Diameter Bored Piles in Non-cohesive Soils. Swedish Geotechnical Institute, Linkoping, Sweden. Report No. 26.
Igoe, D.J.P., Gavin, K.G., O'Kelly, B.C., 2010. Field tests using an instrumented model pile in sand. In: Proceedings of Physical Modelling in Geotechnics. Taylor and Francis Group, pp. 775–780.
Igoe, D.J.P., Gavin, K.G., O'Kelly, B.C., 2011. Shaft capacity of open-ended piles in sand. Journal of Geotechnical and geoenvironmental Engineering (ASCE) 137 (10), 903–913.
Jardine, R.J., Chow, F.C., Overy, R., Standing, J., 2005. ICP Design Methods for Driven Piles in Sands and Clays. Thomas Telford Publishing, London, p. 105.
Kempfert, H.G., Becker, P., 2010. Axial pile resistance of different pile types based on empirical values. In: Proceedings of Geo-Shanghai 2010 Deep Foundations and Geotechnical In-Situ Testing (GSP 205). ASCE, Reston, VA, pp. 149–154.
Kolk, H.J., Baaijens, A.E., Senders, M., 2005. Design criteria for pipe piles in silica sands. In: Proceedings Ofinternational Symposium on Frontiers in Offshore Geomechanics (ISFOG 2005), Perth. Taylor and Francis Group, London, pp. 711–716.
Kulhawy, F.H., Mayne, P.W., 1990. Manual on Estimating Soil Properties for Foundation Design. Electric Power Research Inst., Palo Alto, CA (USA). Cornell Univ., Ithaca, NY (USA). Geotechnical Engineering Group.
Lehane, B.M., Li, Y., Williams, R., 2013. Shaft capacity of displacement piles in clay using the cone penetration test. Journal of Geotechnical and geoenvironmental Engineering (ASCE) 139 (2), 253–266.
Lehane, B.M., Schneider, J.A., Xu, X., 2005. The UWA-05 method for prediction of axial capacity of driven piles in sand. In: Proceedings of International Symposium on Frontiers in Offshore Geomechanics (ISFOG 2005), Perth. Taylor and Francis, London, pp. 683–689.
Meyerhof, G.G., 1976. Bearing capacity and settlement of pile foundations. Journal of the Geotechnical Engineering Division 102, 195–228.
Meyerhof, G.G., 1983. Scale effects of pile capacity. Journal of Geotechnical Engineering, ASCE 108 (GT3), 195–228.

Niazi, F.S., 2014. Static Axial Pile Foundation Response Using Seismic Piezocone Data. Ph.D. Thesis. Georgia Institute of Technology, School of Civil and Environmental Engineering, 926 pp.

Niazi, F.S., Mayne, P.W., 2013. Cone penetration test based direct methods for evaluating static axial capacity of single piles. Geotechnical & Geological Engineering 31 (4), 979–1009.

Niazi, F.S., Mayne, P.W., 2016. CPTu-based enhanced UniCone method for pile capacity. Engineering Geology 212, 21–34.

Nottingham, L.C., 1975. Use of Quasi-Static Friction Cone Penetrometer Data to Predict Load Capacity of Displacement Piles. Ph. D. Thesis. Dept. of Civil Engineering., University of Florida, 553pp.

Paik, K., Salgado, R., Lee, J., Kim, B., 2003. Behavior of open-and closed-ended piles driven into sands. Journal of Geotechnical and Geoenvironmental Engineering 129 (4), 296–306.

Philipponnat, G., 1980. Methode pratique de calcul d'un pieu isole a l'aide du penetrometre statique. Revue Française de Geotechnique 10, 55–64.

Price, G., Wardle, I.F., 1982. A Comparison between cone penetration test result and the performance pf small diameter instrumented piles in stiff clay. In: Proceeding of the Second European Symposium on Penetration Testing, vol. 2. ESOPT-II, Amsterdam, pp. 775–780.

Randolph, M., 2000. Pile-soil interaction for dynamic and static loading. In: Pile-soil Interaction for Dynamic and Static Loading. CRC Press/Balkema, pp. 3–11.

Robertson, P.K., Campanella, R.G., 1989. Guidelines for Geotechnical Design Using the Cone Penetrometer Test and CPT with Pore Pressure Measurement. Soil Mechanics Series, No. 105, fourth ed. UBC, Dept. of Civil Engineering and Hogentogler and Company, Inc. 193 pp.

Schmertmann, J.H., 1978. Guidelines for Cone Test, Performance and Design. Federal Highway Administration, Report FHWA-TS-78209, Washington, 145pp.

Schneider, J.A., 2007. Analysis of Piezocone Data for Displacement Pile Design. Doctoral dissertation, University of Western Australia.

Takesue, K., Sasao, H., Matsumoto, T., 1998. Correlation between Ultimate Pile Skin Friction and CPT Data. Geotechnical Site Characterization (2). Balkema, Rotterdam, pp. 1177–1182.

Togliani, G., 2008. Pile capacity prediction for in-situ tests. In: Proceedings of Geotechnical and Geophysical Site Characterization. Taylor and Francis Group, London, pp. 1187–1192.

Togliani, G., 2010. Pile capacity prediction using CPT—case history. In: Proceedings of the 2nd International Symposium on Cone Penetration Testing, CPT'10. Huntington Beach, CA, Paper No. 3–13. www.cpt10.com.

Tummay, M.T., Fakhroo, M., 1982. Friction Pile Capacity in Cohesive Soils Using Electric Quasi-Static Penetration Tests. Louisiana Department of Transportation and Development Research and Development Section, Baton Rouge, La. Interim Research Rep. No. 1.

Valikhah, F., Eslami, A., Veiskarami, M., 2018. Load-displacement behavior of driven piles in sand using CPT-based stress and strain fields. International Journal of Civil Engineering. https://doi.org/10.1007/s40999-018-0388-7 (Published online: 21 Dec. 2018).

Van Dijk, B., Kolk, H., 2011. CPT-based design method for axial capacity of offshore piles in clays. In: Proceedings of Frontiers in Offshore Geotechnics II. Taylor and Francis Group, London, pp. 555–560.

White, D.J., Bolton, M.D., 2005. Comparing CPT and pile base resistance in sand. Proceedings of the Institution of Civil Engineers-Geotechnical Engineering 158 (1), 3–14.

Wroth, C.P., 1988. Penetration testing-a more rigorous approach to interpretation. In: Proceedings of the 1st International Symposium on Penetration Testing, ISOPT-1, vol. 1. Disney world, pp. 303–311.

Yu, F., Yang, J., 2012. Base capacity of open-ended steel pipe piles in sand. Journal of Geotechnical and geoenvironmental Engineering (ASCE) 138 (9), 1116–1128.

Zhang, G., Robertson, P.K., Brachman, R.W.I., 2002. Estimating liquefaction induced ground settlements from CPT for level ground. Canadian Geotechnical Journal 39 (5), 1168–1180.

CPT and CPTu for foundation settlement and load–displacement (P-Δ) estimation

8

8.1 Introduction

The vast majority of structures are supported by shallow foundations, and that settlement, not bearing capacity and stability, controls most of those designs (Holtz et al., 2011). Therefore, the settlement analysis and the accuracy of its estimation is an important aspect of shallow foundation design.

Despite many advances in foundation engineering, there are still uncertainties which are unavoidable in geotechnical engineering. The source of uncertainty in foundation design can be attributed into four main categories (Whitman, 2000; Wang, 2009):

1. Uncertainties in estimating the loads;
2. Uncertainties associated with variability of the ground conditions at the site;
3. Uncertainties in evaluation of geotechnical material properties;
4. Uncertainties associated with the accuracy of the applied analytical model for representing the actual behavior of the foundation, structure, and ground due to the speculated simplifications and assumptions (Phoon and Retief, 2016).

The nonlinear and time-dependent stress–strain behavior of soils and their elasto-plastic performance under loading and unloading can cause complexity in soil behavior for settlement analysis (Holtz et al., 2011; Fellenius, 2015). Using a single number for soil property as an average measure of the stiffness of all soil layers results in the approximate and incorrect estimation of settlement.

In the other words, in the settlement estimation, the uncertainties associated with the natural variability of ground conditions and the evaluation of the geotechnical properties are usually the greatest issue (Whitman, 2000; Phoon and Retief, 2016). Using the geotechnical parameters obtained by different approaches can result in the great differences in calculated settlement. These differences could be significant in some cases, especially if the proposed structure is particularly sensitive to settlements. Therefore, quantification of the uncertainties in the foundation settlement analysis is worthwhile. The uncertainty in foundation settlement analysis can be reduced by increasing the quality and quantity of site investigation tests. The rapid development of in situ testing techniques along with the limitations and inadequacies of conventional laboratory testing has made in situ tests playing prominent role in geotechnical engineering designs. Among in situ tests, the cone penetration test (CPT) defines the soil profile with great resolution because it continuously retrieves

Piezocone and Cone Penetration Test (CPTu and CPT) Applications in Foundation Engineering
https://doi.org/10.1016/B978-0-08-102766-0.00008-0

data with depth and allows a variety of sensors to be incorporated with the penetrometer (Moshfeghi and Eslami, 2016; Niazi and Mayne, 2016; Eslami et al., 2017; Valikhah et al., 2018).

8.2 Basic equations for settlement estimation

Many methods have been proposed to predict foundation settlement, where most of them are based on typical parameters such as soil stiffness parameters (i.e., elastic modulus, E), foundation width (B), contact pressure (q), and pressure developing in depth (Janbu, 1967; Schmertmann, 1978; Briaud and Gibbens, 1999; Anderson et al., 2007; Briaud, 2007). Immediate and consolidation settlements are separated in all prementioned approaches. In other words, these equations are specialized for only rapid or long-time settlement according to the soil type, course or fine-grained soils, respectively. Hence, a unique equation to predict the settlement for all soil types is rarely found in geotechnical practice. Regardless of immediate or elastic settlements, the general following equations can be used for settlement calculations:

$$s = \frac{kqB}{E_s} \tag{8.1}$$

where k is related to foundation flexibility or rigidity and q, B, and E_s are footing stress, foundation width, and elastic soil modulus, respectively.

While two following equations are commonly used for immediate and consolidation settlements

$$s = \frac{qBI(1 - \nu^2)}{E_s} \tag{8.2}$$

$$s = m_v \Delta H \Delta\sigma'_z \tag{8.3}$$

where I is elastic displacement influence factor and ν is Poisson's ratio. In Eq. (8.3), $m_v = \frac{1}{E_s}$ is coefficient of volume change, ΔH is thickness of layer, and $\Delta\sigma'_z$ is induced vertical effective stress due to applied load from foundation. Realizing Eq. (8.1) and comparing Eqs. (8.2)–(8.3), it can be deducted that the determinant factors are B, E_s, and applied stress in the form of q or $\Delta\sigma'_z$ for settlement calculations.

Therefore, in all the abovementioned equations, settlement is dependent of foundation width, and influenced by depth and soil stiffness. Accordingly, the first step to predict the settlement of a foundation is to recognize the influenced soil zone in an appropriate way. Regarding the changes of the soil type with depth, it is obvious that elastic modulus cannot be specified as a unique and constant value. In addition, in all correlations, soil stiffness parameters, such as E (elastic modulus), M (constrained modulus), and m_v (coefficient of volume change) in the stressing zone, have been applied to estimate the settlement as an average number. In addition, even for a homogenous soil layer, elastic modulus values would gradually increase with depth. Therefore, using a single number for soil property as an average value of the elastic modulus of all soil layers results in approximate and incorrect estimation of settlement amount.

Using a single average number for E and m_v is the common approach, but it is hardly practical to measure the soil elastic modulus in short distances with good accuracy. Hence, the average of two or three different values of E and m_v in affected layers must be considered. The most important step for the settlement estimation is to find precisely the soil elastic modulus values. The more exact soil elastic modulus measuring would result in the more accurate prediction of settlement. Obviously, to estimate the settlement in soil profile, which the layers change frequently, the elastic modulus of each layer must be measured and used just for that specific layer. For this purpose, to determine the stress–strain modulus, the in situ testing records such as SPT, CPT, PMT, and DMT can be realized, representing the actual soil performance in the field under loading. Among them, the CPT is much less prone to error due to differences in equipment and technique. This significant characteristic causes more repeatable and reliable outputs than the SPT, and this fact resulted in a vast amount of CPT-based correlations in foundation engineering practice (Fellenius, 2015).

It has been well recognized in geotechnical practice that soil stiffness modulus is proportional to CPT records as described in Chapter 4.

There are some approaches for foundation settlement estimation which are based on CPT results. The first researches are done by Janbu in the early 1960s, Meyerhof (1976) and Schmertmann (1978).

Meyerhof (1976) correlation (Eq. 8.4) is a simple conservative relation to compute foundation settlements of sandy soils on which CPT results are directly employed.

$$S = \frac{\Delta P.B}{2q_c} \tag{8.4}$$

where ΔP is the net pressure employed to the ground by the foundation, B is the foundation width, q_c is the average of tip resistance for the influenced depth of B, and E is the Young's modulus which considers to be equal to $2q_c$ in this relation.

Schmertmann (1978) present Eq. (8.5) to determine vertical settlement of sandy soils, using CPT results based on elasticity theory.

$$S = C_1 C_2 \Delta P \sum_{1}^{n} \frac{I_z}{C_3 E} \Delta z \tag{8.5}$$

where C_1 indicates the correction rate for the buried depth, C_2 is the correction rate of creep, C_3 is the correction rate for the foundation shape, ΔP is the net pressure employed to the ground by the foundation, I_z is the strain influence factor, ΔZ is the thickness of the examining layers, and E is Young's modulus which is $E = 2q_c$ here. Fig. 8.1 gives the vertical strain influence factor I_z distribution for both square and strip

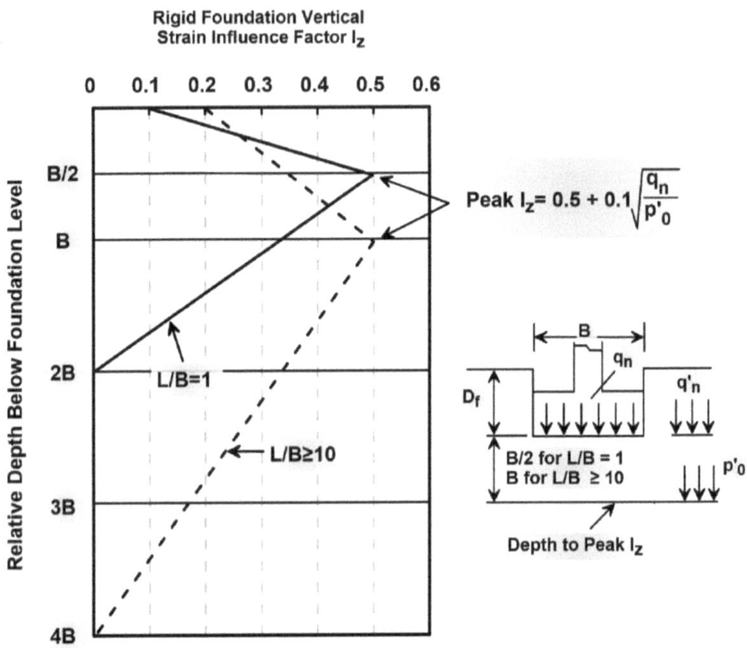

Figure 8.1 Vertical strain influence factor diagrams (Schmertmann, 1978).

foundations if the ratio $L/B \geq 10$. Values for rectangular foundations for $L/B < 10$ can be obtained by interpolation. The depths at which the maximum I_z occurs may be calculated as follows:

$$I_z = 0.5 + 0.1\sqrt{\frac{q_n}{p'_0}} \tag{8.6}$$

where p'_0 = effective overburden pressure at depths B/2 and B for square and strip foundations, respectively. Furthermore, I_z is equal to 0.1 at the base and zero at depth 2B below the base for square footing, whereas for a strip foundation, it is 0.2 at the base and zero at depth 4B.

8.3 Background to Janbu (1967) approach

The Janbu approach, proposed by Janbu in the early 1960s (1963, 1965, 1967) and also 1998, combines the basic principles of linear and nonlinear stress–strain behavior of material. The method is applied to all types of soils. The relation between stress and strain is a function of two nondimensional parameters which are unique for any soil: a stress exponent, "j", and a modulus number, "m", shown in Fig. 8.2.

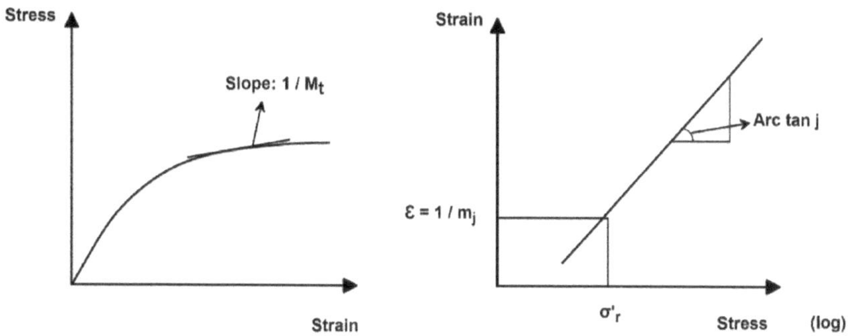

Figure 8.2 Nonlinear stress–strain relation proposed by Janbu (1967).

The Janbu relations are derived from the definition of tangent modulus by the following expression:

$$M_t = \frac{\partial\sigma}{\partial\varepsilon} = m\sigma_r\left(\frac{\sigma'}{\sigma'_r}\right)^{1-j} \tag{8.7}$$

where ε, σ', j, and m are strain induced by increase of effective stress, effective stress, a stress exponent, and a modulus number, respectively, and σ'_r is a constant stress which is equal to 100 kPa. The Janbu expressions for strain in different soil types can be derived from Eq. (8.7). For cohesionless soil, the stress exponent is larger than zero, $j > 0$. Integrating Eq. (8.7) results in Eq. (8.8):

$$\varepsilon = \frac{1}{mj}\left[\left(\frac{\sigma'_1}{\sigma'_r}\right)^j - \left(\frac{\sigma'_0}{\sigma'_r}\right)^j\right] \tag{8.8}$$

where σ'_0 and σ'_1 are original and final effective stress, respectively. In this approach, the settlement in dense coarse-grained soils is assumed to be elastic, which means that the stress exponent is equal to unity and compression is linearly elastic. By inserting $j = 1$, Eq. (8.7) becomes as follows:

$$\varepsilon = \frac{1}{100m}\left(\sigma'_1 - \sigma'_0\right) = \frac{1}{100m}\Delta\sigma' \tag{8.9}$$

For sandy or silty soil, Janbu (1967) assumed that the stress exponent is equal to 0.5. Therefore, Eq. (8.7) is simplified to Eq. (8.10):

$$\varepsilon = \frac{1}{5m}\left(\sqrt{\sigma'_1} - \sqrt{\sigma'_0}\right) \tag{8.10}$$

In cohesive soils, the stress exponent is zero, $j = 0$, and Eq. (6) can be simplified to following equations for normally consolidated and overconsolidated clays, respectively:

$$\varepsilon = \frac{1}{m} \ln \frac{\sigma_1'}{\sigma_0'} \tag{8.11}$$

$$\varepsilon = \frac{1}{m_r} ln \frac{\sigma_p'}{\sigma_0'} + \frac{1}{m} ln \frac{\sigma_1'}{\sigma_p'} \tag{8.12}$$

where m_r is recompression modulus number and σ'_p is preconsolidation settlement. For knowing the original effective stress, the stress increase, and the type of influenced soil by applied stress, only the modulus number, "m," is required. The modulus numbers to use in a particular case can be determined from conventional laboratory testing as well as in situ tests. As a reference, Table 8.1 shows a range of normally conservative modulus numbers, "m," which are typical of various soil types (Proposed by Janbu, 1967 and recommended by CFEM, 2006). The modulus numbers presented in Table 8.1 are approximate and cover wide ranges of values for mixed soils.

Table 8.1 Modulus number (m) and Stress exponent (j) for different soils in Janbu approach (1967).

Soil type		Modulus number	Stress exponent (j)
Till, very dense to dense		300–1000	1
Gravel		40–400	1
Sand	Dense	250–400	0.5
	Compact	150–250	0.5
	Loose	100–150	0.5
Silt	Dense	80–200	0.5
	Compact	60–80	0.5
	Loose	40–60	0.5
Silty clay and clayey silt	Hard, stiff	20–60	0
	Stiff, firm	10–20	0
	Soft	5–10	0
Soft marine clays and organic clays		5–20	0
Peat		1–5	0

Evaluating the settlement of a foundation is a prediction exercise. Its certainty depends on how accurately the soil profile and stress distributions used for analysis and representing parameters for conditions at the site resemble those actually occurring. The accuracy depends also on the quality and quantity of the soil parameters used as input to the analysis, where mainly related to soil stiffness properties.

8.4 Modified stress exponent and modulus number (Malekdoost and Eslami, 2011)

Janbu (1967) has proposed two different equations considering the two major categories of soils: granular and cohesive. This separation is due to different behavior of the granular and cohesive soils in settlement process after surcharging. However, in highly overconsolidated clay, settlement under surcharges is very low, sometimes less than loose sand. It can be stated that the behavior of soft and normally consolidated clays and behavior of overconsolidated clays with low or high OCR values under normal stress are not similar. On the other hand, the large displacements may occur in loose sands due to high vertical pressures. Therefore, it is worth categorizing the soils based on their real condition and applied pressure instead of solely soil type for settlement estimation. Accordingly, based on studies and researches (Robertson et al., 1986; Eslami and Fellenius, 1997), q_t, corrected cone tip resistance in CPT is the valuable parameter for evaluating soil settlement which reflects soil stiffness.

As it is expressed, in Janbu approach for determination of strain, the stress exponent (j) is considered 0.5 for sand and silt in three states of dense, compact, and loose. This parameter is zero for hard, firm, and soft clay. Furthermore, the proposed ranges for modulus number (m) are very wide depending on the soil type as shown in Table 2. This parameter is between 5 and 60 for clays and in the range of 400–800 for gravels. This wide range of the values of "j" and "m" can cause great error and vast predicted values in the calculated strains for different soils. Therefore, the choice of a suitable value for "m" among these wide ranges is one of Janbu method difficulties in settlement estimation. In addition, an appropriate state of soil to presume is not easily recognized, especially in case of hard clay or dense sand. Although several approaches proposed, it is still hard to classify the soil and mainly in strata boundaries. Besides, the settlement for difficult deposits such as sensitive and collapsible soils and peats is significant, where their characteristics are well addressed by CPT and CPTu records.

Alternatively, instead of using the soil gradation classification and the methods based on soil type, like Janbu approach, it is proposed to use the CPT results, i.e., q_t and f_s, for settlement estimation directly. In this case, the high value of q_t, regardless of the soil type which may be dense sand or very overconsolidated clay, indicates an adequate soil resistance against settlement. Expectedly, the settlement in this soil with high value of q_t will be lower than the settlement in the soil with low q_t under the same loading. Therefore, the new equation can be defined based on the load resistance of the soil. As an example, while the value of j for very overconsolidated clay is considered equal to zero and for dense sand 0.5 in Janbu approach, in the proposed method for both will be considered equal to 0.5. This is because of the same behavior of these soils under loading considering stiffness characteristics and regardless the soil type. Hence, via the new approach, instead of using two different equations for granular and cohesive soils, the unique equation can be used for settlement estimation of all types of soils. In addition, with careful study on the range of "m" in Janbu approach, as a matter of fact the values of "m" are in the high ranges for dense and hard soils and in the low ranges for the loose and soft soils like the values of q_t from CPT records. The value of modulus number, "m," can be replaced by applying a coefficient to q_t as follows:

$$m = \alpha q_t \tag{8.13}$$

The calibration studies and recommendations in show that $\alpha = 2$ is the best value to reach the maximum accuracy. To modify Janbu equation so that only the results of CPT will be used for calculations, the appropriate values of "j" for soils must be considered. The new equation for estimation of "j" value, resulted from studies on actual cases for calibration, is as follows:

$$j = \frac{q_t\left[1 + (0.05\, log q_t) \times R_f^2\right]}{5^{log q_t}\left(11\sqrt{R_f} + R_f^2\right)} \tag{8.14}$$

In which, q_t is corrected cone resistance for pore water pressure (from Eq. 8.12) ($\frac{kg}{cm^2}$ or bar) and R_f is friction ratio in percentage. Subsequently, the strain could be obtained from Eq. (8.15) as follows:

$$\varepsilon = \frac{1}{2q_t j}\left[\left(\frac{\sigma'_0 + \Delta\sigma'}{100}\right)^j - \left(\frac{\sigma'_0}{100}\right)^j\right] \tag{8.15}$$

where σ'_0 is initial effective stress in the middle of investigated layer under in situ pressure. $\Delta\sigma'$ indicates the increase of effective stress in the middle of studied layer under the loads of surcharges or applied stress. The stress influenced depth is considered as (2−3)**B* beneath the footings for single foundations, (3−4)**B* for strip and (1−1.5)**B* for mat ones (*B* is the foundation width) (CFEM, 2006).

The value of settlement in the soil can simply be calculated from Eq. (8.16):

$$\varepsilon = \frac{\Delta H}{H} \tag{8.16}$$

In this equation, ΔH is the settlement and H is the thickness of the subject layer. The investigating zone can be divided to several thin layers to increase the accuracy of the calculations. Because CPT records such as q_c and f_s are measured every 20 mm depth intervals, it is suggested to assume the layers have 20 mm thickness.

8.5 CPT-based nonlinear stress–strain approach to evaluate foundation settlement (Valikhah and Eslami, 2019)

Valikhah and Eslami (2017) presented a new relation between the stress exponent, the modulus number, and CPT results with a theoretical basis which can be more reliable to estimate the shallow foundation settlement. Furthermore, the scale effect due to differences between the influenced zone beneath shallow foundation and penetration cone is considered in this proposed approach. The procedure is presented in following sections.

8.5.1 Case study records

A data bank including 46 cases has been compiled from different reliable sources. 16 cases have been used to calibrate the new formulation for settlement estimation (presented in Table 8.2) and the remaining 30 cases employed for evaluation of the new proposed method (presented in Table 8.3). The case histories include 22 square, 17 circular, and 7 rectangular footings of variable embedded depth between 0 and 2.35 m. The settlements for these cases have been measured in the range of 25–260 mm which are about 2%–10% of foundation width (*S/B*). The recorded profiles of q_c in these sites are between 0.5 and 10.70 MPa, and the soil type beneath the foundations is sand and in some cases it contains a combination of silt, clay, and sand.

The CPT profile and pressure versus settlement for two footings (Cases 1 and 2 from Table 8.3) have been shown in Fig. 8.3.

8.5.2 Scale effect realization

The influenced zone beneath the cone and the shallow foundation are shown in Fig. 8.4. The CPT records are founded on the mechanisms of large strain, while the deformation range of the usual foundations are from small to large. Therefore, in using CPT records to estimate the foundation settlement, which may be in small strains, a modification factor is needed.

As shown in Fig. 8.4, the influenced zone beneath the foundations depends on footing width, B. Hence, it can be stated that the soil stiffness is not only related to the soil inherency but also related to the scale of influenced zone. This fact is also expressed by Terzaghi (1955) based on plate load test (PLT) results. He showed that by increasing the plate width, soil reflection factor, k_s, would decrease. k_s for a wide foundation with $B = 2\ m$ which is based on the granular soil is predicted by following expression:

$$k_s = \left(\frac{B + 0.305}{2B}\right)^2 K_{01} \tag{8.17}$$

Table 8.2 Summary of case history records for calibration (Valikhah and Eslami, 2019).

Case No.	Reference	Location	Soil Type	Footing Shape	B (m)	D_f• (m)	GWT(m)	q (kPa)	q_t•(kPa)	R_f	S_m•(mm)
1	Eslami and Gholami (2006)	France	Silt	Square	1	0	-	300	2500	0.5	98
2					1	0	-	325	2800	0.5	97
3		Iran	Silty Sand	Square	0.6	0	-	1260	7000	0.5	55
4					0.6	0	-	1280	10000	0.5	55
5		Brazil	Slity Clay	Circular	0.45	0	-	170	1400	0.6	40
6					0.6	0	-	170	1700	0.6	55
7					0.6	0	-	170	2000	0.6	55
8		USA	Slity Clay	Circular	0.6	1.5	-	520	3100	0.6	60
9					0.6	1.5	-	310	4600	0.6	55
10					0.6	1.5	-	310	5400	0.6	60
11					0.6	1.5	-	690	6000	0.6	60
12	Mayne and Illingworth (2010)	Sweden	Glaciofluvial Sand	Rectangular	0.6	0.4	8	1740	10720	0.51	59
13					1.2	0.6	8	1740	10720	0.51	119
14					1.7	0.8	8	1740	10720	0.51	170
15					2.4	1.1	8	1740	10720	0.51	245.8
16		Australia	Siliceous Dune Sand	Square	0.5	0.5	5.5	480	3440	0.44	51

$\boldsymbol{B}$: foundation width
$\boldsymbol{D_f}$: embedment depth
GTW: ground water table
q: footing stress ($q_{applied}$)
q_t: corrected q_c for pore pressure
$R_f = \frac{f_s}{q_c}$friction ratio
S_m: measured settlement

Table 8.3 Summary of case history records for comparison of settlement prediction methods (Valikhah and Eslami, 2019).

Case No.	Reference	Location	Soil Type	Footing Shape	B (m)	D_f• (m)	GWT(m)	q (kPa)	q_t• (kPa)	R_f	S_m• (mm)
1	Briaud and Gibbens (1999)	USA	Sand, Silty Sand	Square	1	0.76	4.9	1540	7520	0.65	100
2					1.5	0.76	4.9	1540	7520	0.65	154
3	Eslami and Gholami (2006)	France	Silt	Square	1	0	-	375	1700	0.5	115
4					1	0	-	370	2000	0.5	100
5		Iran	Silty Sand	Square	0.6	0	-	1260	3000	0.5	60
6		Brazil	Slity Clay	Circular	0.3	0	-	170	500	0.6	33
7					0.3	0	-	170	900	0.6	25
8		USA	Slity Clay	Circular	0.6	1.5	-	600	1000	0.6	72
9					0.6	1.5	-	600	1700	0.6	72
10					0.6	1.5	-	600	2500	0.6	60
11	Mayne and Illingworth (2010)	South Africa	White Fine Sand	Square	0.69	0	1.2	620	3660	0.54	65
12		Sweden	Glaciofluvial Sand	Rectangular	1	0	3	840	4010	0.63	102.4
13					1	0	3	840	4010	0.63	102.4
14					2.4	1.1	1.5	640	3200	0.63	260
15		Sweden	Compacted Fill	Square	0.46	0	>5	150	880	0.53	47
16					0.63	0	>5	580	3860	0.48	64
17					0.8	0	>5	520	2870	0.58	82

Continued

Table 8.3 Summary of case history records for comparison of settlement prediction methods (Valikhah and Eslami, 2019).—cont'd

Case No.	Reference	Location	Soil Type	Footing Shape	B (m)	D_f•(m)	GWT(m)	q (kPa)	q_t•(kPa)	R_f	S_m•(mm)
18		USA	Alluvial Sand	Circular	2.2	2.2	1.9	1280	6720	0.6	250
19					2.2	2.2	1.9	1280	6720	0.6	250
20					2.35	2.35	1.9	1730	10460	0.52	245
21					2.35	2.35	1.9	1730	10460	0.52	245
22		France	Dune Sand	Square	0.7	0	3	840	4010	0.66	71.7
23					0.7	0	3	840	4010	0.66	71.7
24					1	0	3	840	4010	0.66	102.4
25					1	0	3	840	4010	0.66	102.4
26					1	0	3	840	4010	0.66	102.4
27		Florida	Brown Silty Sand	Circular	1.82	0.6	3	1710	9780	0.55	186
28		Australia	Siliceous Dune Sand	Square	0.5	0.5	5.5	480	3440	0.44	51
29					1	1	5.5	480	3440	0.44	102
30					1	1	5.5	480	3440	0.44	102

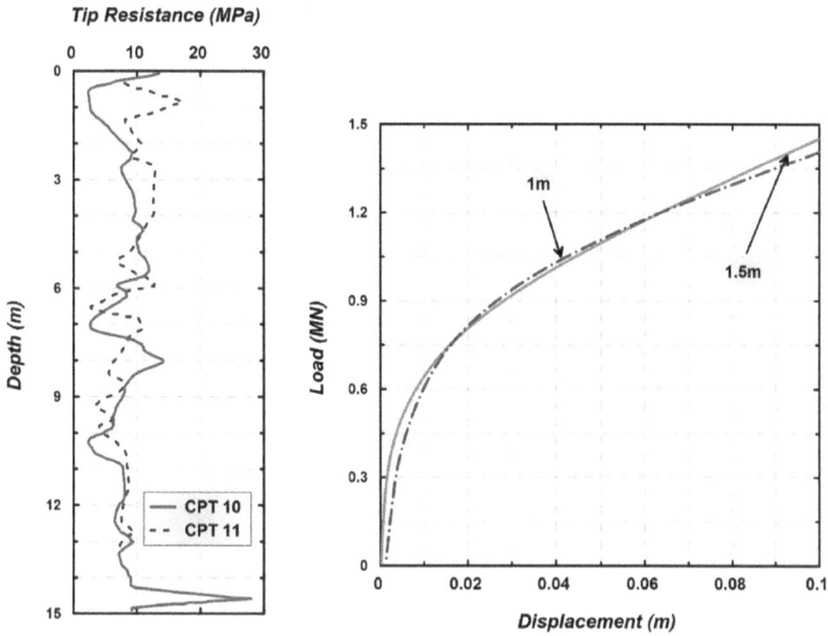

Figure 8.3 CPT records and pressure versus settlement for two cases (Valikhah and Eslami, 2017).

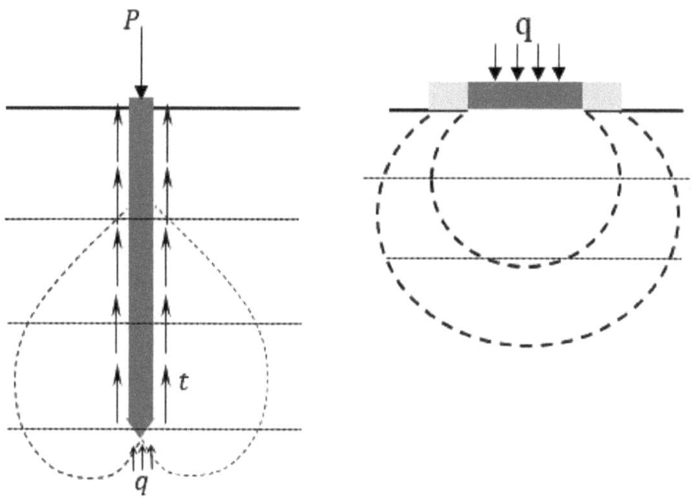

Figure 8.4 CPT records and pressure versus settlement for two cases (Valikhah and Eslami, 2019).

In addition, Vesić (1961) proposed the following equation for k_s and its relation to foundation width and soil properties.

$$k_s = \frac{E_s}{B(1 - \nu_s^2)} \tag{8.18}$$

K_{01}, E_s, and ν_s are reflection factor from PLT, soil elastic modulus, and Poisson ratio, respectively. Therefore, it can be expressed that in using CPT records (with small area) to estimate the foundation settlement, a scale correction factor is necessary.

This scale effect has been considered in the proposed approach to estimate the more realistic settlement. The calibration studies on the cases presented in Table 8.3 show the best relation between α (constant coefficient to estimate modulus number in Eq. 8.13), CPT diameter, b (which is 0.0357 or 0.0437 m for standard electronic penetrometer) and foundation width in maximum accuracy as follows (b and B are in m):

$$\alpha = 0.25b \times \left(\frac{2B + 1}{3B}\right)^3 \tag{8.19}$$

8.5.3 Stress exponent related to CPT results

The stress exponent and the cone tip resistance can be related with any functional relationships. A simple form is found by assuming a hyperbolic relationship between j and q_c. This form is supported using the form of hardening laws presented by traditional soil models such as the hyperbolic soil model (e.g., Lade and Duncan, 1975; Zhang et al., 2014) or the hyperbolic relationship between $sin\phi_{mob.}$ and the maximum shear strain (Clarck, 1998). Therefore, the following equation has been chosen as a basis for the functional dependency of j and q_c:

$$j = \frac{q_c}{x + yq_c} \tag{8.20}$$

In this equation, q_c is in kPa, and two parameters of x and y are found by study on the collected database presented in Table 8.2 as follows:

$$x = 0.02\, R_f + 0.5 \tag{8.21}$$

$$y = 7.53(\sigma'_0)^{-0.25} \tag{8.22}$$

Inspection of the results revealed that these parameters are functions of R_f (friction ratio) and σ'_0 (the initial effective soil stress in kPa at the middle of the target layer at which, the settlement is estimated). This dependency seems to be logical as the stress level effect cannot be disregarded in the settlement estimation.

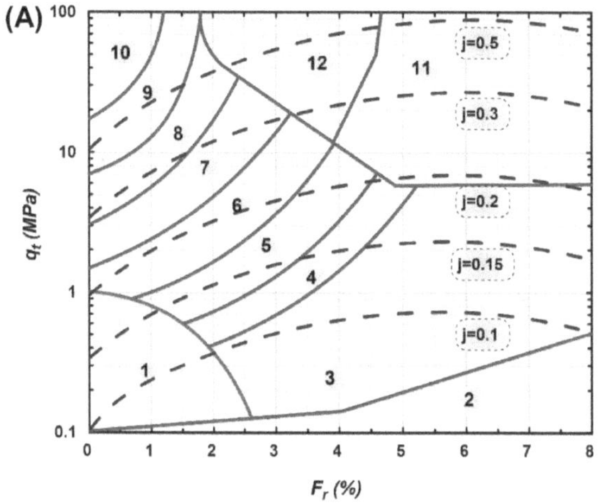

Zone No.	Soil Types
1	Sensitive Fine-Grained Soil
2	Organic Soil
3	Clay
4	Silty Clay to Clay
5	Clayey Silt to Silty Clay
6	Sandy Silt to Silty Clay
7	Silty Sand to Sandy Silt
8	Sand to Silty Sand
9	Sand
10	Sand to Gravelly Sand
11	Very Stiff Fine-Grained Soil
12	Overconsolidated or Cemented Sand to Clayey Sand

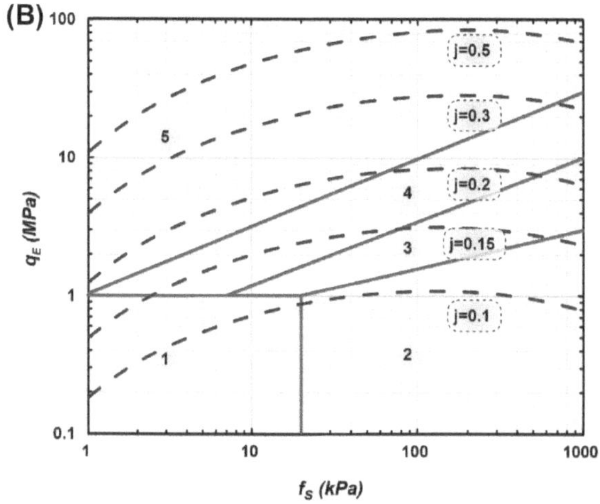

Zone No.	Soil Types
1	Very Soft Clay or Sensitive or Collapsible Soils
2	Clay and/or Silt
3	Clayey Silt and/or Silty Clay
4	Sandy Silt or Silty Sand
5	Sand to Sandy Gravel

Figure 8.5 (A) Campanella et al. (1986) and (B) Eslami-Fellenius (1997) soil behavior classification charts with "j" value proposed by Valikhah and Eslami (2019).

The proposed values of j applying slight changes is employed in Campanella et al. (1986) chart and Eslami-Fellenius (1997) chart as shown in Fig. 8.5. Hence, the parameter of R_f must be replaced by f_s (sleeve friction) and q_E (effective cone resistance in Eslami-Fellenius (1997) method).

Example 8.1. Estimation of Shallow Foundation Settlement:

As an applied example for estimation of shallow foundation settlement using analytical approach proposed by Valikhah and Eslami (2019), Case 2 in Table 8.3 is investigated. The studied case is a 1.5 m square footing with the embedment depth 0.76 m, as shown in Fig. 8.6. The water table is 4.9 m below the ground surface, soil saturated unit weight is 18.5 kN/m^3, and the applied stress is equal to 1540 kPa. The CPT records of this case are illustrated in Fig. 8.3.

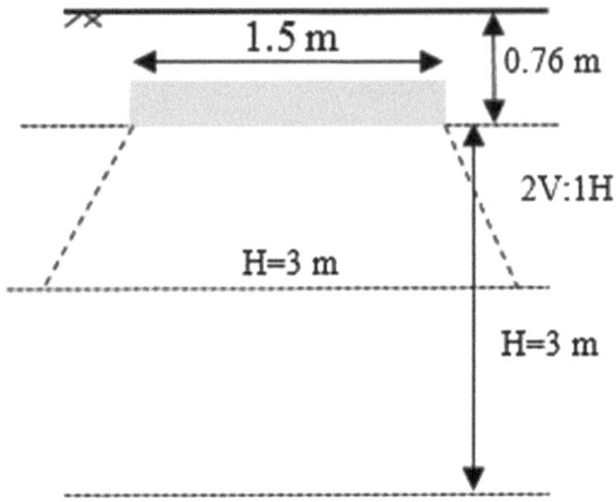

Figure 8.6 Shallow foundation for Example 8.1.

The target layer is assumed H = 2B. The foundation settlement is estimated using proposed approach as below:

The initial stress in the middle of the target layer: $\sigma'_0 = 18.5 \times (0.76 + 1.5) = 41.8\ kPa$.

Applied load: $1540 \times 1.5^2 = 3465\ kN$.

The increase of effective stress under applied pressure by the method of 2V:1H:

$$\Delta\sigma = \frac{3465}{3^2} = 385\ kPa$$

The geometric mean of q_c in the influenced zone: 7520 kPa

$$\text{Friction ratio}: R_f = \frac{f_s}{q_c} = 0.65$$

$$\alpha = 0.25b \times \left(\frac{2B+1}{3B}\right)^3 = 0.25 \times 0.0437 \times \left(\frac{2 \times 1.5 + 1}{3 \times 1.5}\right)^3 = 0.007$$

$$m = \alpha \times q_c = 0.007 \times 7520 = 52.64\ kPa$$

$$x = 0.02\ R_f + 0.5 = 0.02 \times 0.65 + 0.5 = 0.513$$

$$y = 7.53(\sigma'_0)^{-0.25} = 7.53 \times (41.8)^{-0.25} = 2.96$$

$$j = \frac{q_c}{x + yq_c} = \frac{7520}{0.513 + 2.96 \times 7520} = 0.34$$

Table 8.4 Results of different approaches for the foundation settlement of studied example.

Predicted Settlement (mm)					
Janbu (1967)	**Meyerhof (1976)**	**Schmertmann (1978)**	**Malekdoost and Eslami (2011)**	**Valikhah and Eslami (2019)**	**Measured Settlement (mm)**
170.32	126.11	178.93	162.59	150	154

$$\varepsilon = \frac{1}{2q_t j}\left[\left(\frac{\sigma'_0 + \Delta\sigma'}{100}\right)^j - \left(\frac{\sigma'_0}{100}\right)^j\right] = \frac{1}{2 \times 7520 \times 0.34} \times \left[\left(\frac{41.8 + 385}{100}\right)^{0.34} - \left(\frac{41.8}{100}\right)^{0.34}\right] = 0.05$$

$$\Delta H = \varepsilon \times H = 0.05 \times 3 = 150\ mm$$

The measured settlement is 154 mm, so the difference is about 10% which is acceptable. In Table 8.4, results of foundation settlement calculated by different approaches for the studied case are presented.

8.6 Pile group settlement

The proposed approach could be also used for estimation of the pile group settlement. The concept of equivalent footing presented by Terzaghi and Peck (1948) shown in Fig. 8.7A can be used, and the settlement of pile group can be estimated by the proposed approach.

The analysis of the load–transfer curve is illustrated in Fig. 8.7B. Reducing the dead load on the pile has very little effect on the maximum load in the pile, as illustrated in the left side diagram of Fig. 8.7B. The figure also shows a schematic illustration of the settlement in the soil and the downdrag for the pile. The pile cap settlement is the soil settlement at the neutral plane plus the "elastic" compression of the pile for the load in the pile.

8.7 CPT-based load–displacement behavior of foundations (Valikhah et al., 2018; Valikhah, 2019)

Valikhah et al. (2018) and Valikhah (2019) proposed a new analytical–numerical method to estimate the bearing capacity and axial load–displacement behavior of

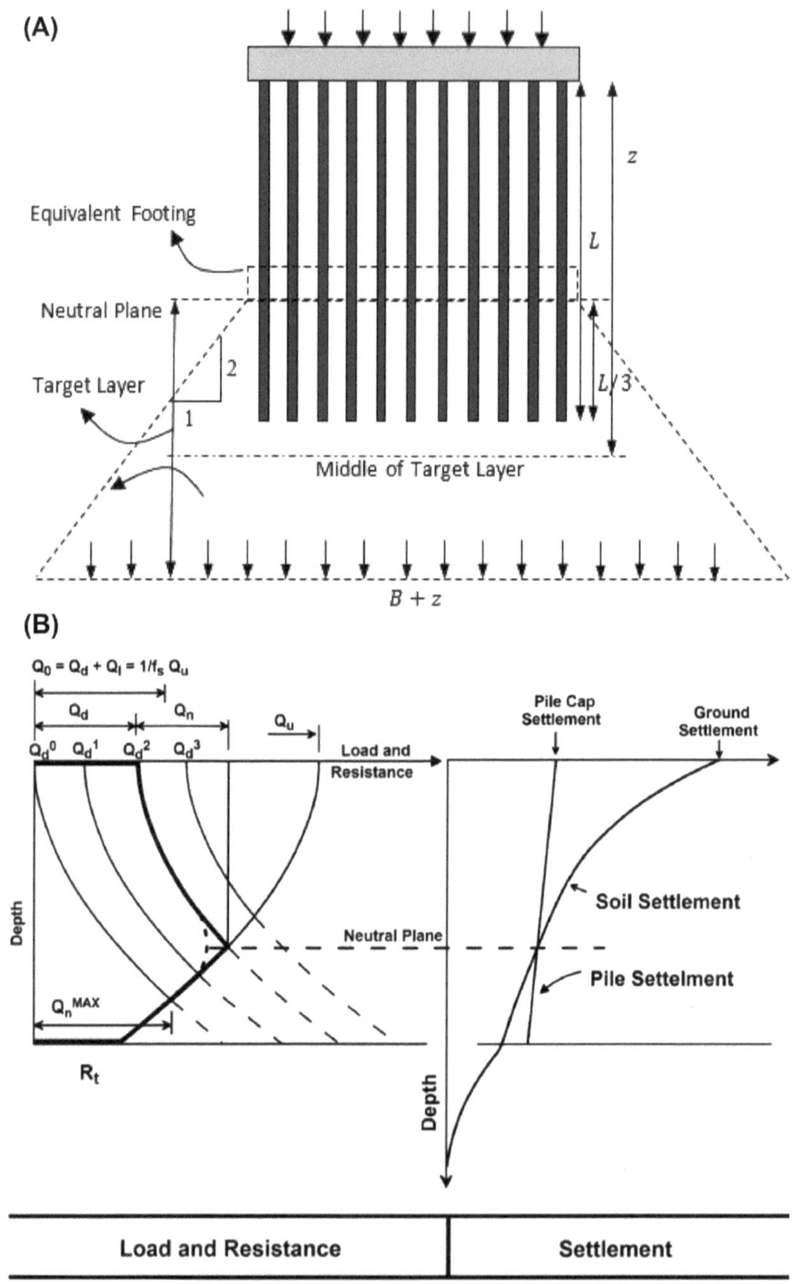

Figure 8.7 (A) Simple model to estimate pile group settlement proposed by Terzaghi and Peck (1948) and (B) load, resistance, and settlement distribution along depth (Fellenius, 2015).

driven piles in granular soils using CPT records. They used the method of stress characteristics to analyze the stress field below and around the pile and, in effect, the failure mechanism. This failure mechanism has been then used by implementation of the kinematic approach of the limit analysis to compute the displacement field. This procedure is employed in a stepwise manner to gradually calculate the stress and displacement field as the pile is assumed to penetrate into the ground. In their proposed method, the mobilization of the friction angle is linked to the gradual increase in shear strains in the field. This is done by making use of the CPT results which are both continuous and reliable in comparison with standard laboratory tests often conducted on disturbed samples at discrete intervals (Eslami and Fellenius, 1997). Hence, the stepwise procedure is expected to give rise to a complete load–displacement behavior of driven piles shown in a practical case.

For this purpose, they collected a database of case histories from the results of 165 full-scale pile load tests with complete information on the soil type and the results of CPT soundings performed close to the pile locations. In the investigated database, all piles were "driven pile" type. Most of the cases were in sand and some in silt and mixed soils.

The same procedure of CPT-based stress and strain field is used to predict the shallow foundation load–displacement behavior which was previously presented by Veiskarami et al. (2014, 2015) to estimate the bearing capacity of shallow foundations. In Fig. 8.8, the stress fields around the pile and beneath the footing are illustrated schematically from the numerical analysis.

The direct CPT-based equations to predict the load–transfer curves (t-z and q-z curves) and the load–displacement behavior of driven piles proposed by Valikhah et al. (2018) are as follows:

$$q = \frac{z}{a_t + b_t z} \tag{8.23}$$

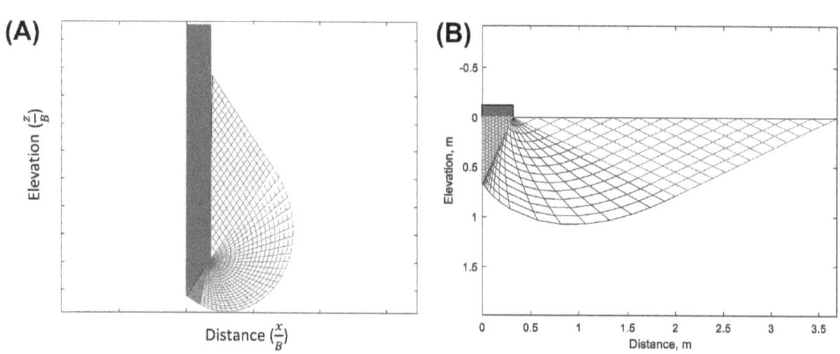

Figure 8.8 Stress field formed (A) around the pile (Valikhah et al., 2018) and (B) beneath the footing.

$$t = \frac{z}{a_s + b_s z} \tag{8.24}$$

$$P = \frac{\Delta}{a + b\Delta} \tag{8.25}$$

In these equations, q is the effective overburden pressure of soil at pile tip, t is shear stress at the pile shaft, z is vertical movement of the pile, P is the pile axial load, and Δ is the movement of the pile head.

Valikhah et al. (2018) with study on the properties of the collected database proposed the relations for a and b parameters for pile tip and shaft based on CPT results as follows:

$$a_t = 0.0009 \ln \frac{1}{\left(\frac{q_c}{\sigma_0}\right)} + 0.0067 \tag{8.26}$$

$$b_t = \frac{0.002}{\sqrt{\sigma_0}} \tag{8.27}$$

$$a_s = 0.0017e^{\left(\frac{97.7}{f_s}\right)} \tag{8.28}$$

$$b_s = 0.0242\left(\frac{1}{f_s}\right)^{0.038} \tag{8.29}$$

$$a = \frac{0.006}{\left(\frac{q_c}{\sigma_0}\right)^{0.4}} \tag{8.30}$$

$$b = \frac{0.089R_f}{B} \tag{8.31}$$

Inspection of the results revealed that these parameters are functions of the ratio q_c/σ_0, σ_0,f_s, R_f, and B, where σ_0 is the initial soil pressure at the depth at which these curves are required. $R_f = \frac{f_s}{q_c}$ is friction ratio, and B is pile width or diameter. This dependency seems to be logical as the stress level effect cannot be disregarded.

Therefore, if the soil conditions and pile properties are provided, by use of above-mentioned relations, the load–transfer curves and load–displacement behavior of piles could be predicted.

In the next section, some examples of estimating the load–displacement behavior of piles and footing using CPT-based stress characteristics method proposed by Valikhah et al. (2018) are presented.

8.7.1 First case study: short driven steel pile (9 m long)

Example 8.2. Prediction of Pile Load–Displacement Behavior:

The steel closed-ended pipe pile with 355 mm wide and 9 m long is investigated. The pile is under 1MN loading (800 kN dead load and 200 kN live load). The CPT profile of the site is shown in Fig. 8.9, and the load–displacement curve of the pile is illustrated in Fig. 8.10.

The step-by-step procedure to predict the load–transfer curves and load–displacement of pile using the analytical–numerical approach proposed by Valikhah et al. (2018) is as below:

The geometric mean of q_c in the influenced zone (4b below and 8b above the pile tip): 7 *MPa*.

The geometric mean of f_s along the pile length: 25 *kPa*.

Unit weight from Mayne (2001) equation: $\gamma_t = 12 + 1.5(ln(f_s+1)) = 16.89$ kPa.

The effective stress at the pile tip level: $\sigma_0 = 2\times16.89 + 7\times6.89 = 82.01$ kPa

$$a_t = 0.0009\ \ln\frac{1}{\left(\frac{q_c}{\sigma_0}\right)} + 0.0067 = 0.0009\ \ln\frac{1}{\left(\frac{7000}{82}\right)} + 0.0067 = 0.0027$$

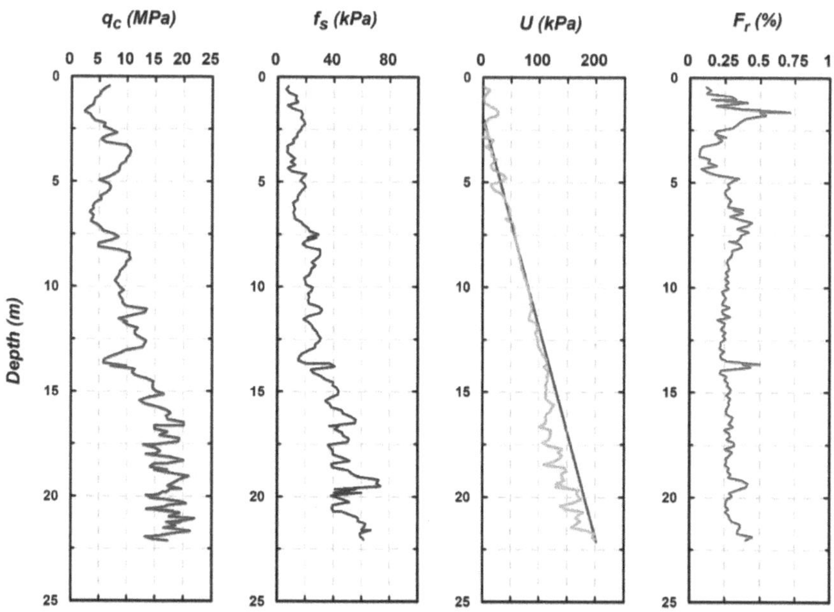

Figure 8.9 CPT results in the first case study site; 9 m pile.
(Data from Fellenius, B.H., April 2015. Basics of Foundation Design. Electronic ed.).

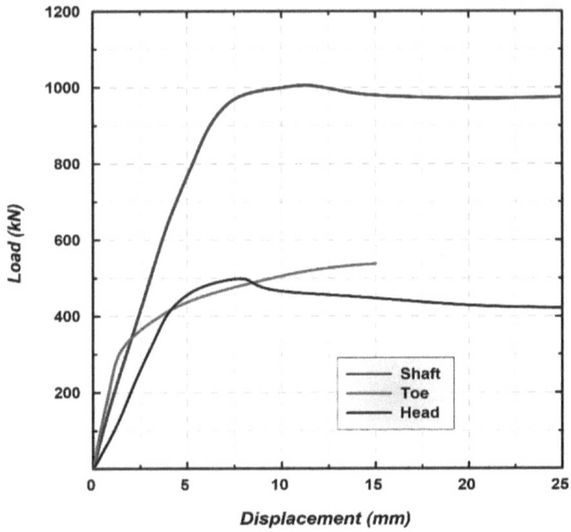

Figure 8.10 Load–displacement of the first case study; 9 m pile.
(Data from Fellenius, B.H., April 2015. Basics of Foundation Design. Electronic ed.).

$$b_t = \frac{0.002}{\sqrt{\sigma_0}} = \frac{0.002}{\sqrt{82}} = 0.00022$$

$$a_s = 0.0017e^{\left(\frac{97.7}{f_s}\right)} = 0.0017e^{\left(\frac{97.7}{25}\right)} = 0.085$$

$$b_s = 0.0242\left(\frac{1}{f_s}\right)^{0.038} = 0.0242\left(\frac{1}{25}\right)^{0.038} = 0.021$$

Friction ratio: Friction ratio : $R_f = \frac{f_s}{q_c} = \frac{25}{7000} = 0.0036$

$$a = \frac{0.006}{\left(\frac{q_c}{\sigma_0}\right)^{0.4}} = \frac{0.006}{\left(\frac{7000}{82}\right)^{0.4}} = 0.001$$

$$b = \frac{0.089R_f}{B} = \frac{0.089 \times 0.0036}{0.355} = 0.0009$$

Using the obtained parameters and Eqs. (8.17)–(8.19), $t-z$, $q-z$, and $P-\Delta$ curves. The results obtained by the proposed approach are shown in Fig. 8.11. Although the proposed approach by Valikhah et al. (2018) is not able to predict the softening behavior of the piles, as shown in the figure, the results are not far from the actual

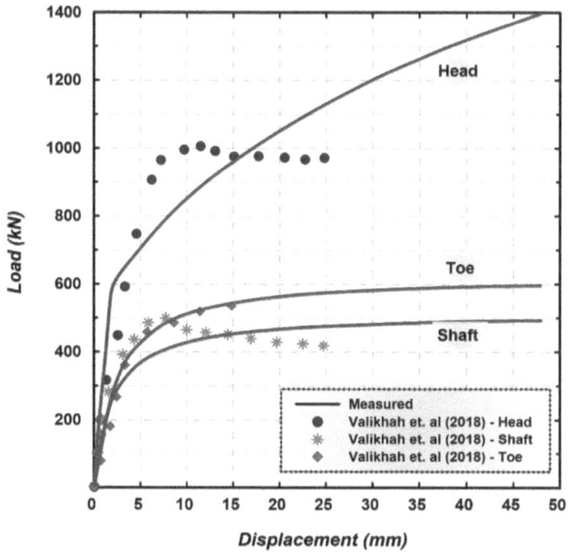

Figure 8.11 Predicted load–displacement responses of the first case study using CPT-based stress characteristics method proposed approach by Valikhah et al. (2018).

behavior of pile and are in acceptable agreement with the measured load–displacement curve for the piles.

8.7.2 Second case study: medium driven concrete pile (15 m long)

The concrete circle pile with 350 mm wide and 15 m long is investigated. The CPT profile of the site is shown in Fig. 8.12, and the load–displacement curve of the pile is illustrated in Fig. 8.13.

The result obtained by the analytical–numerical approach proposed by Valikhah et al. (2018) for this case is shown in Fig. 8.14. As shown, the results are in acceptable agreement with the measured load–displacement curve for the piles.

8.7.3 Third case study: long driven concrete pile (38 m long)

The concrete circle pile with 400 mm wide and 38 m long is investigated. The CPT profile of the site is shown in Fig. 8.15, and the load–displacement curve of the pile is illustrated in Fig. 8.16.

The result obtained by the analytical–numerical approach proposed by Valikhah et al. (2018) for this case is shown in Fig. 8.17. As shown, the obtained results for the long pile are a bit far from the actual behavior of the pile. Therefore, it can be stated that the proposed approach by Valikhah et al. (2018) has acceptable results for short to medium piles.

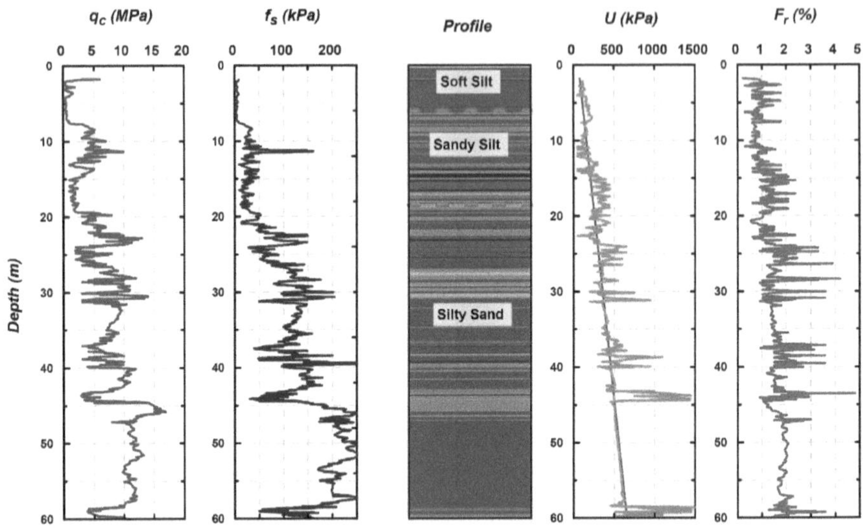

Figure 8.12 CPT results in the second case study site; 15 m pile (Naesgaard et al., 2012).

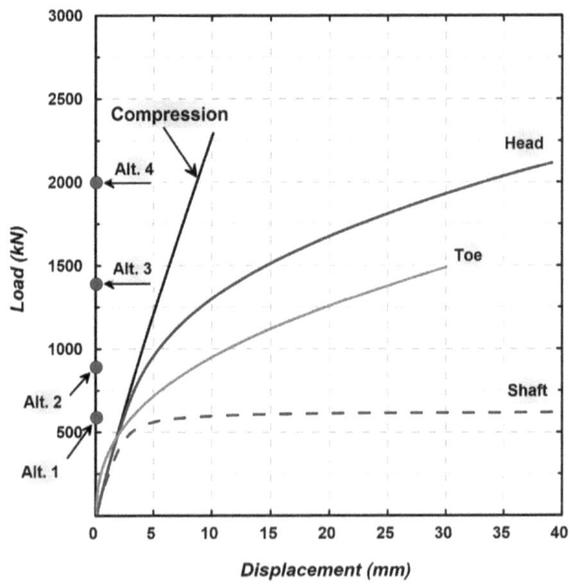

Figure 8.13 Load–displacement of the second case study; 15 m pile (Fellenius, 2019).

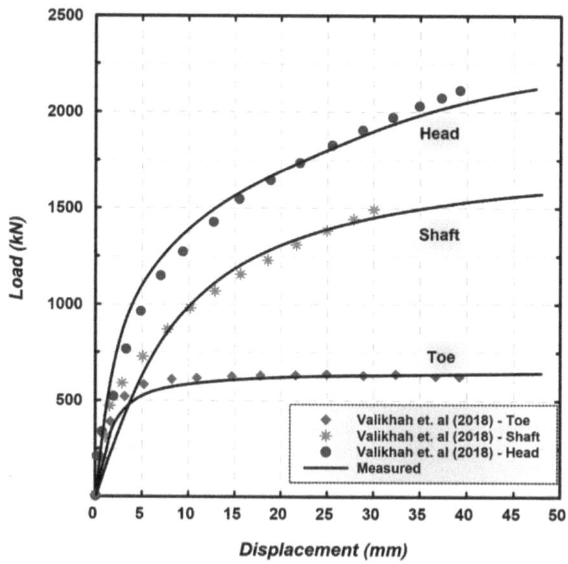

Figure 8.14 Predicted load–displacement responses of the second case study using CPT-based stress characteristics method proposed approach by Valikhah et al. (2018).

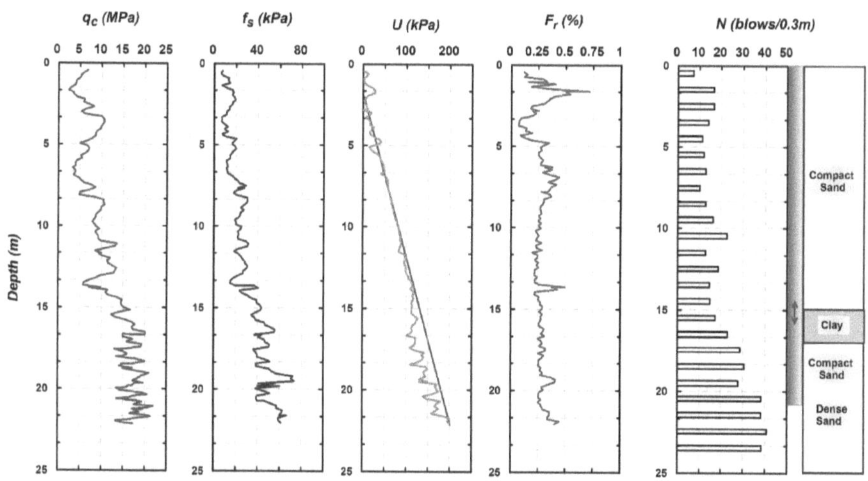

Figure 8.15 CPT results in the third case study site; 38 m pile.
(Data from Fellenius, B.H., April 2015. Basics of Foundation Design. Electronic ed.).

8.7.4 Fourth case study: shallow foundation (1 m width)

As an applied example, the Case 1 in Table 8.3 is investigated. The studied case is a 1 m square footing with the embedment depth 0.76 m. The applied stress is equal to 1540 kPa. The CPT records and the load–displacement curve of the foundation are illustrated in Fig. 8.3. The result obtained by the numerical analysis for this case is shown in Fig. 8.18.

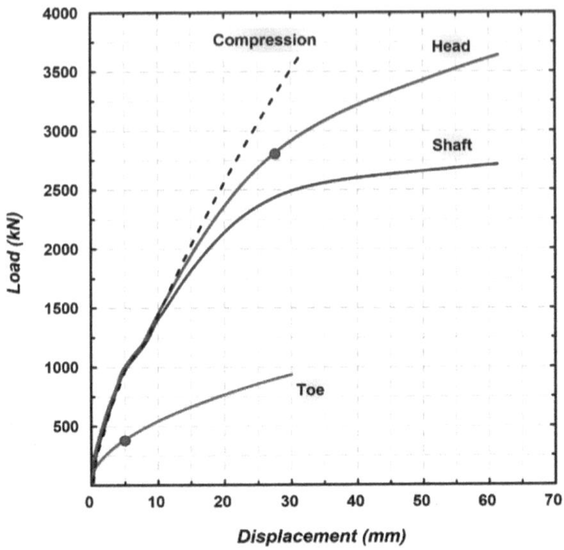

Figure 8.16 Load–displacement of the third case study; 38 m pile.
(Data from Fellenius, B.H., April 2015. Basics of Foundation Design. Electronic ed.).

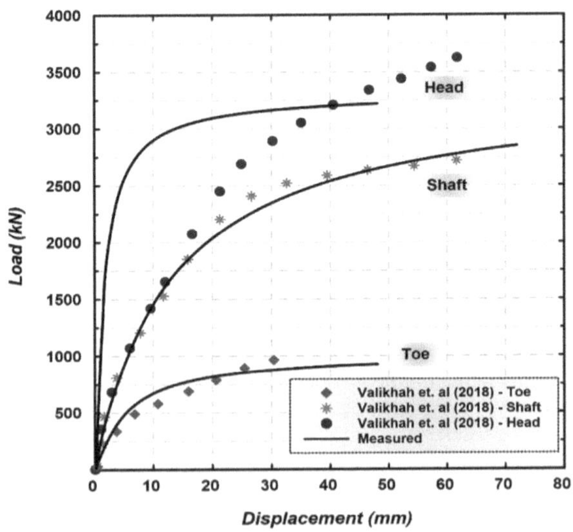

Figure 8.17 Predicted load–displacement responses of the third case study using CPT-based stress characteristics method proposed approach by Valikhah et al. (2018).

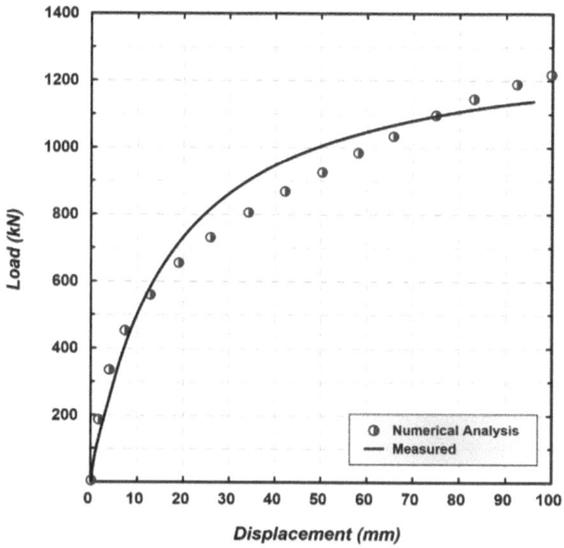

Figure 8.18 Predicted load–displacement responses of the shallow 1 m square footing using CPT-based stress characteristics method.

References

Anderson, B.J., Townsend, F.C., Rahelison, L., 2007. Load testing and settlement prediction of shallow foundation. Journal of Geotechnical and Geoenvironmental Engineering 133 (12), 1494–1502.

Briaud, J.L., 2007. Spread footings in sand: load settlement curve approach. Journal of Geotechnical and Geoenvironmental Engineering 133 (8), 905–920.

Briaud, J.L., Gibbens, R.M., 1999. Behavior of five large spread footing on sand. Journal of Geotechnical and Geoenvironmental Engineering 125 (9), 787–796.

Campanella, R.G., Robertson, P.K., Gillespie, D., Grieg, J., 1986. Use of piezometer cone data. Publication GSP No. 6. In: Proceedings of American Society of Civil Engineering, ASCE, In-Situ 86 Specialty Conference, pp. 1263–1280.

Canadian Foundation Engineering Manual, CFEM, 2006. Bitech Publishers Ltd., British Columbia.

Clark, J.I., 1998. The settlement and bearing capacity of very large foundations on strong soils. 1996 R.M. Hardy Keynote address. Canadian Geotechnical Journal 35, 131–145.

Eslami, A., Fellenius, B., 1997. Pile capacity by direct CPT and CPTu methods applied to 102 case histories. Canadian Geotechnical Journal 34, 886–904.

Eslami, A., Gholami, M., 2006. Analytical model for the ultimate bearing capacity of foundations from cone resistance. International Journal of Science and Technology, Scientica IRANICA 13 (3), 223–233.

Eslami, A., Valikhah, F., Veiskarami, M., Salehi, M., 2017. CPT-based investigation for pile toe and shaft resistances distribution. Geotechnical and Geological Engineering 35 (6), 2891–2905.

Fellenius, B.H., April 2015. Basics of Foundation Design, Electronic ed.

Fellenius, B.H., 2019. Observations and analysis of wide piled foundations. Canadian Geotechnical Journal 56 (3), 378–397.

Holtz, R.D., Kovacs, W.D., Sheahan, T.E., 2011. An Introduction to Geotechnical Engineering, second ed. Prentice Hall.

Janbu, N., 1967. Settlement Calculations Based on the Tangent Modulus Concept. University of Trondheim, Norwegian Institute of Technology. Bulletin (2), 57 pp.

Lade, P.V., Duncan, J.M., 1975. Elastoplastic stress-strain theory for cohesionless soil. Journal of the Geotechnical Engineering Division 101 (10), 1037–1053.

Look, B., 2007. Handbook of Geotechnical Investigation and Design Tables. Taylor and Francis Group.

Malekdoost, M., Eslami, A., 2011. Application of CPT data for estimating foundations settlement. Sharif Civil Engineering Journal 27 (2), 75–85. No. 1.

Mayne, P.W., 2001. Stress-strain-strength-flow parameters from enhanced in-situ tests. In: Proc., International Conference on In-Situ Measurement of Soil Properties and Case Histories, Bali, Indonesia, pp. 27–48.

Mayne, P.W., Illingworth, F., 2010. Direct CPT method for footing response in sands using a database approach. In: Proceedings, 2nd International Symposium on Cone Penetration Testing (CPT'10), 3, Huntington Beach, California, pp. 315–322.

Meyerhof, G., 1976. Bearing capacity and settlement of pile foundations. Journal of the Geotechnical Engineering Division 102 (GT3), 195–228.

Moshfeghi, S., Eslami, A., 2016. Study on pile ultimate capacity criteria and CPT-based direct methods. International Journal of Geotechnical Engineering 12 (1), 28–39.

Naesgaard, E., Amini, A., Uthayakumar, U.M., Fellenius, B.H., 2012. Geotechnical Special Publication, 227. In: Hussein, M.H., Holtz, R.D., Massarsch, K.R., GE, L. (Eds.), Long Piles in Thick Lacustrine and Deltaic Deposits, Two Bridge Foundation Case Histories, Full Scale Testing in Foundation Design, pp. 404–421.

Niazi, F.S., Mayne, P.W., 2016. CPTu-based enhanced UniCone method for pile capacity. Engineering Geology 212, 21–34.

Phoon, K.K., Retief, J.V., 2016. Reliability of Geotechnical Structures in ISO2394. CRC Press.

Robertson, P.K., Campanella, R.G., Gillespie, D., Grieg, J., 1986. Use of piezometer cone data. In: Proceedings of American Society of Civil Engineers, ASCE, In-Situ 86 Specialty Conference, Edited by S. Clemence, Blacksburg, June 23-25, Geotechnical Special Publication GSP No. 6, pp. 1263–1280.

Schmertmann, J.H., 1978. Guidelines for Cone Test, Performance and Design. ReportFHWA-TS-78209. Federal Highway Administration, Washington, 145 pp.

Terzaghi, K.V., 1955. Evaluation of coefficient of subgrade reaction. Géotechnique 5 (4), 297–326.

Terzaghi, K.V., Peck, R.B., 1948. Soil Mechanics in Engineering Practice. John Wiley and Sons, New York, 566 pp.

Valikhah, F., Eslami, A., 2019. CPT-Based Nonlinear Stress–Strain Approach for Evaluating Foundation Settlement: Analytical and Numerical Analysis. Arabian Journal for Science and Engineering 1–6.

Valikhah, F., Eslami, A., Veiskarami, M., 2018. Load-displacement behavior of driven piles in sand using CPT-based stress and strain fields. International Journal of Civil Engineering. https://doi.org/10.1007/s40999-018-0388-7.

Valikhah, F., 2019. Analysis of Bearing Capacity and Axial Load-Displacement of Driven Piles in Sand Using CPT Records (Ph.D. Dissertation). Amirkabir University of Technology.

Veiskarami, M., Kumar, J., Valikhah, F., 2014. Effect of the flow rule on the bearing capacity of strip foundations on sand by the upper-bound limit analysis and slip lines. International Journal of Geomechanics 14 (3), 04014008.

Veiskarami, M., Kumar, J., Valikhah, F., 2015. A practical procedure to estimate the bearing capacity of footings on sand–application to 87 case studies. Iranian Journal of Science and Technology Transactions of Civil Engineering 39 (C2+), 469–483.

Vesić, A.S., 1961. Beams on Elastic Subgrade and the Winkler Hypothesis. 5th ICSMFE, vol. 1, pp. 845–850.

Wang, Y., 2009. Reliability-based economic design optimization of spread foundations. Journal of Geotechnical and Geoenvironmental Engineering 135, 954–959.

Whitman, R.V., 2000. Organizing and evaluating uncertainty in geotechnical engineering. Journal of Geotechnical and Geoenvironmental Engineering 126, 583–593.

Zhang, Q.Q., Li, L.P., Chen, Y.J., 2014. Analysis of compression pile response using a softening model, a hyperbolic model of skin friction, and a Bilinear model of end resistance. Journal of Engineering Mechanics, ASCE 140 (1), 102–111.

CPT & CPTu for ground modification

9

9.1 Introduction

Where a project encounters unfavorable subsurface conditions, possible alternative solutions are avoiding the particular site, modifying the design of the planned structure accordingly, removing and replacing unsuitable soils, or attempting to modify the existing ground.

It becomes necessary to improve geomaterials and geotechnical conditions for many projects to meet the project requirements. To exemplify, ground improvement can be implemented to do the following:

- Increase soil strength
- Mitigate erodibility
- Reduce distortion under stress
- Diminish compressibility
- Control shrinking, swelling, and permeability
- Diminish water pressures
- Redirect seepage
- Prevent the detrimental physical or chemical changes due to environmental conditions
- Mitigate susceptibility to liquefaction and change natural variability of borrow materials or foundation soils

Although almost more than 50 techniques and methods are commonly used in ground improvement practice, they can collectively be classified in five major categories. As stated in Chapter 2, these five categories of soil improvement methods include the following:

1. Earthworks
2. Densification
3. Physical and chemical modification
4. Hydraulic modification
5. Reinforcement

A detailed explanation of approaches and recent developments can be found for ground improvement or soil stabilization as presented by Hausmann (1990), FHWA (2001), and Han (2015).

9.2 General description, function, and application of the methods

Table 9.1 provides the general descriptions, advantages, and applications of the five major categories.

Piezocone and Cone Penetration Test (CPTu and CPT) Applications in Foundation Engineering
https://doi.org/10.1016/B978-0-08-102766-0.00009-2

Table 9.1 Classification of ground improvement methods.

No.	Category	Method	Principle
I	Earthwork (E)	E1: Cut and fill	Remove weak soil by excavation and replace it by strong soil or rocks.
		E2: Shallow soil mixing	Mixing two different soils; clay with sand and sand with clay. Plasticity, gradation, and cohesion can be solved by mixing.
		E3: Unloading and floatation of foundations	Reduce total and differential settlement of foundations by decreasing the total net load using excavation of surface soil. This method is based on floatation principle in physics.
		E4: Lightweight materials	Using some materials with lower density in order to reduce the magnitude of applied loads.
		E5: Preloading and surcharge fills	Preconsolidation of compressible soil by surcharge fill to reduce the compressibility of soil under future applied loadings.
		E6: Deep soil mixing	Lime and lime/cement columns, where quicklime and/or dry cement are mixed in situ with soft soil to stabilize soft clay and silt as well as organic soils.
II	Densification (D)	D1: Static or dynamic rollers	Compaction of fill or ground at shallow depth using a variety of compaction machines such as tampers, classic rollers, intelligence compaction, high-impact energy dynamic compaction (HIEDYC).
		D2: Rapid impact compaction (RIC)	Rapid impact compaction is an intermediate compaction technology between shallow compaction and deep dynamic compaction. It rapidly applies impact on ground surface using a hydraulic hammer.
		D3: Deep dynamic compaction (DDC)	Dropping a heavy weight from air onto ground to densify the granular soil.

Table 9.1 Classification of ground improvement methods.—cont'd

No.	Category	Method	Principle
		D4: Vibro compaction	Vibratory probe inserted to ground to densify the granular soil.
		D5: Sand compaction piles (SCP)	Driving a hollow steel pipe into the ground. The bottom is closed with a collapsible plate down to the required depth and then pipe is filled with sand. The pipe is withdrawn while the air pressure is directed against the sand inside it.
		D6: Rammed aggregate piers (RAPs)	RAPs are a system that uses compacted aggregate to create stiff pier elements and modifies the soil surrounding the piers. Piers are installed by drilling holes into the foundation soils and ramming thin lifts of well-graded aggregate within the holes to form stiff, high-density aggregate piers.
		D7: Blasting	Shock waves and vibrations are generated by explosion to cause the granular soil to settle through liquefaction or compaction.
III	Physicochemical modification (PC)	PC1: Stabilization by admixtures	Treat the weak soil by mixing it with cement, lime, bitumen, or fly ash to increase the soil strength and stiffness, improve volume stability (less moisture sensitivity, control of frost heave), and increase durability.
		PC2: Chemical grouting	Solutions, suspensions, or emulsions of two or more chemicals are injected in soil pores to increase the strength or reduce the permeability of soil or ground.
		PC3: Thermal stabilization	Permanent or temporary changes of the engineering properties of soil by heating or freezing the soil.

Continued

Table 9.1 Classification of ground improvement methods.—cont'd

No.	Category	Method	Principle
		PC4: Electro-osmosis or electrokinetic consolidation	Passing a direct current from anodes to cathodes positioned at predetermined locations in soil which causes water to migrate from the anodes toward the cathodes, then it is collected and removed.
		PC5: Bioengineering	Construction of terraces, berms, fences, catch walls, protective meshes, brushwood fence, shelter fences, and others to protect slopes.
IV	Hydraulic modification (H)	H1: Dewatering by open sumps and ditches	Works well in relatively shallow excavations in dense, well-graded coarse soils, rock and in the case of permeable soils overlying impermeable strata.
		H2: Vertical drains	Vertical drains (sand drains, sand wicks, band drains, wick drains, prefabricated vertical drains (PVD)) are used in conjunction with dewatering systems to lower the groundwater level in stratified soils.
		H3: Vacuum dewatering wells	In fine sands and silts, water does not flow freely under the influence of gravity, due to capillary tension. Therefore, to make dewatering and stabilizing of these soils possible, a vacuum may be applied to the sealed-off filter section of the well.
		H4: Hydraulic fill	The fill material is pumped through pipelines onto the construction site and is discharged with immediate drainage or fed into a pool at the core of the embankment being formed.
		H5: Vibro replacement or stone columns	Hole jetted into soft, fine-grained soil and back filled with densely compacted gravel or sand to form stone columns.

Table 9.1 Classification of ground improvement methods.—cont'd

No.	Category	Method	Principle
V	Reinforcement (R)	R1: Geosystems	A system concluding soil, water, and structural intrusions
		R2: Geosynthetics	Many polymers (plastics) are found in geosynthetics such as polyolefin, polyester, rubber, fiberglass, and natural materials. They may be used to function as a separator, filter, planar drain, reinforcement, cushion/ protection, and/or as a liquid and gas barrier.
		R3: Micropile and helical pile	Use of micromaterials to modify soil to increase strength or reduce permeability.
		R4: Jet grouting	Jet grouting offers a means of forming an impermeable barrier, as well as providing a means for supporting or underpinning structures by high pressure grouting in soil.
		R5: Nailing and anchor	Reinforcing the ground with small inclusions, usually steel bars, where installed horizontally or subhorizontally to improve the shearing resistance of the soil by acting in tension.
		R6: Soil confinement	Increase in soil strength by confining the soil and adding induced cohesion. Some examples of soil confinement are soil bag, crib wall, gabion, and geocell.

Soil type and particle size distribution are essential for preliminary selection of ground improvement methods as shown in Fig. 9.1. This guideline is suitable for ground improvement methods for foundation support.

The thickness and location of problematic geomaterial are also important for the selection of ground improvement methods. For example, when a thin problematic geomaterial layer exists at a shallow depth, the over excavation and replacement method is one of the most suitable and economic methods.

When a relatively thick, loose, cohesionless geomaterial layer exists near ground surface, deep dynamic compaction and vibro compaction methods are suitable. When a relatively thick soft, cohesive, geomaterial layer exists near ground surface, preloading and

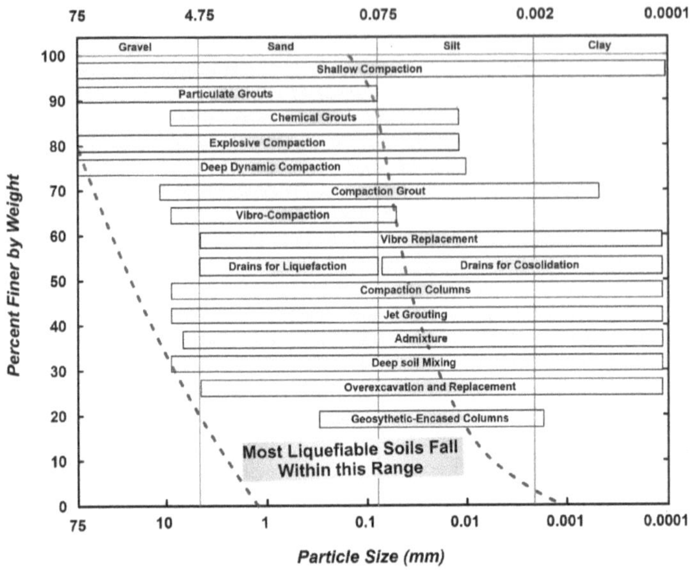

Figure 9.1 Available ground improvement methods for different soil types (Mitchell and Gallagher, 1998).

deep mixing methods may be used. When a site needs to be excavated, the tieback anchors, soil nails, deep mixed columns, and jet-grouted columns may be used. Where a site needs to be elevated, geosynthetic-reinforced slopes and walls can be good choices as supporting systems. The level of groundwater table often affects the selection of ground improvement techniques. For instance, when deep excavation happens in ground with a high groundwater table, deep mixed column walls may be better than soil nailed walls because they not only can retain the geomaterial but also can cut off water flow.

As discussed in Chapters 3–5, cone penetration test (CPT) is recognized as a valuable source of measured data with primary application in soil classification and identification. Hence, in preliminary steps for assessing the necessity of doing soil modification and consequently choosing a method, CPT data can be employed.

9.3 CPT and CPTu application in ground improvement practice

Selection of proper ground improvement method must consider the following aspects: (1) structural characteristics, (2) site geotechnical conditions, (3) environmental constraints, (4) construction procedures, and (5) reliability and durability.

The major question in preliminary step is "Does the project need ground modification?" In situ tests are capable of answering this question by determining the density, water content, strength, and stiffness characteristics of soil layers.

Among in situ tests, CPT is more applicable to be associated with the ground improvement methods. The soil properties can be easily obtained by having variation of tip resistance (q_c) and sleeve friction (f_s) in depth, supplemented by excess pore pressure due to penetration and available formulas that relate these records to soil geotechnical properties.

Usually, soil improvement practice is required in soft, loose, and saturated soils. This feature makes CPT a suitable option for validation of soil improvement activity in different steps, i.e., evaluation, design, and control. The direct CPT data can be used for classification of the soil and interpretation of the ground performance before and after the modification.

Assessment of ground modification effectiveness using laboratory tests is rather a wearing, time-consuming, and expensive approach. Reaching an acceptable insight from geotechnical condition of project needs lots of sampling and laboratory tests. Due to shortage of time in civil projects, the collection of extra samples is rather inappropriate and redundant. Moreover, presence of loose and saturated deposits in ground modification projects make sampling a costly process. Above all, the disturbance of specimens along with the complications in testing process result in more difficulties in utilization of laboratory tests in ground modification projects.

For any stage of ground modification practice, CPT-based approach can be employed via following trends which will be reviewed in detail:

- Quality analysis of soil parameters
- Comparison of records before and after treatment
- Validation by soil behavior classification (SBC)

CPT is used for investigation of soil improvement operations effectiveness in soft, weak, or loose saturated soils due to providing accurate and continuous records, reduction of operator influence on test results, optimized costs, and measurement of excess pore pressure, cone tip, and sleeve resistances in very close depth intervals (Robertson and Wride, 1998).

Also, CPT can obtain more information about the site condition (two or more parameters of soil) (Eslami et al., 2017). Different soil deposit profiles detected by CPT or CPTu logs enable the geotechnical engineer to reach a logical and optimum design.

Both quality control and quality assurance ensure the quality of construction; however, they are done at different stages and by different entities. Quality control refers to procedures, measurements, and observations used by the contractor to monitor and control the construction quality such that all applicable requirements are satisfied. Quality assurance refers to measurements and observations by the owner or the owner's engineer to provide assurance to the owner that the facility has been constructed in accordance with the plans and specifications (Han et al., 2012).

Automatic or computer-controlled installation processes and data collection systems as mentioned in Chapter 1 can reduce the variability of improved geomaterials and avoid human errors. Quality assurance often involves in situ testing and field monitoring, mostly done by penetration testing with the ability of providing continuous geotechnical records.

9.4 Quantitative analysis of soil geotechnical parameters

Application of CPT in soil improvement is providing direct and indirect quantitative parameters which can be used to deduce whether the soil needs any modification work.

Through the indirect approach, the CPT data can be related to soil geotechnical properties such as water content, shear wave velocity, friction angle, relative density, and shear strength by empirical formulas as indicated in Table 9.2. Accordingly, it is possible to decide whether or not ground improvement is required for the encountered soil condition. For this purpose, typical ranges of some geotechnical parameters are shown in Tables 9.3 and 9.4 for sandy and clayey soils.

Some sites are studied for quantitative analysis of CPT data and soil strength parameters before and after ground modification. In the following, the case information, ground improvement practices and data comparison are presented.

9.4.1 Case No. 1: hydraulic modification in Treasure Island—the United States (Rollins et al., 2004)

Treasure Island is a 1.62-km^2 man-made island located in the vicinity of San Francisco. It was built using hydraulic and clamshell dredging. Several drain clusters were installed at the site, and pore pressure and settlement behavior of the site were compared to the parameters after improvement. The water table was approximately about 0.15 m below ground surface. Geotechnical investigations were carried out in the site, and the soil profile. Hydraulically placed fill and native shoal sands to a depth of 4.2 m formed the soil profile of the project. It is usually categorized as SP-SM in the USCS and had a D_{50} between 0.2 and 0.3 mm. Silty sand (SM) to a depth of 7.5 m was situated under this layer, and the SM itself was underlain by Young Bay Mud. Standard penetration test (SPT) and CPT were carried out in the site, and the results were compared before and after installation of the drains.

A few important geotechnical parameters of each soil which were calculated based on CPT soundings are given in Table 9.5. In this table, q_c is cone tip resistance, E is elastic modulus, φ soil is friction angle, V is shear wave velocity, D_r is soil relative density, and m is Janbu modulus number. "a" index in each parameter refers to amount of that parameter after ground modification, and "b" index refers to before ground modification. The changes of the site geotechnical parameters before and after improvement are clear in the figure and table.

Comparison of tip resistance, friction angle, and relative density values before and after ground modification (presented in Table 9.5) with values in Table 9.3 indicates that the site soil was very loose and categorized as a medium dense soil after densification. Besides, average values of Young modulus doubled during soil improvement operation. This matter is coupled with improving Janbu modulus which led to diminishing the soil settlement significantly as a result of hydraulic modification in this site.

Table 9.2 Empirical relationships for estimating soil properties based on cone penetration test records.

Soil type	Parameter	Correlation	Reference
Sandy soils	Elastic modulus (E)	$(2-4)q_c$; normally consolidated sand $(3-7)q_c$; saturated sand $(6-10)q_c$; over consolidated sand $(1-2)q_c$; clayey sand and silty sand	(Bowles, 1996)
	Friction angle $(\varphi)'$	$\varphi' = 17.6 + 11log\frac{q_c/p_a}{\sqrt{\sigma'_{v0}/p_a}}$	(Kulhawy and Mayne, 1990)
	Shear wave velocity (V_s)	$V_s = 277q_c^{0.13}\sigma_{v0}'^{0.27}$	(Baldi et al., 1986)
	Relative density (D_r)	$D_r^2 = \frac{1}{305Q_c}\cdot\frac{q_c}{p_a}\cdot\left(\frac{\sigma'_{v0}}{p_a}\right)^{0.5}$	(Kulhawy and Mayne, 1990)
	Janbu modulus (m)	$m = a\sqrt{\frac{q_c\left(p_a/\sigma'_m\right)^{0.5}}{p_a}}$	(Massarsch, 1994)
		σ'_m can be calculated using the following equation: $D_r = \frac{1}{2.61}\ln\left(\frac{q_c}{181(\sigma'_m)^{0.55}}\right)$	(Baldi et al., 1986)
Clays	Elastic modulus (E)	Same as mentioned above for sandy soils	(Bowles, 1996)
	Shear wave velocity (V_s)	$V_s = 1.75q_c^{0.627}$	(Baldi et al., 1986)
	Undrained shear strength (S_u)	$S_u = \frac{q_c-\sigma_{v0}}{N_k}$	
	Janbu modulus (m)	$M = \alpha_m.q_c,\ M = mp_a\left(\frac{\sigma'_v}{p_a}\right)^{(1-a)} \Rightarrow m = \frac{\alpha_m\cdot q_c}{p_a\left(\frac{\sigma'_v}{p_a}\right)^{(1-a)}}$	(Senneset et al., 1989)
	Over consolidation ratio (OCR)	$OCR = \left(\frac{S_u}{0.4\sigma'}\right)^{\frac{1}{0.85}}$	(Ladd et al., 1977)

Table 9.3 Range of geotechnical parameters in sandy soils (Eslami and Sekhavatian, 2014).

Soil condition / Geotechnical parameters	Very loose	Loose	Medium	Dense	Very dense
N_{SPT}	<4	4−10	10−30	30−35	>50
$q_c\left(\frac{kg}{cm^2}\right)$	<50	50−100	100−150	150−200	>200
$D_r(\%)$	<15	15−35	35−65	65−85	85−100
$\varphi(°)$	<30	30−32	32−35	35−38	>38
$\gamma_{dry}\left(\frac{ton}{m^3}\right)$	1.4	1.4−1.6	1.6−1.8	1.8−2	>2
$\frac{\tau}{\sigma'}$	<0.04	0.04−0.1	0.1−0.35	>0.35	−

Table 9.4 Range of Geotechnical parameters in Clays (Eslami and Sekhavatian, 2014).

Soil condition / Geotechnical parameters	Soft	Medium to stiff	Stiff to hard
N_{SPT}	<5	5−15	>15
$q_c\left(\frac{kg}{cm^2}\right)$	<10	10−50	>50
$S_u\left(\frac{kg}{cm^2}\right)$	<0.5	0−1	1−2
$\gamma_{dry}\left(\frac{ton}{m^3}\right)$	0.9−1.4	1.4−1.7	1.7−2.1
ω (%)	>30	20−40	<20

9.4.2 Case No. 2: compaction modification in map Ta Put Port—Thailand (Bergado et al., 1996)

This port is located in Thailand. The resonance compaction method was carried out in this reclaimed area. Hydraulic fill dredged of clean sands and corals was used to fill up the area. The site is mostly situated on sandy deposits as shown in Fig. 9.2. The compacted area would act as foundation of oil storage tanks, thus evaluating the efficacy of ground improvement is fundamentally important. CPT and SPT were performed to assess the magnitude of modification. The improvement in soil condition is quite significant as shown in Table 9.6. In this table, q_c is cone tip resistance, f_s is sleeve friction, E is elastic modulus, V is shear wave velocity, D_r is soil relative density, and R_f is friction ratio ("a" index refers to after and "b" index to before ground modification).

As shown in Fig. 9.2, N_{60} values before and after ground modification procedure were significantly enhanced, and the cone resistance increased by about 2.5 times.

Table 9.5 Cone penetration test Data and Geotechnical Parameters of Treasure island before and after ground improvement (Eslami et al., 2010).

Depth	q_{cb}	q_{ca}	E_b	E_a	φ_b	φ_a	V_{sb}	V_{sa}	D_{rb}	D_{ra}		
(m)	***(MPa)***		***(MPa)***		(°)		(m/s)		(%)		m_b	m_a
1–2	7.9	10.4	23.6	41.7	42.8	44.1	118.0	124.7	51.4	60.5	204.5	251.7
2–3	6.8	8.1	20.4	32.5	41.2	41.8	129.4	134.6	47.8	54.1	194.7	230.9
3–4	5.5	7.7	16.3	30.6	39.3	40.8	137.2	146.0	43.1	51.7	182.3	223.6
4–5	4.2	6.8	12.7	27.1	37.4	39.6	141.5	153.5	35.1	46.8	163.9	209.8
5–6	3.4	4.2	10.1	16.8	35.9	36.8	145.3	151.8	29.2	31.0	151.1	170.5
6–7	3.5	3.5	10.4	13.9	35.7	35.5	152.5	154.9	31.1	31.0	154.8	168.6
Average	5.2	6.8	15.6	27.1	38.7	39.8	137.3	144.2	39.6	45.8	175.2	209.2

Note: a and b indices represent after and before modification, respectively.

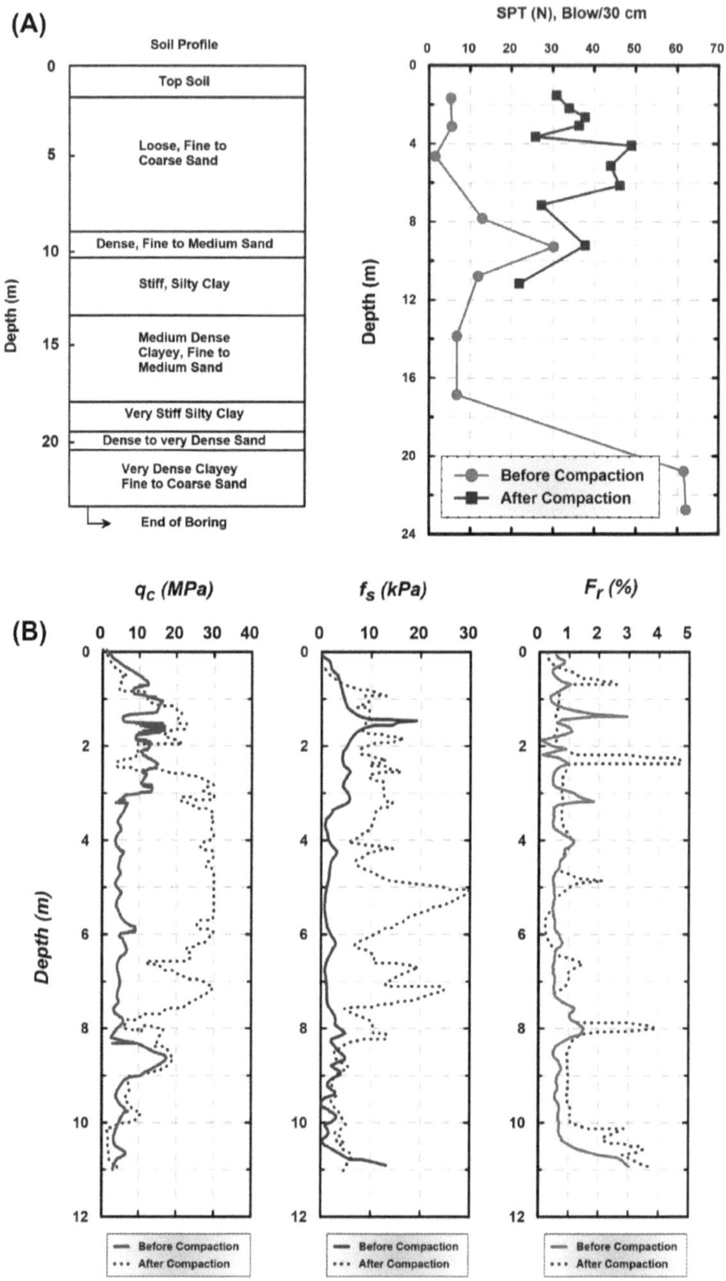

Figure 9.2 Soil profile, standard penetration test (SPT), and cone penetration test (CPT) results before and after resonance compaction at Map Ta Put Port (Bergado et al., 1996).

Table 9.6 Cone penetration test data and geotechnical parameters of Map Ta Put Port before and after ground improvement (Bergado et al., 1996).

Depth	q_{cb}	q_{ca}	f_{sb}	f_{sa}	E_b	E_a	V_{sb}	V_{sa}	D_{rb}	D_{ra}	R_{fb}	R_{fa}
(m)	*(MPa)*		*(KPa)*		*(MPa)*		*(m/s)*		(%)		(%)	
1–2	12.2	17.3	9.8	11.8	36.5	57.4	125.7	141.1	42.0	81.7	8.1	7.1
2–3	12.7	15.0	5.7	12.0	38.0	50.0	142.8	155.6	47.9	60.0	4.5	9.5
3–4	9.0	23.8	5.1	12.9	26.9	79.3	141.6	174.6	41.2	85.6	6.4	5.5
4–5	4.6	29.3	2.6	12.6	13.8	97.6	145.1	198.9	33.1	97.4	5.7	4.3
5–6	5.4	28.9	2.6	13.4	16.2	96.2	155.8	209.1	37.5	99.4	4.9	4.6
6–7	5.3	22.8	2.4	10.2	16.0	76.0	162.5	211.6	38.8	94.0	4.5	4.5
7–8	4.3	15.0	2.8	11.8	13.0	50.0	164.2	207.5	36.3	77.8	6.5	8.1
8–9	8.8	11.3	4.1	7.1	26.5	37.6	185.9	204.8	53.2	69.9	4.8	6.8
9–10	5.2	7.7	2.3	4.1	15.5	25.6	178.9	203.1	42.0	53.7	4.4	5.4
Average	7.5	19.0	4.1	10.6	22.5	63.3	155.8	189.6	41.3	79.9	5.5	6.2

Comparing the results pre- and postmodification operation and based on the soil properties presented in Table 9.3, it can be conceived that the site soil have changed from loose and medium sandy soil to a dense profile. It should be mentioned that dynamic compaction affects deep soil layers more considerably, and the shallow depths are improved insubstantially. For instance, cone tip resistance and sleeve friction within the depth of 4−5 m have improved by nearly six times, leading to improvement of other related geotechnical parameters. It is also obvious that the effect of modification diminishes at depths greater than 6 m. This matter is due to the efficacy of energy transformation down to specific depths.

9.4.3 Case No. 3: vibro replacement in Putrajaya Boulevard—Malaysia (Raju et al., 1998)

Earth embankments to heights of up to 18 m were needed for the construction of the Putrajaya Boulevard. The existence of extremely soft organic clays to a depth of 8−12 m raised a lot of issues of slope stability to the proposed embankment.

To reach sufficient safety factor against slope stability failure, stone columns were installed to improve the shear strength of the soft organic clay. Also, instead of the excavation and replacing option, ground improvement using vibro replacement was chosen as a cheaper and environmental friendly alternative. CPTu sounding data that was performed in the site and related geotechnical parameters before and after ground improvement are shown in Table 9.7. In this table, q_c is cone tip resistance, f_s is sleeve friction, u_2 is excess pore water pressure, E is elastic modulus, V is shear wave velocity, m is Janbu modulus number, and OCR is over consolidated ratio ("a": after and "b": before ground modification).

Considering the loose and organic soil profile of the site, it is noticeable that the applied soil improvement technique has increased the cone sleeve friction by 1.5 times along soil depths. An increase in the cone sleeve friction also enhances other geotechnical properties of soil. Regarding Table 9.4, the applied improvement method has made the loose soil change into the relatively stiff soil, which represents the proper efficiency of the method to improve the geotechnical parameters and resistance of the site soil. Furthermore, with an increase of the over consolidation ratio by approximately two times, the consolidation settlement of the site soil will remarkably reduce.

It can be deduced from the reviewed case studies that due to relatively high expenses of ground modification in depth, appropriate and economic design of the modification method is considerably vital. Consequently, having adequate knowledge about materials, equipment and limitations of various ground improvement methods with regard to soil profile of the project is unavoidable.

It is obvious that sufficient awareness of different soil improvement methods coupled with application of CPT for determination of soil layers profile can lead to an economic and practical method of ground improvement choice and increases its efficacy in improving soil condition.

Table 9.7 Cone penetration test data and geotechnical parameters of Putrajaya Boulevard, Malaysia, before and after ground improvement (Raju et al., 1998).

Depth	q_{cb}	q_{ca}	f_{sb}	f_{sa}	u_{2b}	u_{2a}	E_b	E_a	V_{sb}	V_{sa}				
(*m*)	(*MPa*)		(*kPa*)		(*kPa*)		(*MPa*)		(*m* /*s*)		m_b	m_a	OCR_b	OCR_a
1–2	2.5	3.7	138.2	224.3	−4.2	−8.5	17.6	30.0	129.3	161.5	30.5	44.2	14.3	27.5
2–3	3.5	4.5	197.2	277.2	−5.8	−13.5	26.0	36.4	160.0	179.5	43.3	51.3	13.8	19.9
3–4	3.7	4.8	195.5	315.0	−8.5	−15.0	25.8	38.2	160.0	184.8	41.8	52.3	9.8	14.7
4–5	3.6	5.2	211.8	340.0	−10.0	−21.4	25.1	41.6	157.8	193.6	39.8	55.8	7.4	12.5
5–6	3.7	6.4	211.8	376.5	−13.3	−17.5	26.3	51.4	161.4	217.5	40.8	67.4	6.2	12.5
Average	3.4	4.9	190.9	306.6	−8.4	−15.2	5.6	6.2	24.1	39.5	153.7	187.4	114.2	163.6

9.5 Comparison of pre- and postmodification

As CPT can provide a continuous vertical soil profile, changes of soil conditions in a ground improvement operation can be monitored through comparing CPT records before and after modification procedure. Also, soil improvement operation is a time-dependent process. Hence, it is important to monitor the soil properties after modification operation in different intervals. For instance, in Cases No. 1 and No. 2 in Fig. 9.3, the q_c profile significantly varied for pre- and postground modifications. Therefore, it is suggested to perform CPT some days, weeks, and months after ground improvement procedure to monitor the changes in soil strength and stiffness with time (Massarsch and Fellenius, 2002; Rollins et al., 2004; Eslami, 2015). In this section, a few cases are studied and the monitoring records are reported.

9.5.1 Case No. 4: explosive compaction for liquefaction mitigation (Eslami, 2015; Eslami et al., 2015)

In order to evaluate the performance of explosive compaction (EC) or blast densification (BD) as one of the deep improvement methods, a database of practical EC cases was collected by Eslami (2015) and Eslami et al. (2015) and investigated with complete information including geotechnical conditions of improved deposits, design of EC, and results of monitoring pre- and postexecution of explosion. In all of the case histories, surface settlement caused by EC is measured and the CPT data are reported before and after execution of explosion. The summary of sites information is introduced in Table 9.8. The improved geomaterials often are in the category of sand with small to medium size, hydraulic and alluvial deposits, SMs or mixture of sand and gravel, and fine percentage was about 5%–35%. Maximum underground water table level was 3 m.

For evaluation of EC effects on soil properties, q_c average values have been calculated according to CPT records pre-and postblasting. Fig. 9.3 presents CPT records before and after explosion compaction for all eight case histories.

As illustrated in Fig. 9.3, densification and decreasing liquefaction potential of soils for surface layers (2–3 m) do not happen in some cases and q_c values changed marginally. However, in deeper layers, CPT records after explosion enhanced significantly, and this means that soil in these zones is improved very well. According to Fig. 9.3, EC has been able to increase the average of initial q_c in most of the cases.

Also, corrected cone tip resistance, q_{c1N} (as can be calculated by Eq. 7.43) and normal cone tip resistance (Q_{tn}) which can be obtained by Eq. (9.1), prior and following of explosion were measured:

$$Q_{tn} = \frac{q_t}{\left(\sigma'_v\right)^{0.65}} \tag{9.1}$$

In this equation, q_t is corrected cone tip resistance due to pore pressure and σ'_v is effective vertical stress (bar). These parameters are compared to evaluate the effect

Table 9.8 Summary of case records (Eslami, 2015).

Site No.	Site	References	Soil layer	Layer thickness (m)	GWT (m)
1	Bordeaux (P6A)	Narin (1997)	Sand (SP)	10	1.5
2	Bordeaux (P8A)	Narin (1997)	Sand (SP)	10	1.5
3	Quebec SM-3Dam	Gohl et al. (2000)	Fine sand	2	3
4	South Carolina	Narsilio et al. (2009)	Fine sand	5.5	1
5	Jebba dam (Zone I)	Solymar (1984), Solymar et al. (1984)	Coarse s and—gravel	15	2
6	Jebba dam (test)	Solymar (1984), Solymar et al. (1984)	Coarse sand—gravel	15	2
7	St. Petersburg I	Minaev (1993)	Fine and coarse sand	7–7.5	1.2
8	St. Petersburg II	Minaev (1993)	Fine and coarse sand	7–7.5	1.2

of explosion on soil behavior and decreasing liquefaction potential. Results shown in Fig. 9.4 indicate promising variations due to BD in targeted depth.

In South Carolina site (Site No. 4), settlement occurred after 3 years, although cone tip resistance did not increase significantly. This phenomenon can be related to complex behavior of sands and should be considered in EC assessment and evaluation for settlement issue.

Sequence of blasting with a delay can increase densification and decrease liquefaction potential. Minaev (1993) presents two case histories in Russia (Sites No. 7 and 8). In the first case (Site No. 7) all explosion occurred simultaneously, but in second case (Site No. 8) the time lapse between explosions is between 3 and 5 min. In the first case, cone tip resistance increased 30%. But in the second case, cone tip resistance increased 80%.

Moreover, dividing the deposits thickness to the several sublayers and setting the explosive charges within the several levels will produce a uniform level of compaction comparing to the placement of all designed explosive materials in one borehole in a specific depth. It is observed that a uniform densification occurs by distribution of the explosive charge throughout the target layer.

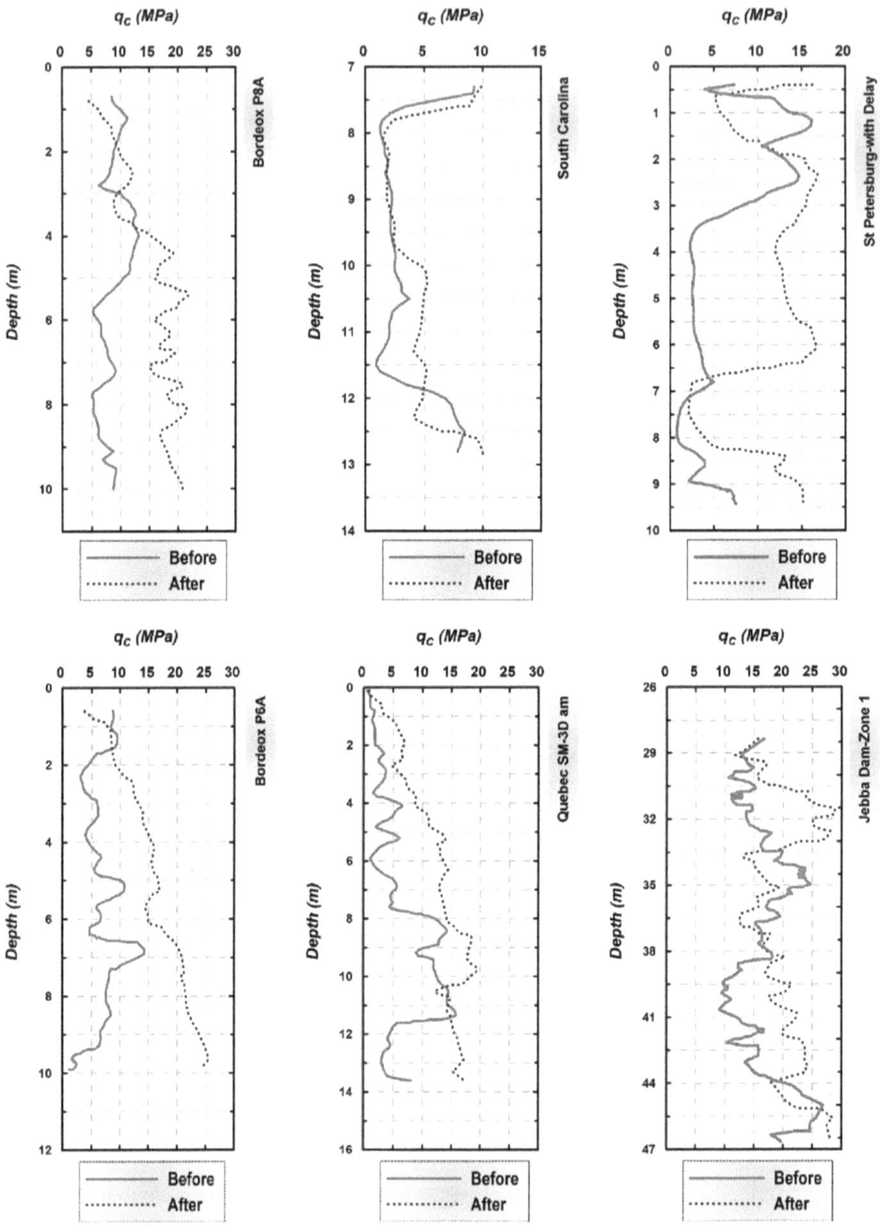

Figure 9.3 Comparison of q_c measurements before and after explosion for cases (Eslami, 2015).

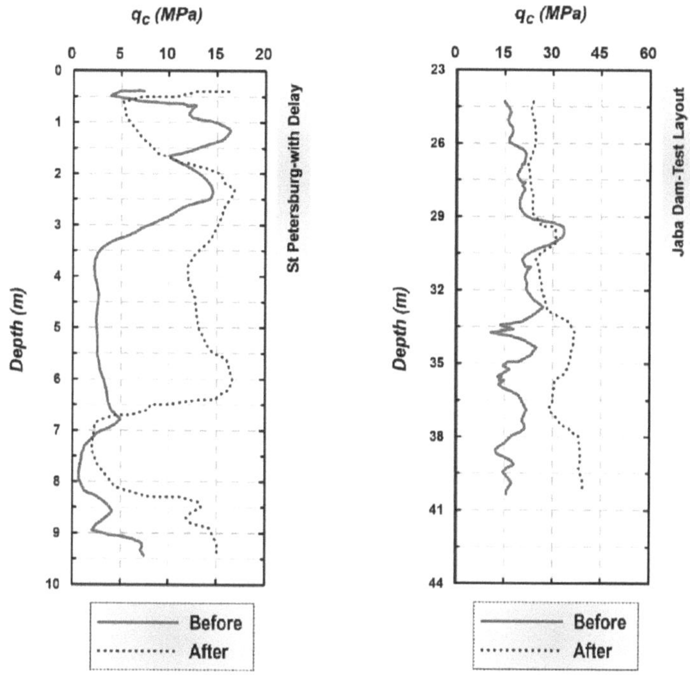

Figure 9.3 cont'd.

9.5.2 Case No. 5: vibro compaction in Hong Kong Airport—Chek Lap Kok (Massarsch and Fellenius, 2002)

The Chek Lap Kok reclaimed land was located along the seashore, prior to the placement of dredged sand fill. The fill consisted partly of calcareous material (fragments of shells and clams) and was placed by bottom dumping, where the water depth exceeded 4 m, and by spraying in shallower water depths. The sand fill was specified to contain less than 10% of silt and clay and the final thickness of the fill prior to compaction was on average about 10 m, but due to the sloping seabed, the thickness varied significantly. The groundwater level before compaction was located about 1 m below the fill surface.

The Müller resonance compaction method is used for the compaction, which incorporates a powerful vibrator that is clamped to the upper end of a flexible compaction probe. The compaction process is monitored and documented using an electronic process control system, which records various parameters for the soil densification process. The CPT results of the site before and after improvement are shown in Fig. 9.5. As it is shown in Fig. 9.5, the cone stress and the sleeve friction have increased in the sand deposit as a result of the resonance vibratory compaction. On average, the increase in cone stress is a doubling or higher, which indicates a definite densification of the sand fill. The specifications requirement of a cone stress of at least 10 MPa was satisfied. The effect of dynamic soil compaction on the stress conditions is also evidenced by the increase in sleeve friction, on average about 2.5 times, which is about the same increase ratio as that of the cone stress.

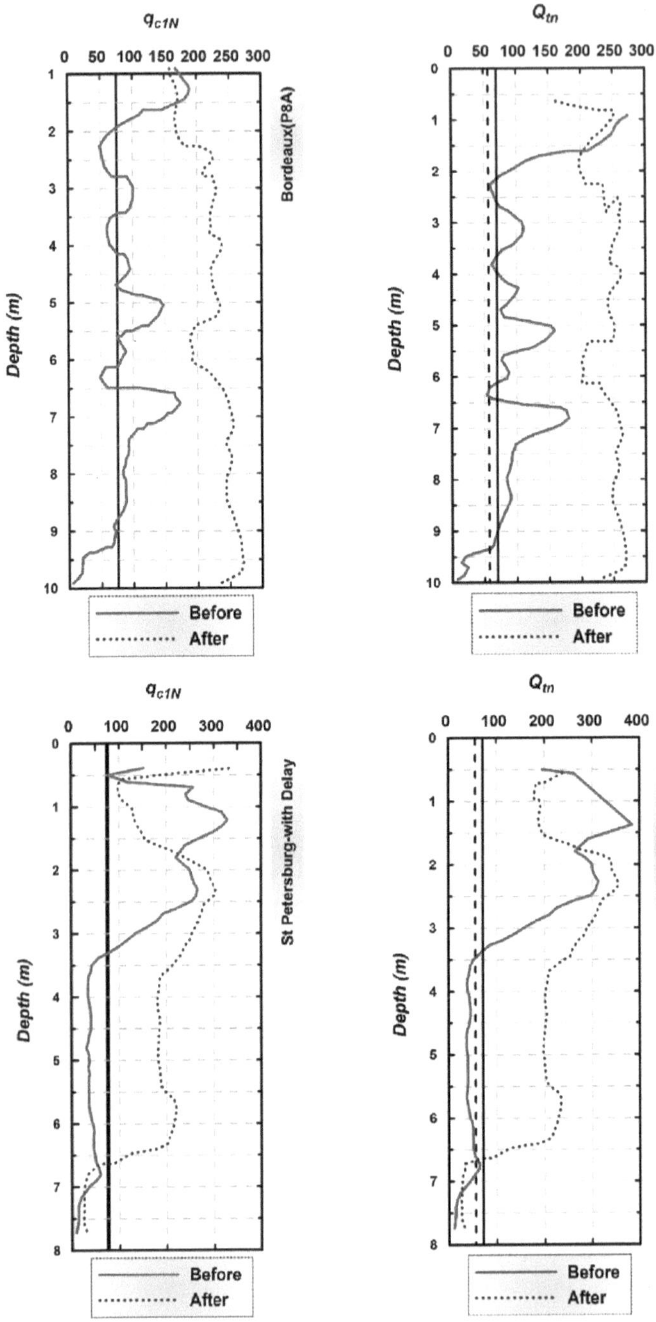

Figure 9.4 Variation of Q_{tn} and q_{c1N} before and after explosive compaction (Eslami, 2015).

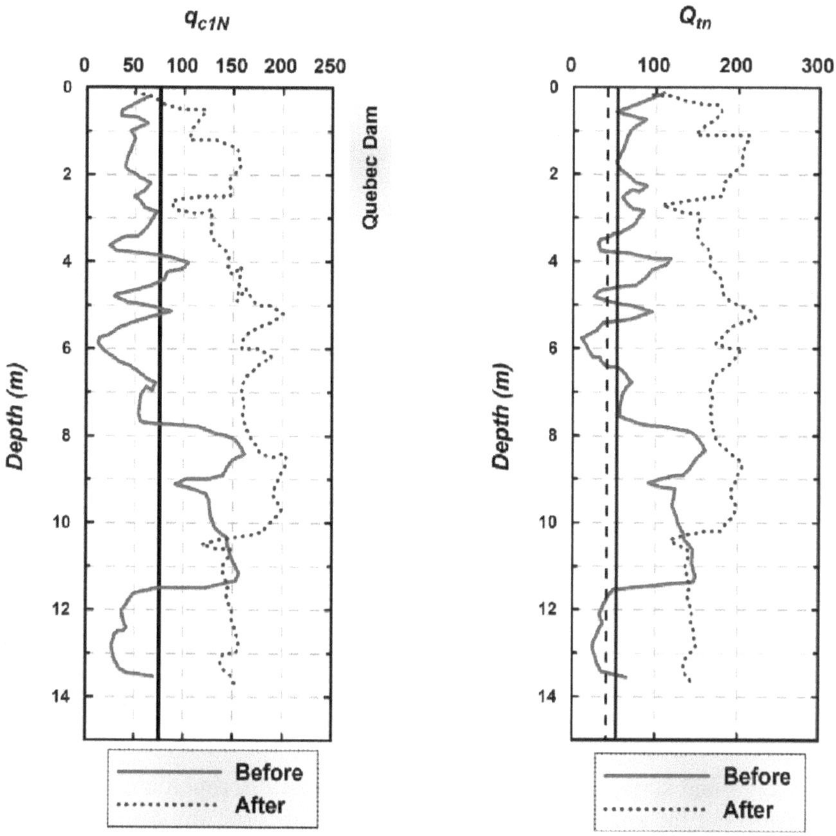

Figure 9.4 cont'd.

The results of cone and sleeve friction measurements are significantly affected by the effective overburden stress (Jamiolkowski et al., 1988). Therefore, it is necessary to consider this effect in interpreting of CPT results. For the depth adjustment of the cone stress, Massarsch (1994) proposed applying a dimensionless adjustment factor, C_M, to the cone stress according to Eq. (9.2) based on the mean effective stress σ'_m:

$$C_M = \left(\frac{\sigma_r}{\sigma'_m}\right)^{0.5} \tag{9.2}$$

where, C_M is the stress adjustment factor ≤ 2.5, σ_r is a reference stress equal to 100 kPa, and σ'_m is the mean effective stress, determined according to Eq. (9.3):

$$\sigma'_m = \frac{\sigma'_v(1 + 2K_0)}{3} \tag{9.3}$$

where, σ'_v is the vertical effective stress and K_0 is the coefficient of horizontal earth pressure (effective stress condition).

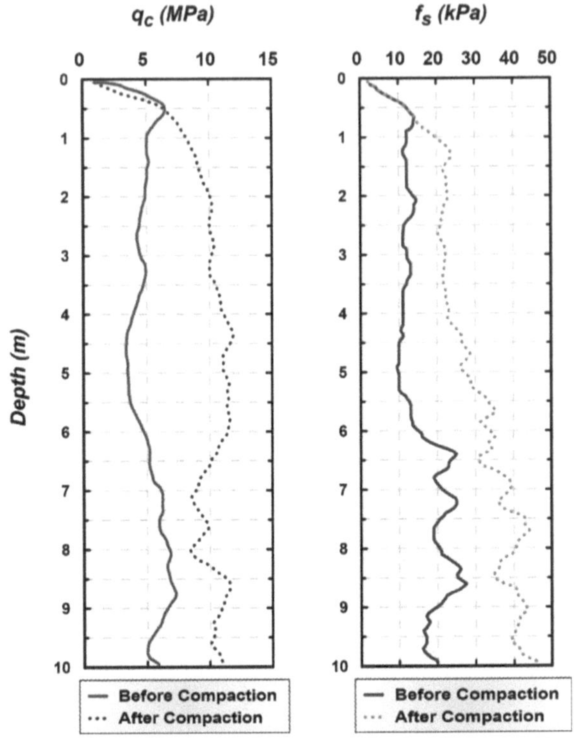

Figure 9.5 Filtered average values of cone stress and sleeve friction from before and after compaction (Massarsch and Fellenius, 2002).

Near the ground surface, values per Eq. (9.2) increase disproportionally and it is necessary to limit the adjustment factor to a value of 2.5. The stress-adjusted cone penetration stress is as Eq. (9.4):

$$q_{tM} = q_t C_M \tag{9.4}$$

where, q_t and q_{tM} are the unadjusted and stress-adjusted corrected cone stress, respectively.

Fig. 9.6 shows the filtered average unadjusted and adjusted cone stress data pre- and postcompaction. The figure clearly demonstrates the results of the extra compaction effort in the upper part of the sand fill. Such excessive compaction results in an upper soil layer that is stiffer than intended, as well as a loss of soil volume which will have to be compensated for by import and placement of additional fill material. This can be an important cost aspect for larger compaction projects.

Massarsch (1994) proposed a semiempirical correlation between the Janbu (1974) modulus number and the adjusted cone stress. The values of cone stress adjusted for depth were inserted in Eq. (9.5):

$$m = a\sqrt{\frac{q_{tM}}{\sigma_r}} \tag{9.5}$$

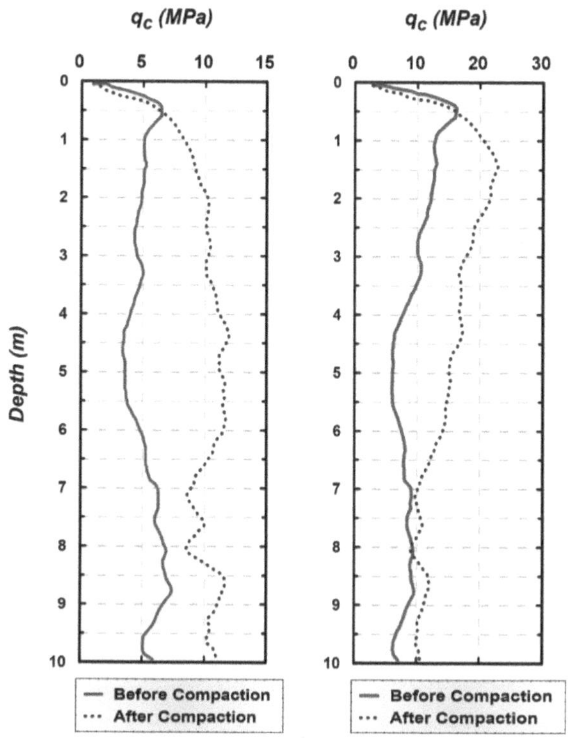

Figure 9.6 Filtered average values of cone stress from before and after compaction; unadjusted in left diagram and adjusted in right diagram (Massarsch and Fellenius, 2002).

where *a* is an empirical modulus modifier, which depends on soil type. The modulus modifier, a, has been determined from the evaluation of extensive field and laboratory data (Massarsch, 1994) and shown to vary within a relatively narrow range for each soil type. Massarsch et al. (1997) proposed the revised values for the modulus modifier between seven for organic soft silt and 45 for dense sand.

The Janbu modulus numbers before and after compaction are plotted in Fig. 9.7. Before compaction, the Janbu modulus number was about 250 near the surface and decreased slightly and approximately linearly to a value of 180 at 10 m depth. After compaction, the modulus number was 450 near the shallow depths, decreasing approximately linearly to 300 at 10 m.

9.6 Soil behavior classification approach

Due to relatively high expenses of ground modification extended in depth, appropriate and economic design of the modification method is considerably vital. Consequently, having adequate knowledge about materials, equipment and limitations of various ground improvement methods with regards to soil profile of the project is unavoidable. It is obvious that sufficient awareness of different soil improvement methods coupled

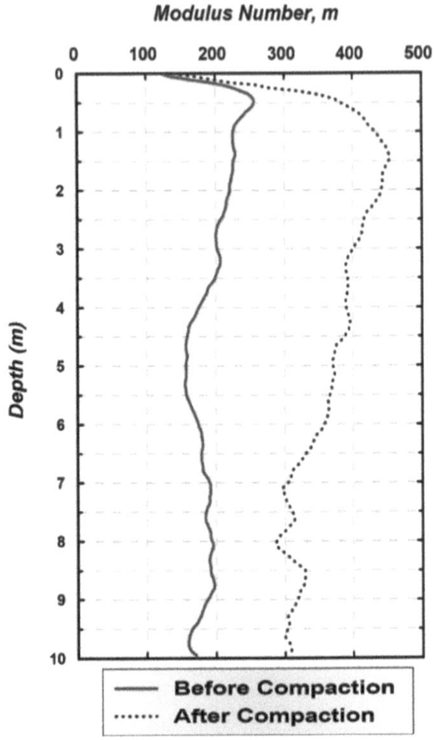

Figure 9.7 Janbu modulus number, m, before and after compaction (Massarsch and Fellenius, 2002).

with application of CPT for determination of soil layers profile can lead to an economic and practical method of ground improvement choice.

Soil behavior classification (SBC) is the third CPT-based approach employed in soil stabilization practice. In evaluation and design procedure, where the CPT data points lie on classification charts is a significant indicator of current soil condition. If in general, Fig. 9.8 is realized as a typical SBC, the soft and weak soils are concentrated in Zones 1 to 3; while Zones 4 and 5 are attributed to stronger or stiffer deposits.

Most of the SBC charts or diagrams described in Chapter 5 are more or less consistent with this trend. However, some charts may have more subdivisions. Hence, having the CPT data pre- and postmodification and comparing them visualize the alternation of soil behavior, both qualitatively and quantitatively. Accordingly, a few cases are reviewed upon this approach to investigate variations of soil properties.

9.6.1 *Case No. 6: Ground improvement and foundation practice for Persian Gulf Bridge—Iran (Asadi et al., 2017)*

In recent years, several investigations have been performed on Persian Gulf subsoil condition and available difficulties for construction of offshore structures. National Persian Gulf Bridge is being constructed to connect the Qeshm Island to main lands,

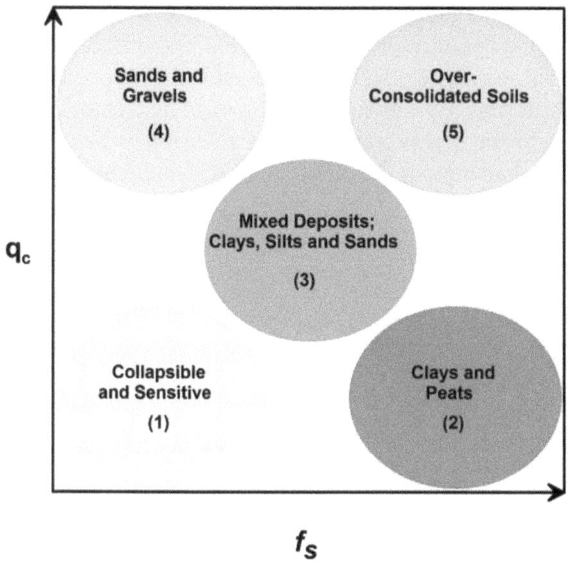

Figure 9.8 Schematic chart for soil behavior classification.

i.e., Bandar Abbass Harbor in Persian Gulf. The bridge length is 3430 m, the height between sea level and lower part of the bridge deck is 25 m. The bridge structure type includes piles, piers, concrete deck, and the parts inside the sea is planned to install on concrete caisson foundation (Petro Sahel Report, 2012).

Eighteen boreholes have been performed in order to investigate the geotechnical conditions of the site through SPT and CPT. The longitudinal profile of the site is shown in Fig. 9.9.

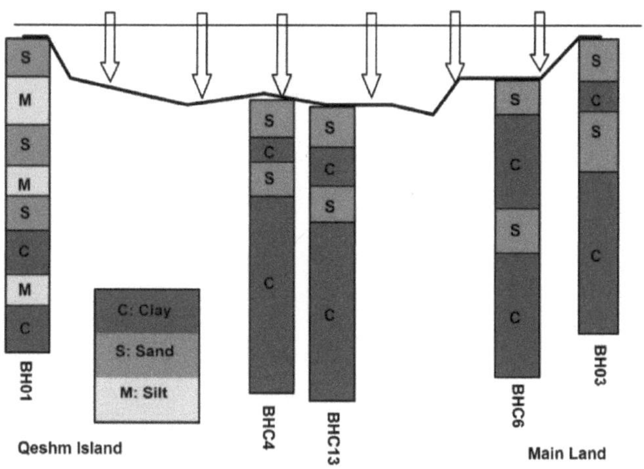

Figure 9.9 Longitudinal profile of the site and the excavated boreholes (Asadi et al., 2017).

The results from the SPT and CPT are indicated in Fig. 9.10. The analyses indicate that the soil layers near the main land coastline are highly variable. Subsurface profile in the upper part 6 m of this area includes a very soft silt layer and in the deeper depths there are layers with variable resistances. By moving from the main land coastline to the deeper parts of the sea, the sediment grains sizes will be greater. They are mostly the loose sands type or silt and in some regions are consolidated clays. In these areas,

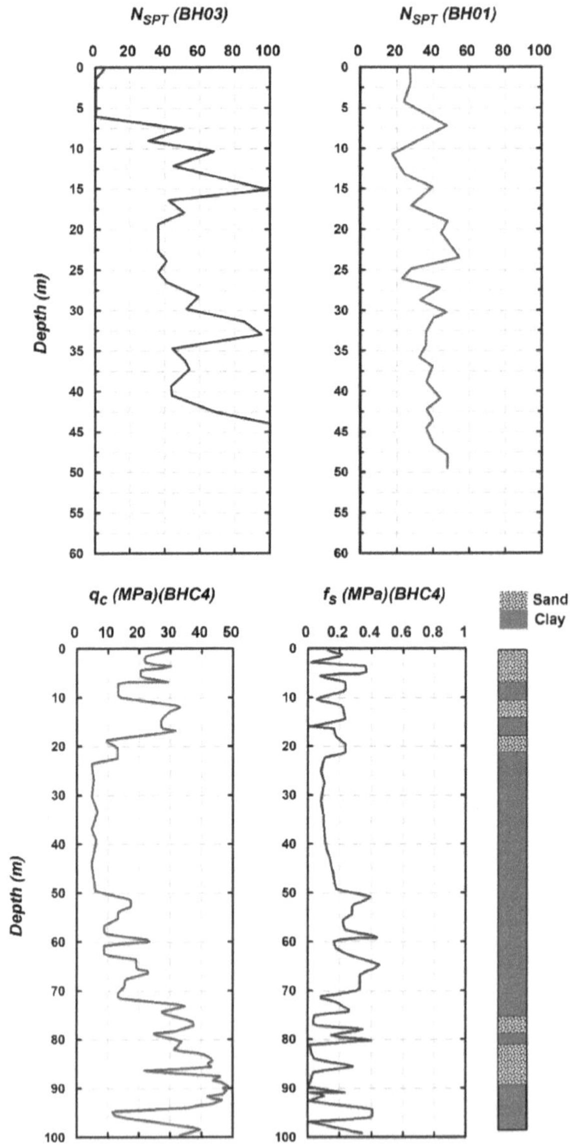

Figure 9.10 Standard penetration test and cone penetration test results close to main land coastline and the Qeshm Island coastline (Asadi et al., 2017).

the resistance of surface sediments is in appropriate range and by moving into the depths the resistance is reduced.

Behavior of subsoil is classified using the CPT records in deep zones of the sea. For this purpose, the SBC charts proposed by Campanella et al. (1986a,b), Eslami and Fellenius (1997), and Robertson (2010) are used, as shown in Fig. 9.11. The studies of Asadi et al. (2017) show that the subsoil between the Qeshm Island and main land is mostly made of sandy materials with varied resistances and in some parts, made of carbonate clays. By moving toward the Qeshm Island, the sediments are more uniform in terms of resistance and mostly clay and layered silt with almost high resistance.

Due to serious loading from the mentioned bridge pronounced by hydrodynamic and earthquake impacts and some subsoil weaknesses, foundation design and construction are important tasks. Choosing appropriate ground improvement methods to stabilize the soil conditions as a supplementary, minimizes the instability problems and prevent the implementation of a large number of deep foundations.

According to the suggested diagram by Mitchell and Gallagher (1998) (Fig. 9.1) and the results of the field studies, the appropriate methods of the ground improvement have been suggested by Asadi et al. (2017) for construction of this bridge. In the zones

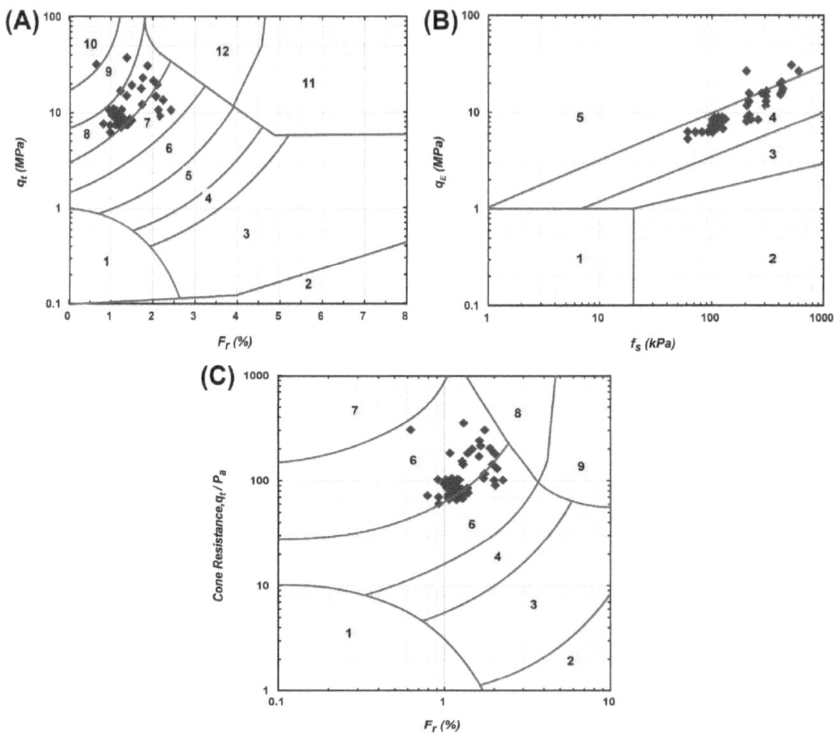

Figure 9.11 Subsoil Behavior Classification in deep parts of the sea using charts of (A) Campanella et al. (1986a,b), (B) Eslami and Fellenius (2004), and (C) Robertson (2010) (Asadi et al., 2017).

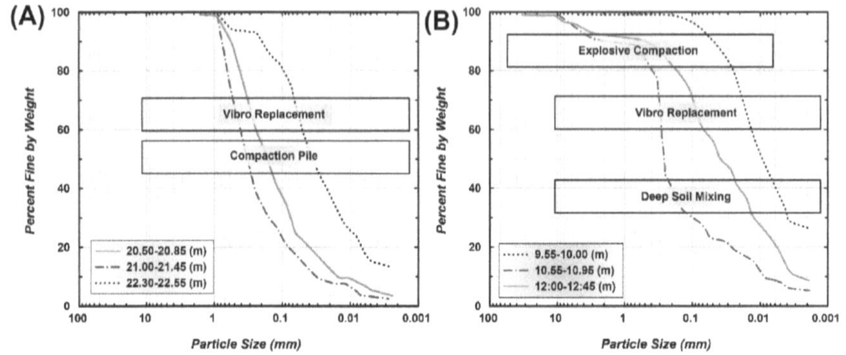

Figure 9.12 Particle size distribution of the soil for BH01 close to the Qeshm coastline and (B) Particle size distribution of the soil for BH03 close to main land (Asadi et al., 2017).

close to the Qeshm Island and in the deeper parts of the sea, the soil is made of loose sands and some depths, with SMs, which is liquefaction prone in depths of 20–40 m (as it is shown in Figs. 9.11 and 9.12). Therefore, vibro replacement, constructing the stone columns, and sand compaction piles can be proposed as appropriate methods to improve the geotechnical characteristics of the areas near to the Qeshm Island. As the areas susceptible to liquefaction close to the main land are almost made of loose sands, EC, deep soil mixing (DSM), and vibro replacement methods seem to be appropriate to compact the subsoil close to the main land.

9.6.2 Case No. 7: explosive compaction project (Eslami, 2015; Eslami et al., 2015)

As presented in Section 9.5, Case No. 4, a database of practical EC cases was collected by Eslami (2015) and Eslami et al. (2015) which the cases information are presented in Table 9.8. In Case No. 4, mentioned sites were studied from the view of pre- and post-modification comparison and in this section, soil classification charts for a better assessment of soil behavior and liquefaction potential of these sites before and after EC are studied. Obtained results are presented in Fig. 9.13.

According to Douglas and Olsen (1981) and Campanella et al. (1986a,b) charts after explosion, soil classification zone changes to an upper one, and this means that soil resistances increase and liquefaction potential decreases. Soil deposits in all of the cases before densification which contain sensitive and noncohesive soils. However, after EC, soils are not in the metastable zones, anymore.

Based on Robertson (1990) and Robertson et al. (1986) methods for soil classification, before explosion soils belong to the Zones 4–6 which include loose sands and after ground modification, soil location shifts to the Zones 7 and 8 which are in the category of dense and very dense sands. This results show that modified soils are excluded from liquefiable soils category.

As presented in Jefferies and Davies (1991) diagram, modified soils are in SMs and dense sands category (Zones 5 and 6) which have less liquefaction potential. Before modification, soils were collapsible and unstable, but after EC instability of soils

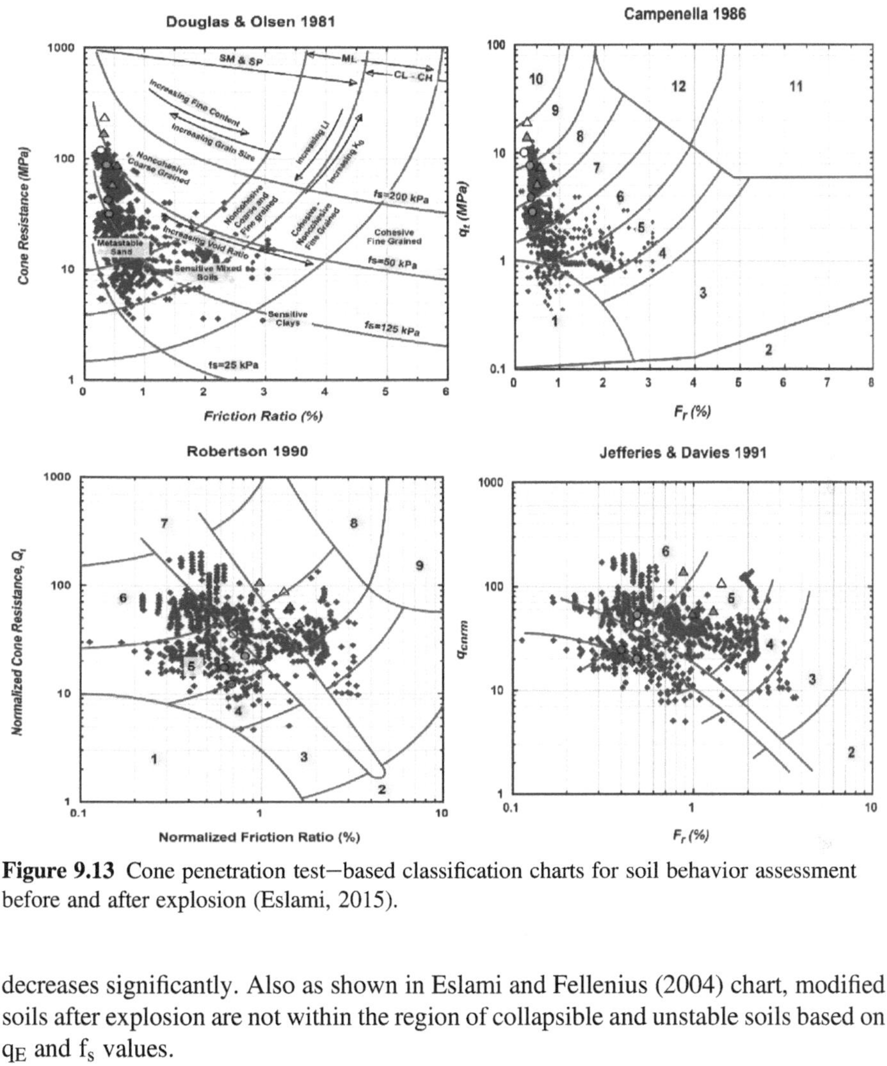

Figure 9.13 Cone penetration test–based classification charts for soil behavior assessment before and after explosion (Eslami, 2015).

decreases significantly. Also as shown in Eslami and Fellenius (2004) chart, modified soils after explosion are not within the region of collapsible and unstable soils based on q_E and f_s values.

9.6.3 Case No. 8: Crude oil contaminated clayey sand behavior using CPT data (Soroush Hagh, 2019)

In order to investigate the behavior of clayey sand contaminated by crude oil, a physical modeling study was performed using CPT data. For this purpose, a CPT system has been constructed in Amirkabir University of Technology (AUT) in 2017 and 2018. The CPT used for testing has 3.57 cm diameter, 15 cm cone height, and 1.21 m total length. Mixed soil (combination of sand and different percentage of clay, i.e., 10%, 30%, and 50% of Kaolinite clay) contaminated by crude oil and without any contaminant materials were studied.

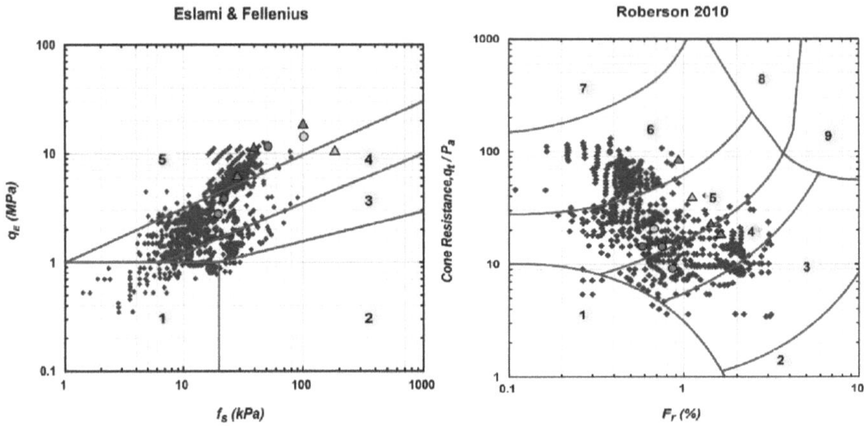

Jebba Dam (Zone 1) after Explosion	▲	Jebba Dam (Zone 1) befor Explosion	●
Jebba Dam (Test) after Explosion	△	Jebba Dam (Test) befor Explosion	○
St. Petersburg I (Without delay) after explosion	▲	St. Petersburg I (Without delay) befor explosion	●
St. Petersburg II (With delay) after explosion	▲	St. Petersburg II (With delay) befor explosion	●

Figure 9.13 cont'd.

Initially, each of the three soil types with the optimum moisture content was tested to determine the maximum strength of the soil in the highest density. Then, each of the samples was mixed with 4% and 8% crude oil. Subsequently, the soil samples with different percentages of the crude oil at their optimum moisture content were investigated to determine the long-term effect of the crude oil penetration into the soil through CPT. In Fig. 9.14, the CPT profiles are presented for the samples with different clay percentages, with and without crude oil.

Also, Eslami and Fellenius (1997) and Campanella et al. (1986a,b) charts have been used to analyze soil behavior according to CPT results as shown in Figs. 9.15–9.17. A number is reported for each data point in these figures which represents clay percentage, optimum water content as well as crude oil percentage. The extra fourth number for some data points indicate the additional water percentage added to the soil sample in order to consider the effect of additional moisture due to precipitation and runoff. For example, the data point in Fig. 9.15 with the number of 10.5.4 indicates the sample of sand mixed with 10% clay, the optimum moisture content is 5%, corresponding to 4% of the crude oil. Similarly, the number 10.5.4.4 represents the same sample with 4% additional water to the soil.

The obtained CPT results indicate that the cone tip resistance and sleeve friction decreased with the increase in the amount of crude oil and/or excess moisture content of the soil sample. On the other hand, the effect of crude oil on the cone sleeve friction

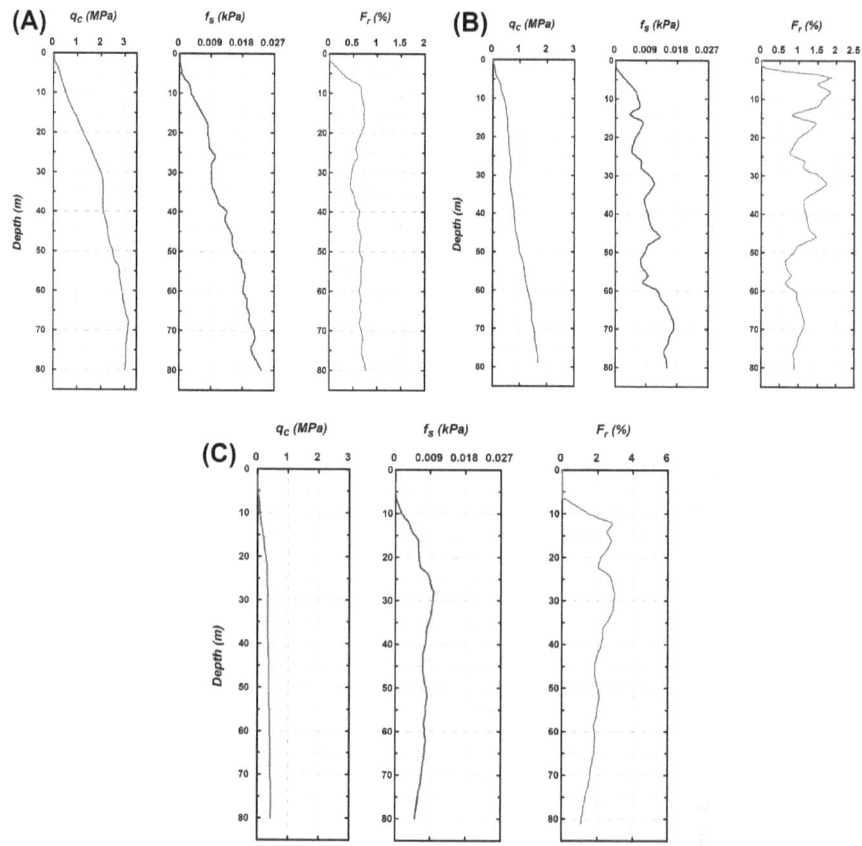

Figure 9.14 Cone penetration test data for sand with (A) 10% clay, 9% water, and 0% (without) crude oil; (B) 30% clay, 12% water, and 4% crude oil; and (C) 50% clay, 8% water, and 8% crude oil (Soroush Hagh, 2019).

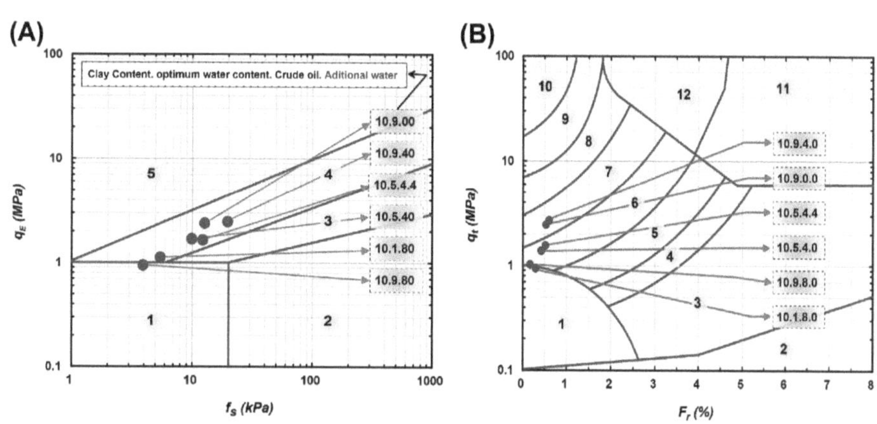

Figure 9.15 The soil behavior classification changes trend of sandy soil mixed with 10% clay based on (A) Eslami and Fellenius (1997) chart and (B) Campanella et al. (1986a,b) chart (Soroush Hagh, 2019).

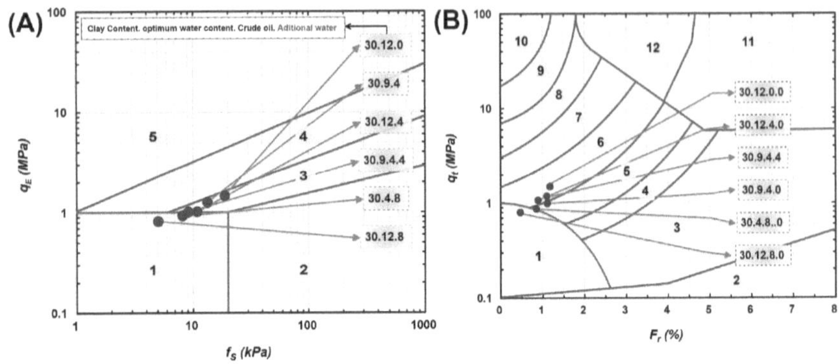

Figure 9.16 The soil behavior classification changes trend of sandy soil mixed with 30% clay based on (A) Eslami and Fellenius (1997) chart and (B) Campanella et al. (1986a,b) chart (Soroush Hagh, 2019).

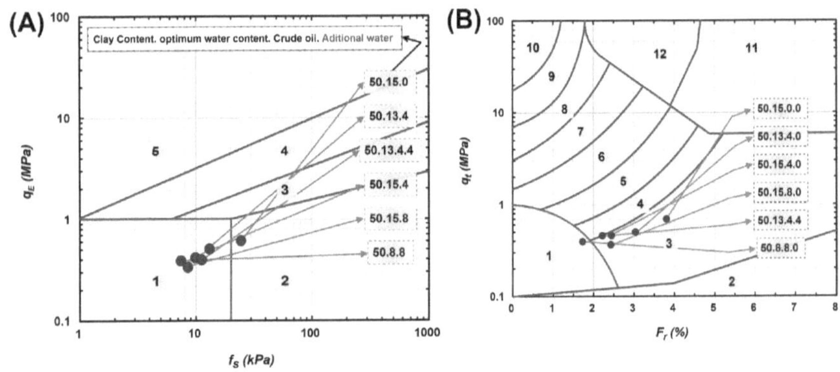

Figure 9.17 The soil behavior classification changes trend of sandy soil mixed with 50% clay based on (A) Eslami and Fellenius (1997) chart and (B) Campanella et al. (1986a,b) chart (Soroush Hagh, 2019).

values is greater than those of cone tip resistance. Also, this decrease in cone sleeve friction becomes more significant as the clay percentage in samples increase, which is due to crude oil effect on soil cohesion.

References

Asadi, F., Eslami, A., Valikhah, F., 2017. Ground improvement and foundation practice for Persian Gulf bridge (causeway); Bandar Abbas Harbor–Qeshm island. Marine Georesources & Geotechnology 35 (4), 538–547.

Baldi, G., Bellotti, R., Ghionna, V., Jamiolkowski, M., Pasqualini, E., 1986. Interpretation of CPT's and CPTU's. 2nd part: drained penetration. In: Proceedings of 4th International Geotechnical Seminar, Singapore, pp. 143–156.

Bergado, D.T., Anderson, L.R., Miura, N., Balasubramaniam, A.S., 1996. Soft Ground Improvement in Lowland and Other Environments. American Society of Civil Engineers, USA.

Bowles, J.E., 1996. Foundations Analysis and Design, fifth ed. McGraw- Hill, New York.

Campanella, R.G., Robertson, P.K., Gillespie, D., 1986a. Seismic cone penetration test. In: Proceeding of in Situ 86, ASCE Specialty Conference. Blacking, Virginia, pp. 116–130.

Campanella, R.G., Robertson, P.K., Gillespie, D., Grieg, J., 1986b. Use of piezometer cone data. In: Proceedings of American Society of Civil Engineering, ASCE, In-Situ 86 Specialty Conference, pp. 1263–1280. Publication GSP No. 6.

Douglas, B.J., Olsen, R.S., 1981. Soil classification using electric cone penetrometer. In: Symposium on Cone Penetration Testing and Experience, ASCE. Geotechnical Engineering Division, St. Louis, pp. 209–227.

Eslami, A., 2015. Investigation of explosive compaction (EC) for liquefaction mitigation using CPT records. Bulletin of Earthquake Engineering 13 (11), 3239–3257.

Eslami, A., Fellenius, B., 1997. Pile capacity by direct CPT and CPTu methods applied to 102 case histories. Canadian Geotechnical Journal 34, 886–904.

Eslami, A., Fellenius, B.H., 2004. CPT and CPTu data for soil profile interpretation: review of methods and a proposed new approach. Iran Journal of Science and Technology, Transection B 28 (B1), 69–86.

Eslami, A., Mehrdad, M.A., Khosravi, M.T., 2010. Qualitative control of geotechnical parameter of improved soil by CPT and CPTu. Sharif Journal Vol. 51, pp. 49–59.

Eslami, A., Pirouzi, A., Omer, J.R., Shakeran, M., 2015. CPT-based evaluation of Blast Densification (BD) performance in loose deposits with settlement and resistance considerations. Geotechnical & Geological Engineering 33 (5), 1279–1293.

Eslami, A., Sekhavatian, A., 2014. Geotechnical engineering: principles, investigations and interpretations. Amirkabir University of Technology Publication, In Persian.

Eslami, A., Valikhah, F., Veiskarami, M., Salehi, M., 2017. CPT-based investigation for pile toe and shaft resistances distribution. Geotechnical & Geological Engineering 35 (6), 2891–2905.

FHWA-SA-98–086 R, 2001. Ground Improvement Technical Summaries. In: Prepared for US Department of Transportation, Federal Highway Administration, Vol: vols. I and II.

Gohl, W.B., Jefferies, M.G., Howie, J.A., Digge, D., 2000. Explosive compaction: design, implementation and effectiveness. Géotechnique 50 (6), 657–667.

Han, J., 2015. Principles and Practices of Ground Improvement. John Wiley & Sons.

Han, J., Bhandari, A., Wang, F., 2012. DEM analysis of stresses and deformations of geogrid-reinforced embankments over piles. International Journal of Geomechanics 12 (4), 340–350.

Hausmann, M.R., 1990. Engineering Principles of Ground Modification. McGraw Hill International Edition, 632pp.

Jamiolkowski, M., Ghionna, V.N., Lancelotta, R., Pasqualini, E., 1988. New correlations of penetration tests for design practice. In: DeRuiter, J., Balkema, A.A. (Eds.), Proceedings of the 1st International Symposium on Penetration Testing, ISOPT-1, Orlando, Fla., 20–24 March, ISBN 90 6191 801 4, pp. 263–296. Rotterdam.

Janbu, N., 1974. Relationship between tangent modulus and cone resistance. In: Proceedings of the European Symposium on Cone Penetration Testing, vol. 2 (1). ESOPT, pp. 132–134.

Jefferies, M.G., Davies, M.O., 1991. Soil classification by the cone penetration test: discussion. Canadian Geotechnical Journal 28 (1), 173–176.

Kulhawy, F.H., Mayne, P.W., 1990. Manual on Estimating Soil Properties for Foundation Design. Electric Power Research Inst., Palo Alto, CA (USA). Cornell Univ., Ithaca, NY (USA). Geotechnical Engineering Group.

Ladd, C.C., Foott, R., Ishihara, K., Schlosser, F., Poulos, H.G., 1977. Stress-deformation and strength characteristics. In: S.O.A. Report, 9th ICSMFE, Tokyo, vol. 2, pp. 421–494.

Look, B., 2007. Handbook of Geotechnical Investigation and Design Tables. Taylor and Francis Group.

Massarsch, K.R., 1994. Settlement analysis of compacted fill. In: Proceedings of the 13th International Conference on Soil Mechanics and Foundation Engineering, vol. 1. ICSMFE, New Delhi, pp. 325–328.

Massarsch, K.M., Fellenius, B.H., 2002. Dynamic compaction of granular soils. Canadian Geotechnical Journal 39 (3), 695–709.

Massarsch, K.R., Westerberg, E., Broms, B.B., 1997. Footings supported on settlement-reducing vibrated soil nails. In: Proceedings of the 15th International Conference on Soil Mechanics and Foundation Engineering, vol. 97 (1). ICSMFE, Hamburg, pp. 1533–1539.

Minaev, O.P., 1993. Effective compaction of water-saturated soils by blasting translated from Russian. Osnovaniya Fundamenty Mekhanika Gruntov 30 (2), 17–19.

Mitchell, J.K., Gallagher, P.M., 1998. Engineering Guidelines on Ground Improvement for Civil Works Structures and Facilities. U.S. Army Corp of Engineering Division Directorate of Civil Works, Washington, DC.

Narin, W.A., 1997. Investigation of the mechanisms and predictive methodologies for explosive compaction. In: Civil and Environmental Engineering. Univ. of California at Berkeley, Berkeley, Calif.

Narsilio, G.A., Carlos Santamarina, J., Hebeler, T., Bachus, R., 2009. Blast densification: multi-instrumented case history. Journal of Geotechnical and Geoenvironmental Engineering 135 (6), 723–734.

Petro Sahel Report, 2012. Analysis and Design of Approach Bridges-Roadway Substructure. Petro Sahel Persian Gulf Development (letter No. 10105/72 Amirkabir University of Technology).

Raju, V.R., Wenger, R., Godenzie, D., 1998. Ground improvement using vibro techniques-case histories from S.E. Asia. In: Ground Improvement Conference, Singapore, 1998.

Robertson, P.K., Campanella, R.G., Gillespie, D., Grieg, J., 1986. Use of piezometer cone data. Proceedings of American Society of Civil Engineers, ASCE, In-Situ 86 Specialty Conference. In: Clemence, S. (Ed.), Geotechnical Special Publication GSP No. 6, Blacksburg, pp. 1263–1280.

Robertson, P.K., 1990. Soil classification using the cone penetration test. Canadian Geotechnical Journal 27 (1), 151–158.

Robertson, P.K., 2010. Soil behavior type from CPT. In: 2nd International Symposium on Cone Penetration Testing, Huntington Beach, CA, USA, May 2010.

Robertson, P.K., Wride, C.E., 1998. Evaluating cyclic liquefaction potential using the cone penetration test. Canadian Geotechnical Journal 35, 442–459.

Rollins, K., Anderson, J.K.S., Goughnour, R.R., Mccain, A.K., 2004. Liquefaction hazard mitigation using vertical composite drains. In: 13th World Conference on Earthquake Engineering, Vancouver, B.C., Canada, August 1-6, 2004. No. 2880.

Senneset, K., Sandven, R., Janbu, N., 1989. Evaluation of Soil Parameters from Piezocone Tests. Transportation Research Record 1235. National Research Council, Washington, DC, pp. 24–37.

Solymar, Z.V., 1984. Compaction of alluvial sands by deep blasting. Canadian Geotechnical Journal 21, 305–321.

Solymar, Z.V., Ioabachie, B.C., Gupta, R.C., Williams, L.R., 1984. Earth foundation treatment at Jebba dam site. Journal of Geotechnical and Geoenvironmental Engineering 110 (10), 1415–1430.
Soroush Hagh, A., 2019. The Effect of Crude Oil Contamination on Soil Behavior Classification (SBC) of Clayey Sand Based on Cone Penetration Test (CPT). PhD Dissertation. Amirkabir SSUniversity of Technology (AUT), School of Civil and Environmental Engineering.

CPT-based soft computing implementations in foundation engineering

10

10.1 Introduction

In geotechnical practice, some laboratory tests are accompanied by several difficulties during their performance including sample preparation and transportation, size limitations, inconsistency of outputs, and keeping real field stresses unchanged during soil sampling. Therefore, in situ tests, such as the cone penetration test (CPT), which do not involve these constraints, can be a more straightforward alternative to determine soil properties. Indeed, both laboratory and in situ procedures can be employed to provide complementary considerations in geotechnical and, particularly, foundation engineering practice.

The main objective of this chapter is to present methods that can predict the more common geotechnical parameters in foundation engineering more precisely. A significant advantage of the soft computing techniques is the nonlinear behavior of multivariate dynamic system which can be efficiently computed with no a priori assumption with regard to the distribution of the data (Narimanzadeh et al., 2003).

Soft computing techniques, such as artificial neural networks (ANNs) and support vector machines (SVMs), can be regarded as efficient substitutes for the traditional regression methods previously used in data handling (empirical correlations).

Such methods have been used in geotechnical engineering with various degrees of success. Although soft computing techniques can overcome this problem, their main drawback is their disability to present a convenient relation so as to be useable by others. However, this problem is resolved with the proposed group method of data handling (GMDH)-type neural network (NN) (Atashkari et al., 2007; Jamali et al., 2009; MolaAbasi et al., 2019). GMDH and SVM approaches as candidates for soft computing techniques are briefly presented as follows.

10.2 Review of GMDH-type neural network method

GMDH is a self-organizing approach that can help complicated models to be regularly generated depending on their performance analysis on multiinput data ($X_{i,1}$, $X_{i,2}$, $X_{i,3}$, ...) single-output records (Y_1, Y_2, Y_3, ...). *GMDH* can model complex systems without needing to recognize them. The main hint of GMDH as a feed-forward network is to present an analytical function based on a quadratic node transfer function whose coefficients are acquired by multiregression method (Sheikholeslami et al., 2014).

Piezocone and Cone Penetration Test (CPTu and CPT) Applications in Foundation Engineering
https://doi.org/10.1016/B978-0-08-102766-0.00010-9

Regarding CPT records, *GMDH* aims at distinguishing functional structure of a model with hidden layer. When using *GMDH* method, a model will be consisting of a set of neurons, in a way that different pairs distributed in each layer are connected through a quadratic polynomial. Hence, new neurons will be produced in the next layer. Such a model can be used to map inputs to outputs (Madandoust et al., 2010).

The following stages encapsulate the training processes of GMDH method (MolaAbasi et al., 2019):

- Create the first layer of the network. The number of neurons is determined using $\frac{n(n-1)}{2}$, each of which is a partial description. A layer created for a four-input network is depicted in Fig. 10.1A.
- Carry out the performance evaluation and remove the neurons providing deficient results. Fig. 10.1B illustrates the removed neurons in a lighter color.
- Create one more layer which is considered as the inputs of the outputs of the previous layer. The number of neurons on this layer is also computed using $\frac{n(n-1)}{2}$ (Fig. 10.1C).
- Train and select the neurons on the new layer. After selection, add another layer (Fig. 10.1D).
- If any layer remains with just one neuron, after the selection process, stop training (Fig. 10.1E).
- If in any layer, except layer one, the best performance value is poorer than that of the previous layer, also stop training. In this case, the algorithm chooses the best neuron in the remaining last layer and removes its other nodes.
- Remove all neurons in all previous layers which do not contribute to the output of the network (Fig. 10.1F).

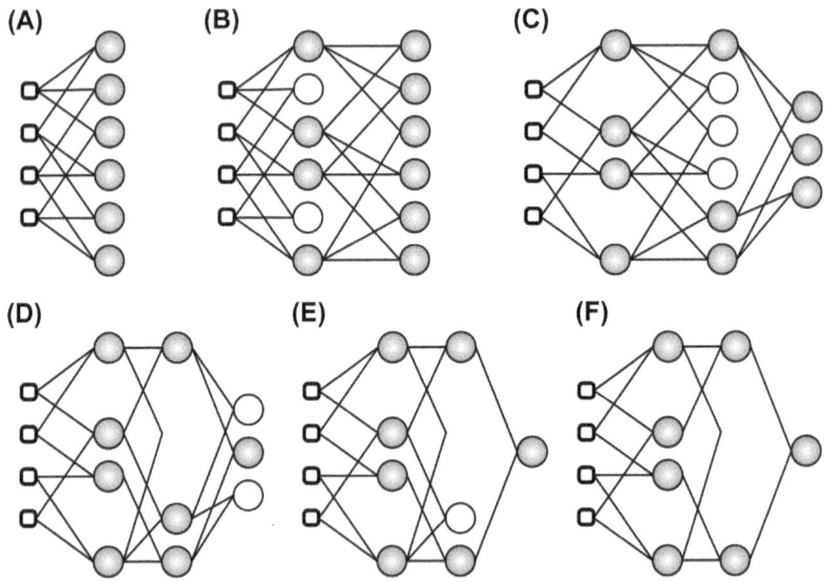

Figure 10.1 Graphical training process of group method of data handling.

10.3 Review of support vector machine method

The SVM, which combines the standards of structural risk minimization and the statistical learning theory, has been fruitfully employed in solving classification and functional regression problems in geotechnical applications. SVMs are useful approaches to solve classification and regression problems. They are rooted in the statistical learning theory. This idea was first proposed by Vapnik (1995), on the basis of structural risk minimization, and then developed in more details. Structural risk minimization mainly aims at selecting the optimum model for the existing training patterns (Scholkopf and Smola, 2001).

One of the main disadvantages of most neural networks like multilayer perceptron is their low generalizability when accessing high accuracy. Whereas the SVM has a high generalization capability and can balance the accuracy and generalization (Schölkopf et al., 1999). In addition, the SVM does not involve the learning algorithm and convergence problems which are associated with ordinary neural network.

In the SVM, the learning result will be a hyperplane by which the data will be divided with the maximum safety margin and minimum error. Solving the optimization Eq. (10.1) is equal to obtaining the mentioned hyperplane for the classes ω_1 and ω_2.

$$\begin{aligned} &\min_{W,\xi} \frac{1}{2}\|W\|^2 + C\sum_{i=1}^{n} \xi_i \\ &s.t. d_i.\left(W^T x_i + b\right) \geq 1 - \xi_i \quad , \ where\, d_i = \begin{cases} 1 for\, \omega_1 \\ -1 for\, \omega_2 \end{cases} \\ &\xi_i \geq 0\, \forall i \end{aligned} \tag{10.1}$$

where x_i is ith pattern, C indicates the coefficient of determining soft margin (box constraint), W is the weight vector, and b indicates the hyperplane bias. Moreover, ξ_i indicates the slack variable corresponding to ith sample which is considered for handling noisy or linearly nonseparable data. To solve Eq. (10.1), its dual equation considering the Lagrange function L (λ) and the Karush–Kuhn–Tucker conditions is formulated as follows (Abe, 2005):

$$\begin{aligned} &\max_{\lambda} L(\lambda) = \sum_i \lambda_i - \frac{1}{2}\sum_i \sum_j \lambda_i \lambda_j d_i d_j x_i^T x_j \\ &s.t.\, 0 \leq \lambda_i \leq C,\ \forall i \quad , \ where\ d_i = \begin{cases} 1 for\, \omega_1 \\ -1 for\, \omega_2 \end{cases} \\ &\sum_i \lambda_i d_i = 0 \end{aligned} \tag{10.2}$$

where λ_i is the Lagrange coefficient corresponding to the constraint of ith pattern. Solving the above dual equation, the optimum solution of Eq. (10.1) will be achieved. The obtained decision surface from the aforementioned equation is a linear decision surface. This equation has been also extended for the nonlinear decision surfaces using kernel functions; and in this way, decision surfaces for the linearly nonseparable equations is obtainable (Schölkopf et al., 1999). Considering the mapping $x_i \rightarrow \varphi(x_i), \forall i$, the optimization Eq. (10.1) can be rewritten for the kernel space. If the kernel function is $K(x_i, x_j) = \varphi(x_i)^T \varphi(x_j)$, then the dual of this equation in the kernel space can be formulated as follows:

$$\max_{\lambda} L(\lambda) = \sum_i \lambda_i - \frac{1}{2} \sum_i \sum_j \lambda_i \lambda_j d_i d_j K(x_i, x_j)$$

$$s.t. \; 0 \le \lambda_i \le C, \; \forall i \qquad \sum_i \lambda_i d_i = 0 \; , \; where \; d_i = \begin{cases} 1 \, for \, \omega_1 \\ -1 \, for \, \omega_2 \end{cases} \tag{10.3}$$

By solving the optimization Eq. (10.3), the solution of SVM equation can be obtained in the kernel space, and the nonlinear decision surface for the considered patterns can be determined.

It is worthy to note that the function $\varphi(x_i)$ can be either specified or unspecified, but the kernel function $K(x_i, x_j)$ should be specified to solve the equation as described in Khan et al. (2016). Selection of the appropriate kernel function is important, and it can affect the accuracy and the generalization performance of the classifier. Some of the most well-known kernel functions are provided in Abe (2005).

Basic SVM is a binary classifier and designed for classifying a two-class problem. However, it can be modified to solve multiclass and regression problems. SVMs designed for regression problems are usually called support vector regression (SVR) (Drucker et al., 1997).

In the following sections, pile-bearing capacity, shallow foundation settlement, internal friction angle of marine deposits, shear wave velocity, and liquefaction potential predicted by GMDH-type NN and SVM have been presented.

10.4 Case Study: pile-bearing capacity

The relation between total and shaft pile capacity and CPT/CPTu records has been investigated and presented via two case studies, namely GMDH and SVR approaches.

10.4.1 Case Study No. 1: pile-bearing capacity—GMDH method

Hassani (2011) applied GMDH-type NN to predict pile-bearing capacity (R_u) in a research work, based on CPT records. To validate the ability of GMDH models to

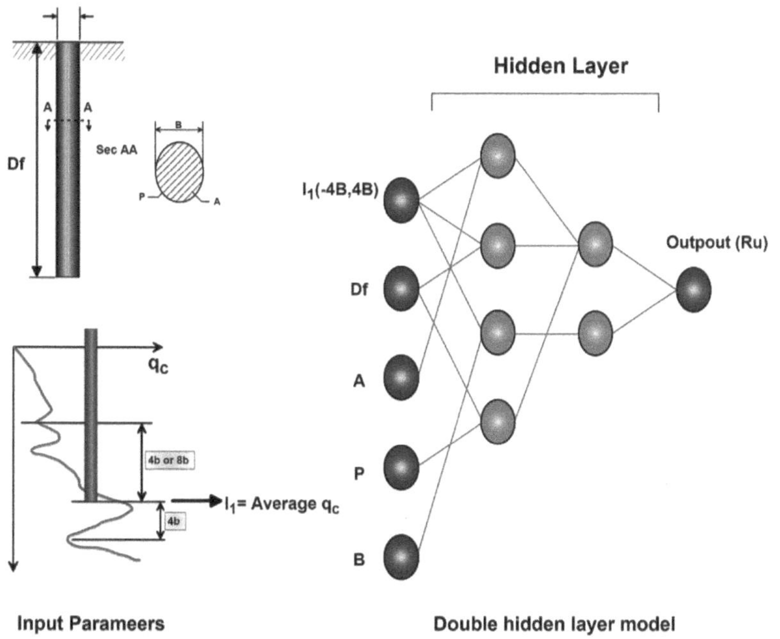

Figure 10.2 Double hidden layer group method of data handling structure for R_u with input parameters (Hassani, 2011).

predict R_u, the compiled dataset has been arbitrarily divided into training and testing datasets, statistical characteristics of which were the same. Among 90 collected data, 20 have been randomly applied in the testing stage and the remaining datasets (70 records) have been selected in the training stage. Several hidden layers and q_c average versus pile depth (-4—4B and -8—4B; B is pile diameter) parameter were considered for pile-bearing capacity model. As an example, the double hidden layer model is presented in Fig. 10.2. The suitable prediction of such GMDH method is also shown in Fig. 10.3 (training set). In addition, to examine the performance of the different hidden layer model, the testing series of case records are used in Fig. 10.4, for the model development. The obtained results of predicting R_u were in good agreement with the measured ones ($R^2 = 0.96$). Thus, it can be concluded that the proposed method possessed trustworthy predicting capabilities.

10.4.2 Case Study No. 2: pile-bearing capacity—support vector regression method

Kordjazi et al. (2014) examined the potential of SVR in prediction of R_u based on CPT records. To develop the SVR models, a databank including pile geometry, CPT records and static pile load tests extracted from renowned references is used. In the Kordjazi et al. (2014) research, 83 and 25 cases are applied for training and testing series, respectively. Statistical input and output parameters used in Kordjazi et al. (2014)

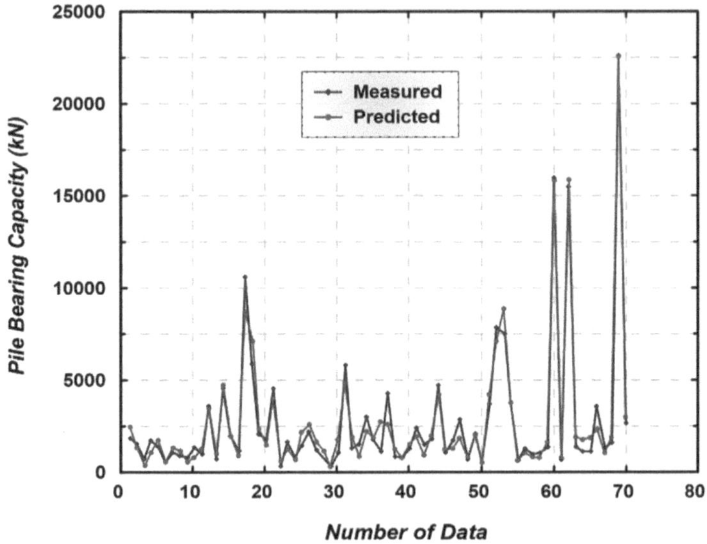

Figure 10.3 Estimated versus measured pile capacity (training set) (Hassani, 2011).

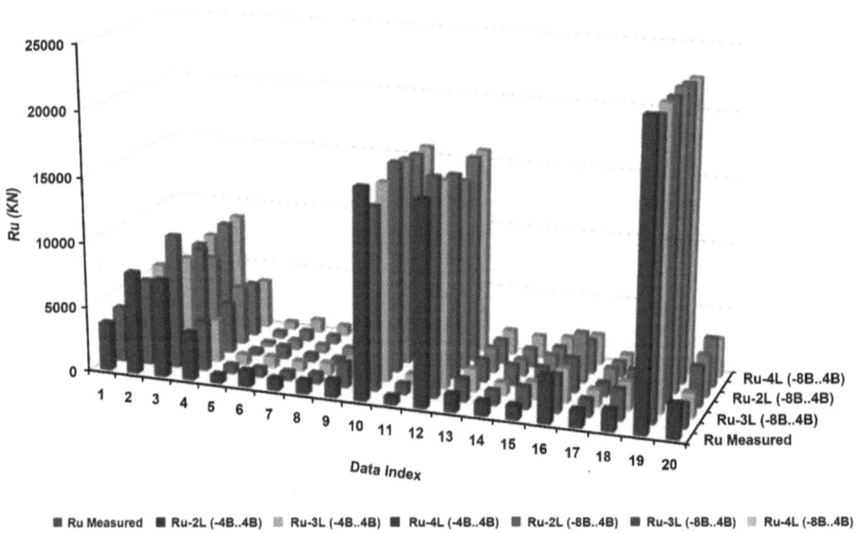

Figure 10.4 Estimated versus measured pile capacity (testing set) for different models (Hassani, 2011).

have been presented in Table 10.1. The results of the best values of SVR tuning parameters regarding the CPT data and considering three types of kernel functions are shown in Table 10.2. In Table 10.3 and Fig. 10.5, the efficiency of the SVR model for predicting R_u is presented. From the output of the testing series, it can be concluded that the proposed SVR model could successfully predict R_u.

Table 10.1 Statistical input and output parameters (Kordjazi et al., 2014).

Model variable sand datasets	Statistical parameters				
	Mean	Standard Deviation	Minimum	Maximum	Range
A_{tip} (m^2)					
Training	0.1827	0.1783	0.008	0.7854	0.7774
Testing	0.1519	0.1176	0.008	0.503	0.495
O (mm)					
Training	1625.12	866.92	585	7341.3	6756.3
Testing	1425.68	416.67	858	2510	1652
L_{embed} (m)					
Training	15.7	10.36	5.5	67	61.5
Testing	16.17	8.47	6.5	36.5	30
q_{c1} (MPa)					
Training	4.13	3.17	0.02	15.07	15.05
Testing	3.76	2.85	0.24	11.72	10.74
f_{s1} (kPa)					
Training	74.17	59.03	0.73	283.94	283.21
Testing	69.49	43.67	10.76	176.75	159.7
q_{c2} (MPa)					
Training	6.21	5.86	0.32	30.71	30.39
Testing	5.73	4.09	1	18.42	17.43
f_{s2} (kPa)					
Training	103.52	96.91	2.12	618.66	616.54
Testing	100.65	63.69	15	303.08	288
q_{c3} (MPa)					
Training	7.36	6.07	0.27	32.59	32.32
Testing	6.63	4.76	0.7	23.05	22.36
f_{s3} (kPa)					
Training	126.57	95.58	7.99	396.57	388.58
Testing	139.18	100.87	25	388	363

Continued

Table 10.1 Statistical input and output parameters (Kordjazi et al., 2014).—cont'd

Model variable sand datasets	Statistical parameters				
	Mean	Standard Deviation	Minimum	Maximum	Range
q_{ctip} (MPa)					
Training	8.75	6.28	0.25	27.11	26.86
Testing	9.03	6.03	1.15	22.3	21.15
P_u (kN)					
Training	1988.2	1811.6	60	10,910	10,850
Testing	1891.68	1302.78	520	5850	5330

A_{tip}, Pile area; O, Pile diameter; L_{embed}, Length of pile; f_{si}, Average of shaft resistance along ith segment; q_{ci}, Average of cone tip resistance along ith segment; q_{ctip}, average of q_c below the pile; Pu, Pile capacity.

Table 10.2 Best values of support vector regression tuning parameters (Kordjazi et al., 2014).

Type of kernel	Parameter	Optimum value of C	Optimum value of ε	Number of vectors
Rbf	$\sigma = 0.9$	1.1	0.02	57
Pearson VII	$\sigma = 0.9$; $w = 7$	0.95	0.02	59
Polynomial	$d = 1.95$	0.3	0.005	78

Table 10.3 Statistical results of the support vector regression models (Kordjazi et al., 2014).

	Training		Testing	
Type of kernel	R^2	RMSE	R^2	RMSE
Rbf	0.979	394.38	0.972	318.85
Pearson VII	0.982	372.25	0.972	315.86
Polynomial	0.966	471.41	0.936	469.62

10.4.3 Case Study No. 3: pile shaft bearing capacity

A polynomial model based on GMDH-type NN was used for correlating the shaft bearing capacity of piles with CPT data by Ardalan et al. (2009). The database includes the results of static loading tests performed on 40 piles. The pile unit shaft resistance (r_s) formula corresponding to the *GMDH* model is as follows:

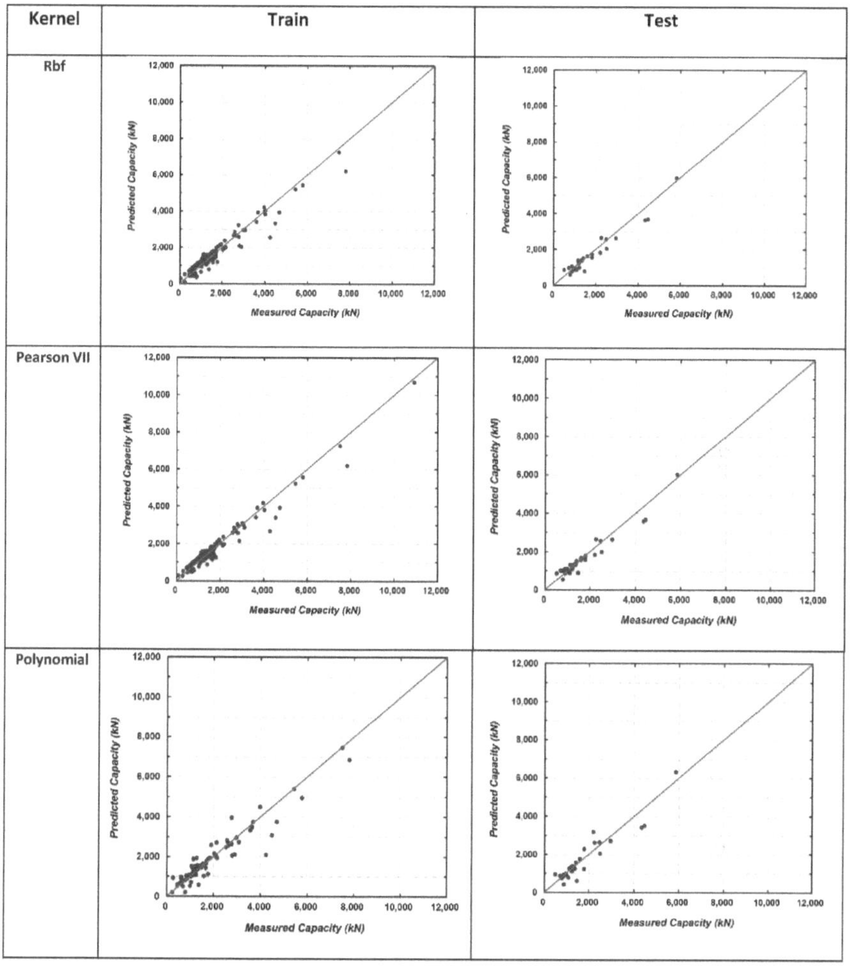

Figure 10.5 Comparison between the measured and predicted values of R_u using the several support vector regression kernels (Kordjazi et al., 2014).

$$r_s = -0.0083 + 1.6206Y_2 - 0.00212\, q_E - 9.404Y_2^2 - 0.00002\, q_E + 0.047335\, Y_2 q_E$$

$$Y_1 = 0.017464 + 0.00231 q_E - 0.4253\, f_s - 0.000243\, q_E - 1.654 f_s^2 + 0.02066\, q_E f_s$$

$$Y_2 = -0.68397 - 1.1222 f_s + 5.829306 Y_1 - 3.928 f_s^2 - 80.2061 Y_1^2 + 36.11058 Y_1 f_s \tag{10.4}$$

where q_E is the geometric average of the effective cone point resistance and f_s is shaft friction resistance. Based on Eq. (10.4), the variation of r_s versus input parameters is illustrated in Fig. 10.6. The results of Ardalan et al. (2009) indicated that the proposed

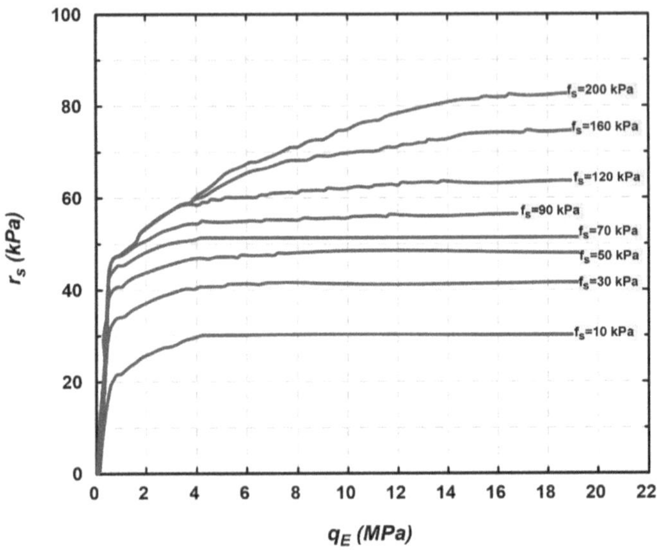

Figure 10.6 Results of the group method of data handling model for r_s (Ardalan et al., 2009).

GMDH method with the average values of $R^2 = 0.96$ demonstrated better agreement with the measured pile shaft resistance.

10.5 Case Study No. 4: shallow foundation settlement

A few relations and equations have been proposed to predict shallow foundation settlement (Sm), most of which are based on different typical parameters, such as elastic modulus (E_s), foundation width (B), contact pressure (q), and subsoil resistance as discussed in Chapter 8. In other words, these equations are to estimate only rapid or long-time settlements based on soil type. Hence, a unique equation to predict the settlement for all soil types is rarely found in geotechnical practices. MolaAbasi et al. (2019) applied the optimum hidden layer (single) GMDH to predict S_m/B. It is worth noting that the single hidden layer was the best and optimum layer in the modeling. In applying a single hidden layer GMDH, the connection of input parameters q (kPa), R_f (%), and q_t (MPa) gives the best correlation as follows:

$$\frac{\mathrm{Sm}}{\mathrm{B}} = 0.93 - 0.00003\mathrm{q} + 0.0095\mathrm{Y} - 0.00000001q^2 + 0.001Y^2 + 0.0006Y.\mathrm{q}$$

$$\text{Where,} \quad Y = 0.18 - 0.012q_t - 0.217R_f + 0.00025q_t^2 + 0.148R_f^2 + 0.017q_t.R_f \tag{10.5}$$

where q, B, R_f, and q_t are foundation contact pressure, foundation width, friction ratio, and cone resistance, respectively.

As can be realized from Eq. (10.5), the relation consists of two polynomial functions. It may seem that the equation is lengthy and needs too many calculations, while in reality, by means of computational software, the equation can be considered to be a concise formula. Besides, in comparison with the previous methods proposed by Janbu (1967) and Valikhah and Eslami (2017), as presented in Chapter 8, where the calculations are separately conducted for each layer, Eq. (10.5) does not have such limitations.

Based on Eq. (10.5), settlement is affected by the soil strength properties such as R_f and q_t (corresponding to relation Y) and the loading on the foundation (q). Therefore, the parameter Y has to be firstly computed in the first layer by using the strength properties of the soil, and then the parameter Sm/B would be calculated in the next stage using parameters Y and q.

Moreover, based on Eq. (10.5), the variations of input parameters are illustrated in Fig. 10.7. As shown in Fig. 10.7, by increasing q_t up to 4 MPa, the Sm/B value increases. Then, a higher increase of q_t leads to a decrease in the Sm/B. Such a trend can be attributed to the fact that the parameters of B and R_f were considered 1 and 0.4 for plotting Fig. 10.7, which resulted in a variance of about 2 mm. In addition, by increasing the load on the foundation, the Sm/B value increases.

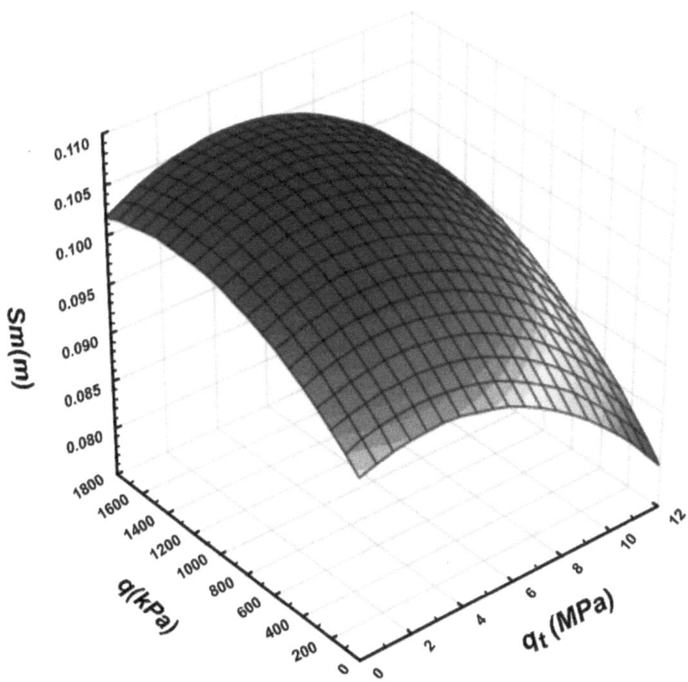

Figure 10.7 Correlation of Sm/B using Eq. (10.5) and considering B = 1m and Rf = 0.4 (MolaAbasi et al., 2019).

10.6 Case Study No. 5: friction angle, cohesion, and shear stress

Mola-Abasi and Eslami (2018) employed a GMDH method to predict friction angle (φ) of the marine deposits based on a dataset (50 case records) from a geotechnical investigation in south shoreline of Iran. They assumed q_C, R_f, and σ' as input parameters to evaluate φ. Using an optimum single hidden layer GMDH model, the combination of input variables R_f, σ', and $\ln(q_c)$ gives the suitable result. The GMDH formula corresponding to φ prediction is as follows:

$$\varphi = -90.6 + 0.83ln(q_c) + 55.02Y^2 - 8.64\ln(q_c)^2 + 0.019Y.\ln(q_c)$$

$$Y = 1000(0.04 - 0.00005\sigma' - 1.07R_f + 0.00000007\sigma'^2 - 0.023R_f^2 + 0.0015\sigma'.R_f \quad (10.6)$$

The graphical presentation of Eq. (10.6) is illustrated in Fig. 10.8.

Additionally, using the GMDH algorithm, the shear strength (τ) based on Mohr–Columb failure relation ($\tau = C + \sigma'\tan(\varphi)$) and also cohesion of the fine grained soils (C) can be predicted. Therefore, based on the same database, τ and C were modeled via the same structure of single hidden layer GMDH for estimating τ and C. The polynomials corresponding to τ and C model are presented in Eqs. (10.7) and (10.8), respectively:

$$\begin{aligned} \tau &= 555.1 + 24\text{Y} + 0.3\ln(q_c) + 6.1Y^2 + 9.64\ln(q_c)^2 + 1.1Y.\ln(q_c) \\ Y &= 11.2 - 0.0085\sigma' + 3.07R_f + 0.00004\sigma'^2 + 0.023R_f^2 + 0.015\sigma'.R_f \end{aligned} \quad (10.7)$$

$$\begin{aligned} \text{C} &= 241.2 - 2\text{Y} - 149.8\ln(q_c) + 0.0003Y^2 + 23\ln(q_c)^2 + 9Y.\ln(q_c) \\ Y &= -11.3 + 0.25\sigma' + 374.5R_f + 0.00001\sigma'^2 + 8.2R_f^2 + 7.7\sigma'.R_f \end{aligned} \quad (10.8)$$

The results obtained from GMDH in predicting C, φ, and τ are in good agreement with the measured values as presented in Fig. 10.9. Thus, reliable predicting capabilities were obtained.

10.7 Case Study: liquefaction potential assessment

Liquefaction is one of the serious phenomena in geotechnical earthquake engineering and primarily associates with saturated cohesionless soil displacements under undrained conditions during earthquake. Building up pore water pressure, as drainage cannot occur, is a striking indication of soil liquefaction phenomenon that results in loss of strength and stiffness. To assess the potential for "triggering" or beginning

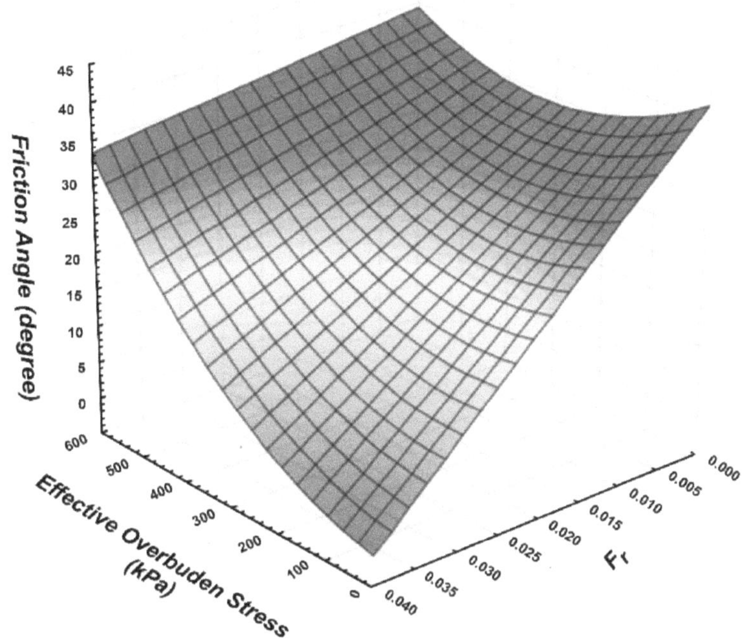

Figure 10.8 Correlation of φ using Eq. (10.6) (MolaAbasi and Eslami, 2018).

of seismically caused liquefaction, CPT records are employed. In this part, several soft computing techniques in liquefaction prediction are presented as provided in the following subsections.

10.7.1 Case Study No. 6: CRR prediction using polynomial model

Eslami et al. (2014) applied the GMDH approach in the field of geotechnical earthquake engineering and proposed a new polynomial model to predict CRR based on CPT parameters, applied on a total of 182 datasets, including field investigation records from 18 earthquakes.

To predict the liquefaction potential, several iterations of the new polynomial model were performed. The evolved model results in a simple polynomial equation for the CRR (refer to Chapter 4 for more information). The corresponding GMDH representation for CRR is as follows:

$$\begin{aligned} CRR = 0.075 + 0.002q_{C1} - 0.0005\,f_s + 0.00018q_{C1}^2 - 0.0001\ f_s^2 \\ + 0.0016q_{C1}\,f_s \end{aligned} \tag{10.9}$$

The isometric view of CRR variation with respect to f_s and q_{c1} is shown in Fig. 10.10.

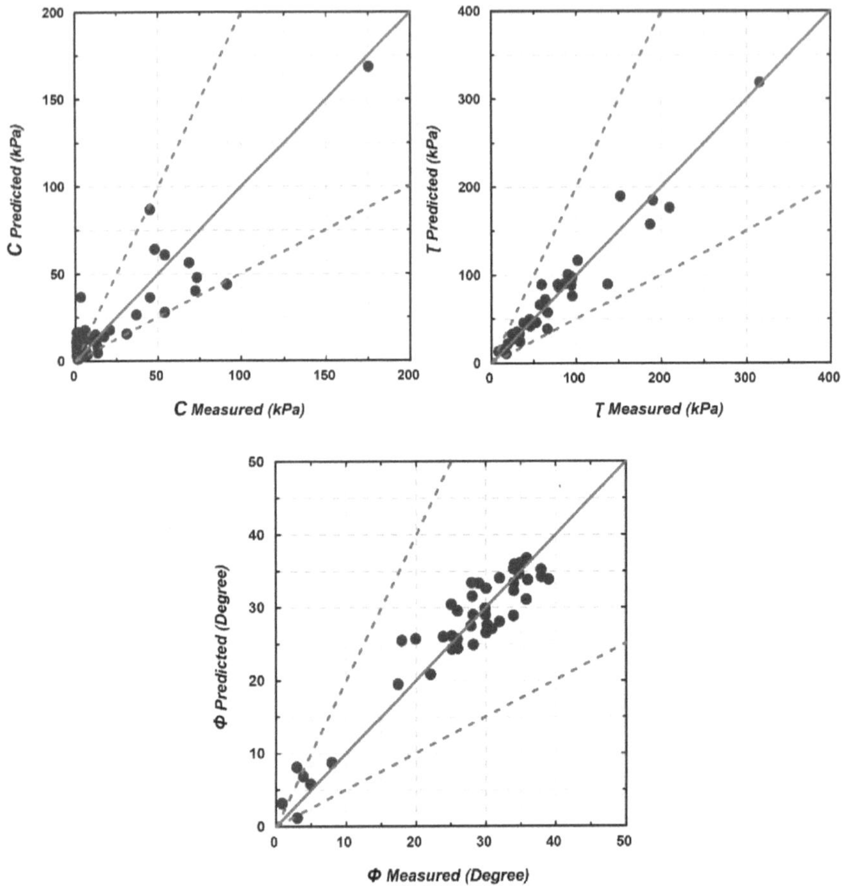

Figure 10.9 Comparison between the estimated and obtained values for C, φ, and τ using group method of data handling (MolaAbasi and Eslami, 2018).

In addition, the performance of Eslami et al. (2014) (Eq. 10.9) with high accuracy (number of correctly predicted cases/total number of cases) 0.901 for liquefied and nonliquefied conditions is presented as in Fig. 10.11.

10.7.2 Case Study No. 7: evolutionary-based approach

Rezania et al. (2010) proposed a new approach as per evolutionary polynomial regression (EPR) for the assessment of the liquefaction potential as a result of earthquakes. For such evaluation, the EPR models were trained and tested separately using two different datasets of field case histories. Moreover, the data were collected from different sources in the literature and include 420 field cases. After analysis of different

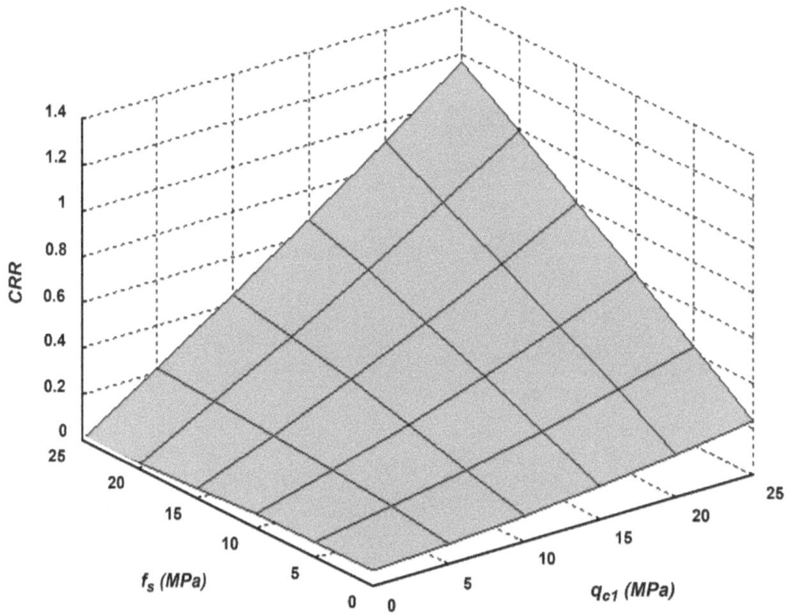

Figure 10.10 Variation of CRR versus q_{c1} and f_s (Eslami et al., 2014).

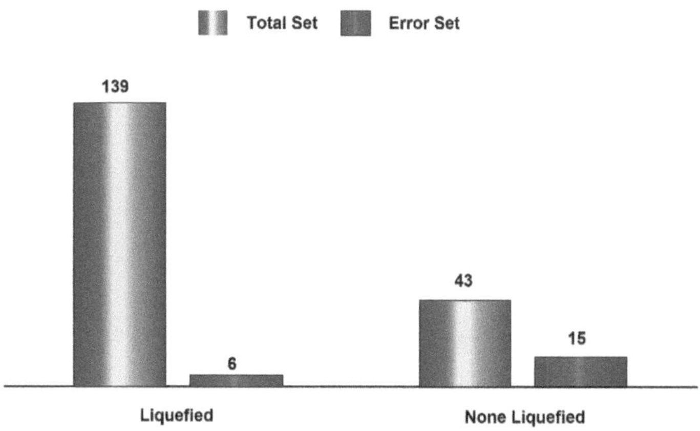

Figure 10.11 Liquefaction prediction over the datasets (Eslami et al., 2014).

alternative models, they came to the conclusion that Eqs. (10.10) to (10.12) are the most robust and practical models for clean sand, silty sand, and silty sand to sandy silt, respectively. For each case, "1" represents the occurrence, while "0" denotes nonoccurrence of liquefaction.

$$Y = \mathrm{ROUND}\left\{8133.187\frac{Z^2a_{max}^2}{M^2\sigma q_{c1}} - 35.778\frac{z^2\sigma' D_{50}^2 a_{max}^2}{M^2\sigma} + 0.423\right\} \tag{10.10}$$

$$Y = \mathrm{ROUND}\left\{-4.718\frac{Z^2D_{50}^2q_{c_1}^2}{M^2\sigma\sigma\, a_{\max}^2} + 979702.139\frac{a_{\max}}{M^2\sigma^2\sigma^{\,2}D_{50}^2q_{c_1}}\right.$$

$$-0.001\frac{Z^2\sigma^{\,2}D_{50}^2q_{c_1}a_{\max}^2}{M\sigma} - 208166757.628\frac{ZD_{50}^2a_{\max}^2}{M\sigma^2\sigma\, q_{c_1}^2} \tag{10.11}$$

$$\left.+4544.990\frac{MZD_{50}a_{\max}}{\sigma^{\,2}q_{c_1}} + 0.319\right\}$$

$$Y = \mathrm{ROUND}\left\{-6.660E-10\frac{Z^2\sigma^2}{\sigma' D_{50}^2 q_{c1}a_{max}} + 6.417E-9\frac{\sigma\sigma'^2 a_{max}^2}{M^2D_{50}^2} - 9.684E\right.$$

$$-6\frac{\sigma^2 q_{c1}}{M^2\sigma' D_{50}a_{max}^2} - 3.438\frac{q_{c1}^2}{M^2Z\sigma' D_{50}} + 5.704E-4\frac{\sigma}{M\sigma' D_{50}q_{c1}^2a_{max}}$$

$$\left.-2.709\frac{Z^2a_{max}}{\sigma'} + 1.330\right\}$$

(10.12)

Rezania et al. (2011) chose the following formula among all the relationships generated by the EPR process, in accordance with its performance on both training and testing sets. The opted model is as follows:

$$Y = \mathrm{ROUND}\left\{\frac{0.00777 \times I_C^2}{CSR} - \frac{2.5\times 10^{-5}q_{c1N}}{CSR^2} - 0.00963q_{c1N}\right.$$

$$-\frac{0.000712I_c\sigma_v'}{CSR_{7.5}} - 1.43\times 10^{-5}q_{c1N}^2 - \frac{951I_c}{q_{c1N}\sigma_v^2} + \frac{4.6\times 10^{-9}.q_{c1N}.\sigma'^2}{CSR_{7.5}^2}$$

$$\left.+1.547\right\}$$

10.13

A surface fitting program developed in MATLABs was used to fit a 3D surface through the generated virtual marginal points. The best surface passing through artificial points was as Eq. (10.14) and is also presented in Fig. 10.12.

$$CRR = \exp\left[-6.994 + 7.9\times 10^{-6}q_{CN1}^{2.5} + 1.115\times Ln(\sigma_v')\right] \tag{10.14}$$

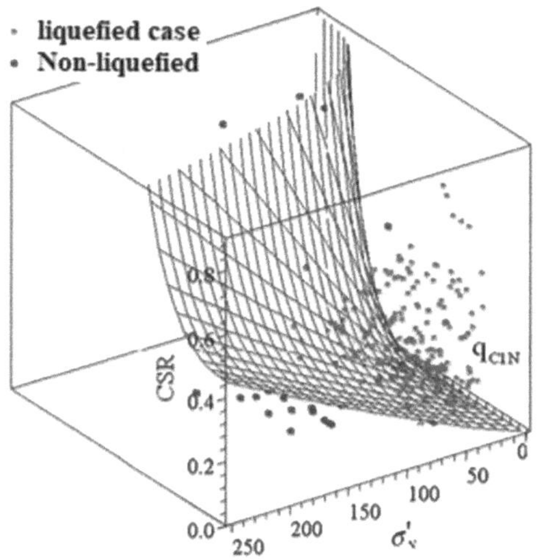

Figure 10.12 Cone penetration test–based case histories and virtual 3D separator surface (Rezania et al., 2010).

10.7.3 Case Study No. 8: Triangular chart mapping and SVM

The main idea of triangular chart is depicted from soil classification chart as presented in Chapter 5. MolaAbasi et al. (2019) have tried triangular chart to derive a more accurate chart for liquefaction prediction based on CPT records (Fig. 10.13). As illustrated in Fig. 10.15, q_{C1}, R_f, and $CSR_{7.5}$ variations are between 0 and 1. Therefore, as presented by MolaAbasi et al. (2019), q_{C1}, R_f, and $CSR_{7.5}$ should be normalized by 30 MPa, 4%, and 0.6 respectively.

In the proposed approach, SVM classifiers present suitable decision surfaces with a minimum classification risk and a maximum generalizability for the CPT data, based on the two features obtained from the triangular chart. The general procedure of the approach is shown in Fig. 10.14.

SVM classifiers are formed considering all the CPT data and based on the kernel functions and the best tuning parameters selected through the cross-validation process.

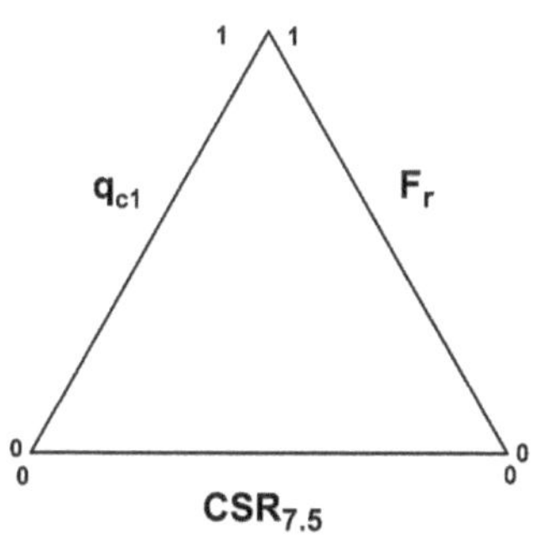

Figure 10.13 Triangular chart for liquefaction via cone penetration test data (MolaAbasi et al., 2019).

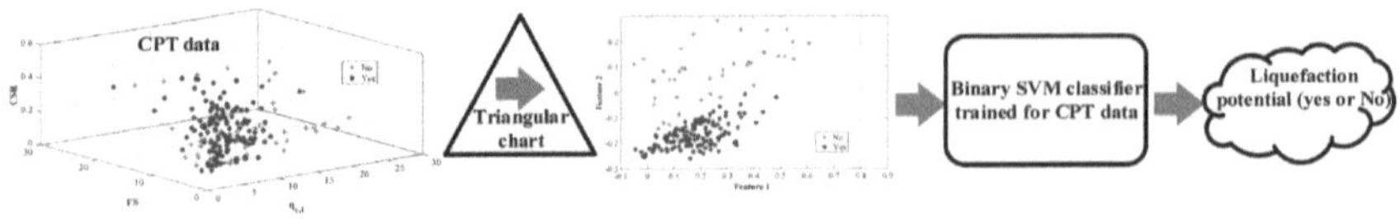

Figure 10.14 General procedure of the MolaAbasi et al. (2019) approach based on the triangular chart identification (MolaAbasi et al., 2019).

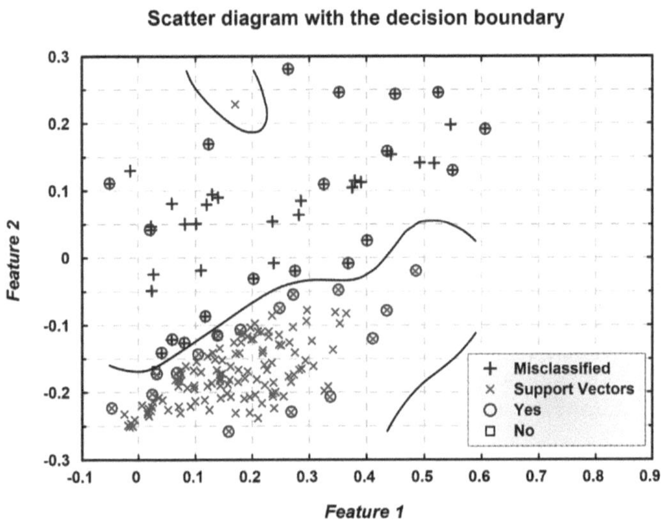

Figure 10.15 Support vector machine decision boundary based on the Gaussian kernel function with all cone penetration test samples obtained from the triangular chart, support vectors, and misclassified samples (MolaAbasi et al., 2019).

Fig. 10.15 shows the SVM decision boundary based on the Gaussian kernel function (best model) with all the CPT samples obtained from the triangular chart, the support vectors, and the misclassified samples. As it can be realized from Fig. 10.15, the data of two classes are completely separated and only two samples are misclassified.

References

Abe, S., 2005. Support Vector Machines for Pattern Classification, vol. 53. Springer.

Ardalan, H., Eslami, A., Nariman-Zadeh, N., 2009. Piles shaft capacity from CPT and CPTu data by polynomial neural networks and genetic algorithms. Computers and Geotechnics 36 (4), 616–625.

Atashkari, K., Nariman-Zadeh, N., Gölcü, M., Khalkhali, A., Jamali, A.J.E.C., 2007. Modelling and multi-objective optimization of a variable valve-timing spark-ignition engine using polynomial neural networks and evolutionary algorithms. Energy Conversion and Management 48 (3), 1029–1041.

Drucker, H., Burges, C.J., Kaufman, L., Smola, A.J., Vapnik, V., 1997. Support vector regression machines. In: Advances in Neural Information Processing Systems, pp. 155–161.

Eslami, A., Mola-Abasi, H., Shourijeh, P.T., 2014. A polynomial model for predicting liquefaction potential from cone penetration test data. Scientia Iranica. Transaction A, Civil Engineering 21 (1), 44–52.

Hassani, A., 2011. Pile Bearing Capacity Prediction Based on CPT Records Using GMDH Type Neural Network. MSC thesis, Amirkabir University of Technology.

Jamali, A., Nariman-Zadeh, N., Darvizeh, A., Masoumi, A., Hamrang, S., 2009. Multi-objective evolutionary optimization of polynomial neural networks for modelling and prediction of explosive cutting process. Engineering Applications of Artificial Intelligence 22 (4–5), 676–687.

Janbu, N., 1967. Settlement Calculations Based on the Tangent Modulus Concept. University of Trondheim, Norwegian Institute of Technology, Bulletin (2), p. 57.

Khan, F., Enzmann, F., Kersten, M., 2016. Multi-phase classification by a least-squares support vector machine approach in tomography images of geological samples. Solid Earth 7 (2), 481–492.

Kordjazi, A., Nejad, F.P., Jaksa, M.B., 2014. Prediction of ultimate axial load-carrying capacity of piles using a support vector machine based on CPT data. Computers and Geotechnics 55, 91–102.

Madandoust, R., Ghavidel, R., Nariman-Zadeh, N., 2010. Evolutionary design of generalized GMDH-type neural network for prediction of concrete compressive strength using UPV. Computational Materials Science 49 (3), 556–567.

Mola-Abasi, H., Eslami, A., 2018. Prediction of drained soil shear strength parameters of marine deposit from CPTu data using GMDH-type neural network. Marine Georesources & Geotechnology 1–10.

MolaAbasi, H., Saberian, M., Kordnaeij, A., Omer, J., Li, J., Kharazmi, P., 2019. Predicting the stress-strain behaviour of zeolite-cemented sand based on the unconfined compression test using GMDH type neural network. Journal of Adhesion Science and Technology 1–18.

Nariman-Zadeh, N., Darvizeh, A., Ahmad-Zadeh, G.R., 2003. Hybrid genetic design of GMDH-type neural networks using singular value decomposition for modelling and prediction of the explosive cutting process. Proceedings of the Institution of Mechanical Engineers – Part B: Journal of Engineering Manufacture 217 (6), 779–790.

Rezania, M., Faramarzi, A., Javadi, A.A., 2011. An evolutionary based approach for assessment of earthquake-induced soil liquefaction and lateral displacement. Engineering Applications of Artificial Intelligence 24 (1), 142–153.

Rezania, M., Javadi, A.A., Giustolisi, O., 2010. Evaluation of liquefaction potential based on CPT results using evolutionary polynomial regression. Computers and Geotechnics 37 (1–2), 82–92.

Schölkopf, B., Burges, C.J.C., Smola, A.J., 1999. Advances in Kernel Methods: Support Vector Learning. MIT press.

Schölkopf, B., Smola, A.J., 2001. Learning with Kernels: Support Vector Machines, Regularization, Optimization, and beyond. MIT press.

Sheikholeslami, M., Sheykholeslami, F.B., Khoshhal, S., Mola-Abasi, H., Ganji, D.D., Rokni, H.B., 2014. Effect of magnetic field on Cu–water nanofluid heat transfer using GMDH-type neural network. Neural Computing & Applications 25 (1), 171–178.

Valikhah, F., Eslami, A., 2017. Foundations settlement estimation by non-linear stress-strain approach using CPT records. In: DFI 42nd Annual Conference on Deep Foundations, October 24–26, 2017, New Orleans, LA.

Vapnik, V.N., 1995. The Nature of Statistical Learning Theory. SpringVerlag, New York, p. 188.

Uncertainty and reliability in foundation engineering: CPT-Based approach

11

11.1 Introduction

Geotechnical design or especially foundation engineering is the art of implementation of site investigations, design models, codes, and computation. This is accompanied by making decision over prevailing conditions to have an optimum design based on the importance of the project as illustrated in Fig. 11.1.

In conventional design issues, engineers consider material properties such as friction angle, soil undrained shear strength, or uniaxial compressive resistance of concrete as a definite parameter, that is, the mentioned property, if properly specified, has a certain value. However, with an insight into a borehole or even the results of the SPT or CPT records, one can notice the latent uncertainty in soil properties. Similarly, uncertainties are not limited solely to material properties, and simplified

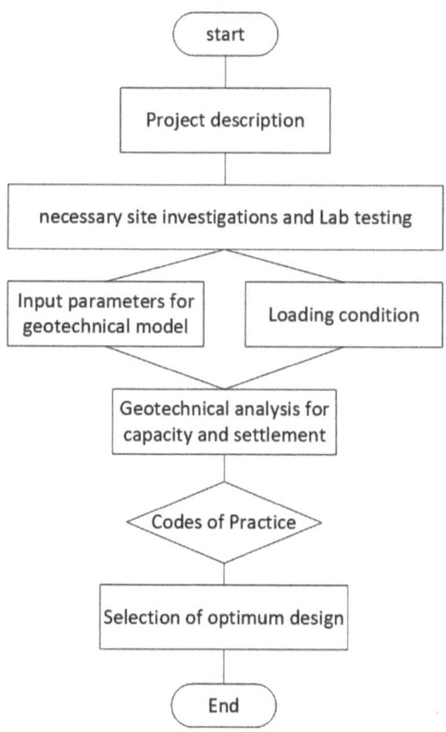

Figure 11.1 Foundation design algorithm.

Piezocone and Cone Penetration Test (CPTu and CPT) Applications in Foundation Engineering
https://doi.org/10.1016/B978-0-08-102766-0.00011-0

constitutive models, and even estimation of material characteristics insert great amount uncertainties in design process. Observations in this field confirm the prevailing differences between measured and predicted design values.

The uncertainty can be divided into two general categories: the first category comes from the inherent random nature of materials. This type of uncertainty exist invariably and cannot be eliminated, but it can be better described and considered by doing more tests or using databases. Parameters such as undrained shear strength is of this kind, and its value differs spatially and over time. In fact, shear strength is estimated with a limited number of samples and the corresponding tests; while being used for design at another point other than where the sample was taken from. The difference between the estimated value of resistance at that point and its actual resistance is the inherent uncertainty. Inherent soil variability is due to complex process of geomaterial formation such as sedimentation, weathering, stress history, and time.

The second category of uncertainty is related to our inadequate and limited knowledge of a system or process. Generally, this kind of uncertainty can be categorized into three main types named model uncertainty, transformation uncertainty, and measurement/random noise error.

The measurement errors are caused by equipment, operator, and random testing effects during the measurements. In addition, empirical methods or other models that incorporate laboratory or field measurements into soil design characteristics are of this category. Therefore, this kind of uncertainty can be reduced by increasing the quantity and quality of experiments or using better models. Statistical uncertainty due to the limited amount of information can be reduced by testing more samples. However, errors caused by the sampling or measurement process cannot be eliminated by increasing sampling.

Furthermore, there is a great deal of uncertainty in the estimation of live and dynamic loads such as earthquakes, winds, floods, and so on. Uncertainties associated with static loads, as opposed to uncertainties in the material parameters, can be ignored. It should also be noted that under dynamic load, the value of material parameters may have significant variations that these uncertainties may not be visible. For example, under the cyclic loading, the shear strength of the soil decreases, resulting in greater uncertainty about the performance of this geotechnical parameter.

11.2 Uncertainty sources in geotechnical engineering

CPT with continuously recording outputs makes it possible for geotechnical engineers to be more confident about inherent soil variability. Soil properties are variable with both depth and location. Fig. 11.2 depicts a typical record of a CPT sounding. As can be seen, soil properties are different in depth. In conventional design methods, a parameter named "trend" which is the average value of recorded parameters is used for design purposes. To express soil spatial variability, random field and probabilistic approach are feasible tools.

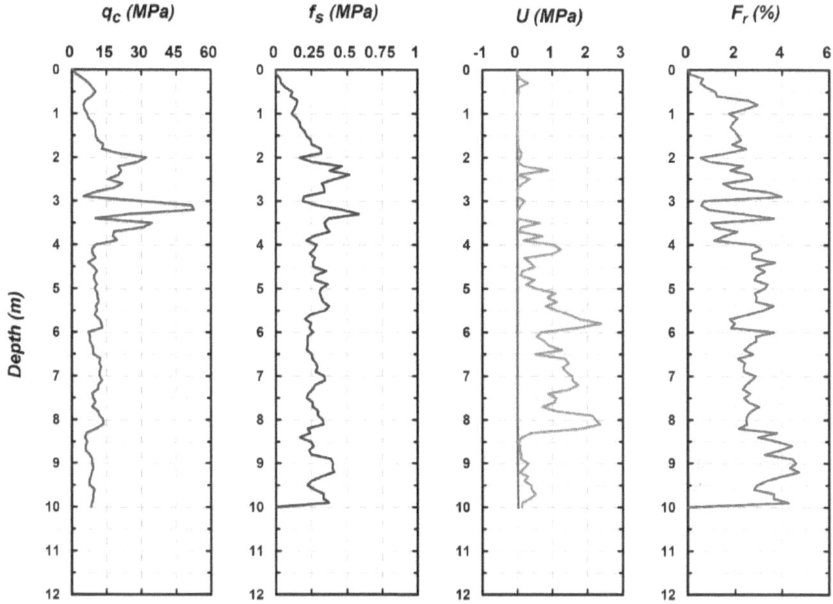

Figure 11.2 An example of CPT records and the variation of soil properties among the depth (Eslami, 1996).

To consider soil as a random field, it is necessary to have the trend or average value, coefficient of variation (COV), and correlation length for that special soil characteristic. The COV is determined as the ratio of standard deviation to mean of data. Correlation length is the length that soil parameters are strongly correlated, and it can be calculated through several ways such as variance reduction method, direct integration of sample autocorrelation function, Bartlet's limit method, and autocorrelation model fitting (Kenarsari et al., 2011).

As illustrated in Fig. 11.3, a property, named Ʒ, is composed of two parts. The first is a straight line called the trend, t(z). Extracting the trend value from the rest of data,

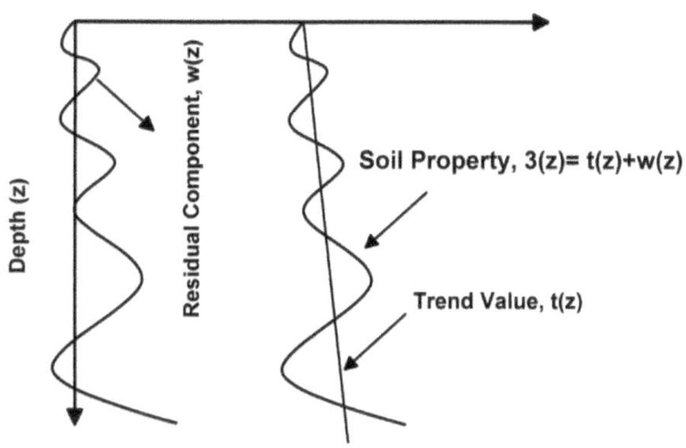

Figure 11.3 Soil inherent variability with depth.

the remaining parts show variation around the zero line and is called the residual components or w(z). This residual component is an indicator of soil spatial variability, and in cases it is an indicator of other accompanied uncertainties, being elaborated in the following paragraphs (Phoon and Kulhawy, 1999).

The statistical uncertainty in the estimation of parameters is due to the limited set of data and the technique used to estimate them. In general, the measurement uncertainty is due to three main sources: (1) equipment error, (2) errors caused by operator, and (3) the effect of random tests during measurements. The amount of this type of uncertainties can be determined by manufacturer's information about equipment and the operator for laboratory and field tests. Random test errors are attributed to the dispersion in the results of the experiments, which cannot be distinguished from the inherent soil variability for the studied parameter.

Tables 11.1 and 11.2 summarize the results of the measurement error for field and laboratory tests. What is clear is that the measurement error will be greater in practice

Table 11.1 Summary of measurement error for common in situ tests (Kulhawy and Trautmann, 1996).

	Coefficient of variation, COV (%)				
Test	**Equipment**	**Procedure**	**Random**	**Total(1)**	**Range(2)**
Standard penetration test (SPT)	5–75(3)	5–75(3)	12–15	14–100(3)	15–45
Mechanical cone penetration test (MCPT)	5	10–15(4)	10–15(4)	15–22(4)	15–25
Electric cone penetration test (ECPT)	3	5	5–10(4)	7–12(4)	5–15
Vane shear test (VST)	5	8	10	14	10–20
Dilatometer test (DMT)	5	5	8	11	5–15
Pressuremeter test, prebored (PMT)	5	12	10	16	10–20(5)
Self-boring pressuremeter test (SBPMT)	8	15	8	19	15–25(5)

[1] $COV(Total) = [COV(Equipment)^2 + COV(Procedure)^2 + COV(Random)^2]^{0.5}$
[2] Because of limited data and judgment involved in estimating COVs, ranges represent magnitudes of field test measurement error.
[3] Best to worst case scenarios, respectively, for SPT.
[4] Tip and side resistance, respectively, for CPT.
[5] It is likely that results may differ for p_0, p_f, and p_L, but the data are insufficient to clarify this issue.

Table 11.2 Summary of total measurement error of some laboratory tests (Phoon and Kulhawy, 1999)

Test	Property	Soil type	Coefficient of variation (%)	
			Range	Mean
Atterberg tests	Plasticity index	Fine grained	5-51	24
Triaxial compression	Effective angle of friction, $\overline{\varphi}$	Clay, silt	7-56	24
Direct shear	Shear strength, Cu	Clay, silt	19-20	20
Triaxial compression	Shear strength, Cu	Clay, silt	8-38	19
Direct shear	Effective angle of friction, $\overline{\varphi}$	Sand	13-14	14
Direct shear	Effective angle of friction, tan $\overline{\varphi}$	Clay	6-22	14
Direct shear	Effective angle of friction, $\overline{\varphi}$	Clay, silt	3-29	13
Atterberg tests	Plastic limit	Fine grained	7-18	10
Triaxial compression	Effective angle of friction, tan $\overline{\varphi}$	Sand, silt	2-22	8
Atterberg tests	Liquid limit	Fine grained	3-11	7
Unit weight	Density	Fine grained	1-2	1

because the results presented in this table are related to controlled conditions. According to the table, it can be concluded that the maximum measurement error is related to the SPT test and the lowest measurement error is related to the DMT and ECPT tests.

Transformation error is another source of uncertainty accompanied by limited number of case records. Each experimental correlation is developed based on a database which can be restricted to a place or special soil type, etc. Furthermore, it is a correlation, using assumptions and special computational model, etc. In other words, an error is implicitly imported to the soil property estimation.

Model error, especially in geotechnical engineering, include many uncertainties for a variety of reasons such as simplifications related to equilibrium and deformation analysis, ignoring the effects of the three-dimensional modeling, etc. The model uncertainty is defined as the ratio of real quantity to predicted quantity with a model. The uncertainty of the model is characterized by the mean and COV. The difference between the mean value of the unit number represents a systematic error in the model, while the standard deviation indicates the variability in the predicted values of the model.

11.3 Principles: random analysis techniques in geotechnical engineering

Random analyzes provide suitable tools for handling variability of soil properties, and by developing rational algorithms for estimating soil design parameters based on probabilities, they can determine the risk levels associated with the design. Depending on the level and type of reliability analysis, these analyzes provide one or more of the following results:

- Probability of failure (or the possibility of undesirable performance)
- Reliability index (β)
- The most likely combination of parameters that lead to failure
- Sensitivity of the results to any change in the parameters

The Target reliability index for each geotechnical design should be determined according to the importance of supported structure. There are many quantified and not quantified sources of uncertainties in geotechnical engineering. As a result, a safety factor higher than one cannot guarantee the safety purposes.

Many of the reliability methods have been developed by researchers in structural engineering, such as first- and second-order methods. Because considering uncertainty in geotechnical designs is somehow complicated, it is preferable to not apply them. Therefore, it is important to make geotechnical engineers familiar with this issue. In structural engineering reliability–based design has been a common topic because the constitutive models for materials such as steel and concrete are linear and these materials are prepared under controlled conditions. However, everything is different in geotechnical engineering; soil, itself, has inherent spatial variability, and its characteristics are accompanied with statistical uncertainty, measurement error, and transformation uncertainty. Besides, constitutive models are nonlinear and have more complexity. Based on what has been explained, it can be admitted that reliability methods are a very useful tool for investigating geotechnical engineering issues. One of the most commonly used cases is the determination of the probability of failure and the reliability index of the proposed geotechnical design.

11.3.1 Methods for determining the reliability index

For each geotechnical problem, there should be a performance function which is called limit state function or margin of safety (M) as well. The main inputs for this function are resistance (R) and loading (Q) as its random variables. Despite parameters such as mean, standard deviation, scale of fluctuation, determined based on a database, each random variable should be identified by a distribution function such as normal, beta, lognormal distributions, etc. Thereafter, the performance function will have a distribution function. Based on central limit theorem, this function tends to have a normal distribution.

According to this performance function failure occurs when $M < 0$ as shown in Fig. 11.4 and Eq. (11.1). For precise and accurate determination of the probability

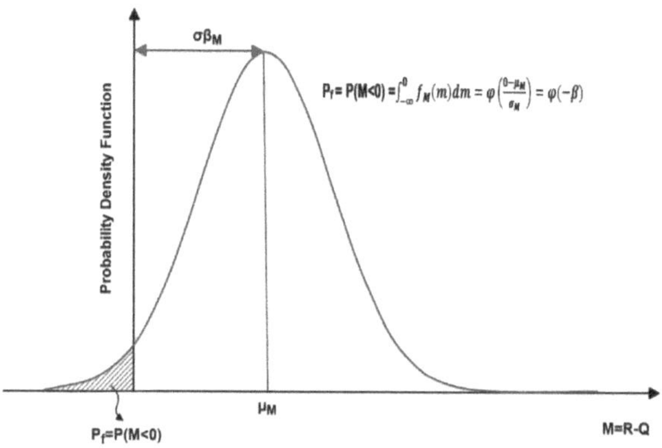

Figure 11.4 Definition of probability of failure and reliability index (β).

of failure for a nonlinear performance function, it is necessary to compute a multiple integral which is somehow complicated. In this regard, several methods such as first-order second moment method (FOSM), first-order reliability method (FORM), second-order reliability method, point estimation method, and Monte Carlo simulation (MCS) have been developed for determining probability of failure or in other words reliability index in geotechnical designs (Beacher and Christian, 2003; Ang and Tang, 2007; Harr, 1987).

$$P_f = \iiint_{M<0}^{1} \cdots \int f_x(X)dx \tag{11.1}$$

where x is the vector of the input parameters of the problem, f_x is the joint probability density function of the random variables, and M is the limit sate function, which can be defined in different forms and represents safety.

11.3.2 First-order second moment method

This approach is based on Taylor expansion theory. It is called first order because it includes only first-order terms in a Taylor expansion. Accordingly, it approximates a nonlinear function as a linear function in the vicinity of the mean value. Besides, this method is also called second moment because it enables estimates to be made of the variance, i.e., second moment of the function.

In this method, the reliability index is expressed as a function in terms of random input parameters. This method calculates the mean and standard deviation of the function in question based on the mean and standard deviation and the correlation coefficient of the input parameters.

Geometrically, the reliability index is the shortest distance between the mean, i.e., the most probable point and the failure point (Fig. 11.4). In this case, Φ is the standard normal distribution function and β is the reliability index, which is calculated as the ratio of average to standard deviation of the performance function.

$$\beta = \frac{\mu_M}{\sigma_M} \tag{11.2}$$

Let the performance function be f(X) and its expansion around the mean is as follows:

$$f(X) = f(\mu_X) + (X - \mu_X)\frac{df}{dx} + \ldots \tag{11.3}$$

Hence, the expectation of a nonlinear function of a single random variable is

$$E[f(X)] \approx E\left[f(\mu_X) + (X - \mu_X)\frac{df}{dx}|_{X=\mu_X}\right] = f(\mu_X) \tag{11.4}$$

And its variance is

$$Var[f(X)] \approx var\left[f(\mu_X) + (X - \mu_X)\frac{df}{dx}|_{X=\mu_X}\right] = \left(\frac{df}{dx}\right)^2 Var[X] \tag{11.5}$$

Table 11.3 shows the reliability index for the performance functions, mechanically equivalent. But due to their different appearances, their first derivatives will be different, resulting in different reliability indices when the random variables are considered the same for all these three cases and this is one major shortcoming of this method. Furthermore, application of first derivatives for nonlinear functions do not lead to accurate and precise approximation of reliability index.

Table 11.3 Reliability index for mechanically equal performance functions.

Performance function (M)	Reliability index (β)
R-Q	$\frac{\mu_R-\mu_Q}{\sqrt{\sigma_R^2+\sigma_Q^2}}$
(R/Q)-1	$\frac{\mu_R-\mu_Q}{\mu_R\left(\sqrt{\frac{\sigma_R^2}{\mu_R^2}+\frac{\sigma_Q^2}{\mu_Q^2}}\right)}$
Ln(R/Q)	$\frac{Ln\left(\frac{\mu_R}{\mu_Q}\right)}{\sqrt{\frac{\sigma_R^2}{\mu_R^2}+\frac{\sigma_Q^2}{\mu_Q^2}}}$

11.3.3 First-order reliability method

To overcome the deficiencies of the FOSM approach, mentioned above, the FORM analysis was developed. Apparently, the first iteration algorithm of FORM was supposed by Hasofer and Lind (1974), lacking several shortcomings due to its assumptions (Beacher and Christian, 2003). Thereafter, Rackwitz and Fiessler (1978) developed an iteration process to overcome these deficiencies (Haldar and Mahdavan, 1999).

11.3.4 Monte Carlo simulation

The Monte Carlo method is a very powerful and flexible simulation tool that can be used in a wide range of issues. The application of this method is simple, and if the simulations are complete, the results of this method can be fairly reasonable.

The advantages of this method can be high flexibility in the combination of data with a wide range of probabilistic distributions without creating high estimates and the ability to model the correlation between variables. In this method, based on the actual number of realizations, the frequency of the negative event for the performance function can be used as an indicator in calculating the probability of failure. However, in the first- and second-order methods, the probability of failure can be estimated by assuming a normal distribution for the performance function.

The optimal number of simulations leads to a proper estimation of the statistical moments of the joint distribution function. In this way, information can be obtained such as failure probability. In this method, the value of the reliability index is calculated according to Eq. (11.6).

$$R = 1 - \frac{N_f}{N} \tag{11.6}$$

where N is the total number of simulations performed, and N_f is the number of failed simulations. In this way, the probability of failure is computed as follows:

$$P_f = \frac{N_f}{N} \tag{11.7}$$

For example,
Estimate a simple integral of $\int x^2 dx$ in MCS.

```
Solution: one can run the following MATLAB code:
sum = 0;
% "n", shows the dimension of the matrix which contains generated
random numbers.
%"k" shows the number of iterations which is considered to estimate
this integral.
n = 100;
k = 100;
```

```
for kk=1:k.

hit=0;
miss=0;
X=rand(n);
Y=rand(n);

  for i=1:n
    for j=1:n

      W(i,j)=X(i,j)^2;
    if Y(i,j)<W(i,j);
        hit=hit+1;

    else

        miss=miss+1;

    end
  end
end
        integration=hit/(n*n);
        sum=sum+integration;

end

%the number which is called 'result' is the mean of iterations which is
done above and also is the final result.

result=sum/k
```

11.4 Case studies

In this section, different aspects of uncertainty in geotechnical engineering will be investigated as follows. The first example is about loading uncertainty effect on a relic bridge in Babolsar, Iran, as elaborated by Eslami et al. (2016). The second is about model uncertainty of various axial pile–bearing capacity predictive methods (Heidarie Golafzani et al., 2019a,b). Then, an algorithm was developed to reproduce CPT data based on their inherent soil variability (Jamshidi Chenari et al., 2018). Finally, for a specific method, the effect of inherent soil variability, transformation error, and model uncertainty were exhaustively investigated (Heidarie Golafzani, 2018).

11.4.1 Case study No. 1: loading uncertainty effects on shallow foundation failure (application of MCS)

Eslami et al. (2016) investigated the effect of loading uncertainty in design of a historic bridge located in Babolsar, the North of Iran. This pedestrian bridge experienced a flood in 1964 and was damaged due to geotechnical, hydraulic, and structural factors. The structural parts of the bridge were reinforced concrete and they included four piers in the river and five spans with a hinge support of almost equal lengths. Under each of these four piers, there was a shallow foundation. The width of this foundation was 1.4 m and its length was 4 m. In addition, the embedment depth was 2m before the flood. The width of the bridge was 2 m and its length was 101 m. By raising water level due to flooding in 1964 and creation of whirl water streams, scouring happened around the foundation and the piers of the bridge settled and consequently its decks subsided in the supports. Fig. 11.5 illustrates this historical bridge over time.

As time passed, the settlement increased to the extent that the historic bridge of the city was damaged. The bridge was scrutinized in three loading conditions, i.e., (1) before occurrence of the flood; (2) during the flood event; and (3) after the flood. In deterministic analyses, it is assumed that the failure occurs when FS is less than two. The ratio of ultimate bearing capacity to the bearing stress (i.e., the ratio of

Figure 11.5 Babolsar pedestrian bridge over time: (A) original bridge after 1964 flood, (B) damaged bridge (2012), and (C) retrofitted bridge (2015) (Eslami et al., 2016).

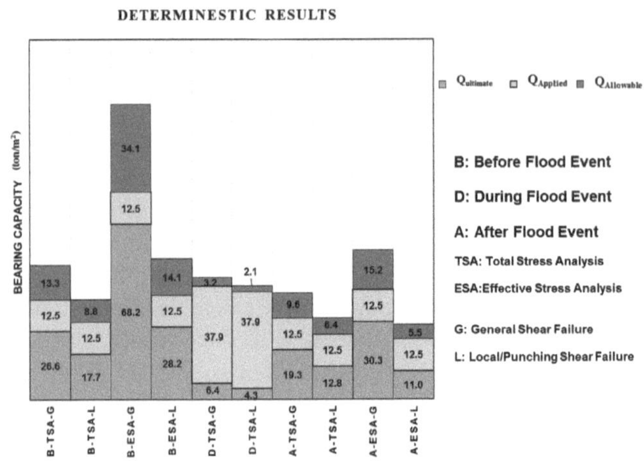

Figure 11.6 Results of deterministic solutions (Eslami et al., 2016).

applied vertical load to the area of footing) is considered as factor of safety. The results of deterministic analysis are shown in Fig. 11.6. The symbols B, D, and A stand for before, during, and after flood conditions. TSA and ESA present total stress analysis and effective stress analysis. G and L substitutes for general shear failure and local/punching shear failure, respectively.

As shown in Fig. 11.6, under condition of before flood loading, local shear/punching failure occurrence is probable for total stress analysis and the bridge did not satisfy the capacity requirements. For other cases, the bridge remained safe. In addition, under during flood condition, due to changes in loading and foundation conditions, the performance of the bridge became unsatisfactory, and for all cases q_{ult} is extremely lower than q_{app}. Furthermore, under after flood condition, analyses have been done with the assumption that no failure had been occurred and only the embedment depth became zero during flood. The results show that, with the reduction of embedment depth, at least three of four cases show failure occurrence. In reliability-based analyses, soil shear strength parameters, i.e., c (cohesion) and ϕ (friction angel), are considered as random variables, distributed normally in two conditions: uncorrelated and correlated. The factor of safety, FS is defined as the ratio of ultimate bearing capacity to the applied stress. The probability of failure P_f is defined as the ratio of the frequency of FS less than unity to the total number of FS.

In Table 11.4, the results of reliability analysis are shown. In addition, the ultimate bearing capacity for both methods was compared and concluded that the amounts of bearing capacity are close to each other. But reliability analysis shows that there is always a probability of failure for all cases. The probability of failure before flood is lower than after flood condition, and this is as a result of decrease in foundation embedment depth. During flood, probability of failure for both cases were 100%, due to extreme loading and reduction in foundation bearing capacity.

Because of lateral applied pressures induced by flood, eccentric loading is unavoidable. When eccentricity exceeds certain limits, tension in foundation base (uplift)

Table 11.4 MCS results compared with deterministic results (Eslami et al., 2016).

			Pf (%)	FS		qult		
Condition			**MCS**	**Mean**	**SD**	**Mean**	**SD**	**q_{ult}(Deterministic)**
Before flood	TSA	General shear failure	1.68	2.13	0.53	26.63	6.66	26.56
		Local shear failure	11.83	1.42	0.36	17.74	4.44	17.68
	ESA	General shear failure	0	5.70	1.66	71.17	20.77	68.18
		Local shear failure	0.06	2.18	0.45	27.29	5.6	28.18
During flood	TSA	General shear failure	100	0.07	0.02	2.7	0.67	6.37
		Local shear failure	100	0.05	0.01	1.8	0.45	4.25
After flood	TSA	General shear failure	8.1	1.54	0.39	19.26	4.8	19.25
		Local shear failure	45.8	1.03	0.26	12.8	3.2	12.8
	ESA	General shear failure	0.25	3.3	1.08	41.67	13.55	30.3
		Local shear failure	27.27	1.2	0.32	15.05	4.06	11.01

*** q_{ult} is in $^{ton}/_{m^2}$***.

develops. Because the tensile strength of soil is approximately zero, part of the footing will not transmit loads to the soil. To avoid this, the footing should be designed such that $e_B < \frac{B}{6}$ and $e_L < \frac{L}{6}$ (B and L are footing width and length, respectively; e_B, e_L are eccentricity in these directions). In the case of Babolsar bridge, by changing the bearing width from 2 to 3m, the eccentricity (i.e., e_L) changes from 1.67 to 2.5m. Because the minimum length should satisfy the inequality of $6e_L < L$, this footing requires at least a length of 10−15 m. In comparison to the present length, it was clear that the bridge would collapse during flood event.

Investigations demonstrated the importance of soil investigations, soil tests, and the effect of loading combinations especially in extreme events. Because of deterministic and reliability analyses, it is evident that shallow foundation is not appropriate in this case and as a result of occurrence of flood or other dynamic loads, the pier foundation of this bridge is supposed to be constructed deeper or at least semideep accompanying with bridge superstructure.

11.4.2 Case study No. 2: CPT production considering soil inherent variability

As piles have key role in geotechnical engineering, the estimation of their bearing capacity is of great importance. In this regard, various methods with different theoretical assumptions and bases have been developed. In situ-based methods, performed under field stresses and boundary conditions, are considered as complementary tools to static analysis and seem to be reasonable and cost-effective enough to estimate pile-bearing capacity. Therefore, reproducing CPT records based on existing CPT are different realizations for a specific site and can improve pile design scheme, significantly (Jamshidi Chenari et al., 2018).

Although many investigations have been conducted to extract the statistical properties of soil characteristics and to model the variability of soil specifications in geotechnical problems by presuming a constant mean and variance through the field, random field theory has been developed to model soil variability, especially in recent years (Jamshidi Chenari and Kamyab Farahbakhsh, 2015; and Ching and Phoon, 2017). However, the statistical uncertainty is inevitable due to the lack of collected amount of information, and this causes the parameters, applied to characterize random field, not to be estimated with complete certainty (Ching and Phoon, 2017).

Furthermore, the stationary random field theory is not capable of simulating the CPT profile accurately. The variability in CPT data arises from two sources: (A) lithological variability which is due to different soil types and (B) the inherent heterogeneity derived from different depositional conditions of similar soil particles.

Most of studies in random field generation of CPT profiles consider the stationary residual CPT profile resulting in constant mean, variance, correlation structure, and probability distribution function throughout the depth. Jamshidi Chenari and Kamyab Farahbakhsh (2015) showed the difference between stationary and nonstationary realization approaches, efficiently. They demonstrated that the proposed nonstationary approach had better performance in realization of CPT profiles than the stationary approach. However, the nonstationary algorithm adopted by Jamshidi Chenari and Kamyab Farahbakhsh (2015) still needs modifications, i.e., considering the soil stratification to enhance its capability to simulate the lithological heterogeneity.

11.4.2.1 CPT-based soil classification

A simple mathematical procedure, categorized into several steps: (1) finding the soil classification index of each sample at the given depth in the CPT profile based on Eslami-Fellenius (1997) soil classification system; (2) strata detection by using a simple mathematical definition (i.e., uniformity index [UI] herein); (3) merging the neighboring primary layers with the same names and indices; and (4) lens detection, was adapted (Jamshidi Chenari et al., 2018). The uniformity or the variability of each soil types can be assessed by the definition of a variable called UI as following equation:

$$UI = \frac{Number\ of\ a\ Specific\ Soil\ Type\ in\ a\ Window}{Total\ Number\ of\ Data} \times 100 \tag{11.8}$$

A moving window is defined to sweep through the CPT dataset and calculate the UI of each consisting soil type at each depth. The UI of each soil type through the depth is stored in a matrix with [N×D] dimensions where N denotes the number of sampling during the test and D the number of consisting soil types. A sample utilization of the proposed procedure is illustrated in Fig. 11.7 for a sample CPT profile.

Comparison between the Fuzzy soil classification method (Zhang and Tumay, 1999) and the method proposed in current study is possible. In this regard, the Eslami–Fellenius soil type indices of 1 and 2 are merged together and named EFSG1,

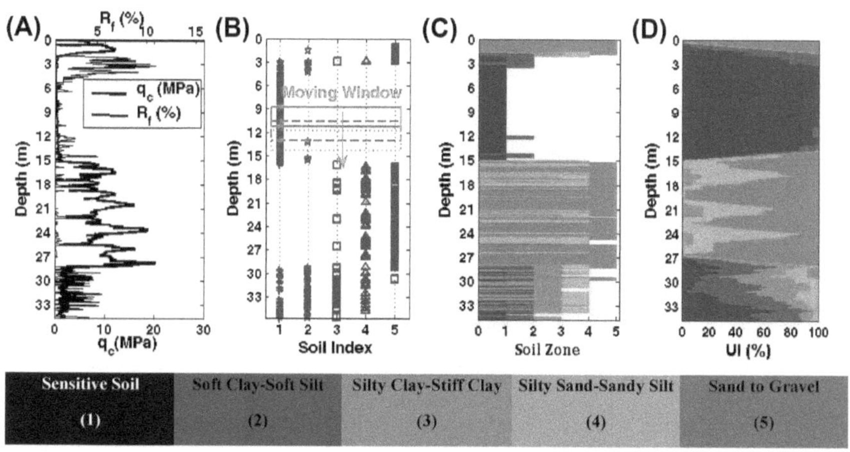

Figure 11.7 Uniformity index definition and calculation for a sample CPT data (case no. 1) through the whole depth: (A) CPT profile; (B) soil index estimation; (C) soil zones identification; and (D) uniformity index (Jamshidi Chenari et al., 2018).

representing a clayey soil group; combination of soil type indices of 3 and 4 are named EFSG2, representing a mixture soil group, and soil type index 5 is renamed as EFSG3, representative of a sandy soil group. For the CPT sounding of Fig. 11.7, these two classification methods are applied and shown in Fig. 11.8.

Comparisons show that in spite of differences between the two classification schemes, they closely resemble each other, e.g., the segment of depth, classified as *EFSG*1 based on the proposed method, agrees well in the same length with the *HPC* designation in the Fuzzy approach (Zhang and Tumay, 1999). Because the procedure adopted in current study consists of 5 soil-type groups, it could classify the subsoil more accurately compared with the Fuzzy classification method.

11.4.2.2 Soil stratification

Fig. 11.9 depicts the presented flowchart by Jamshidi Chenari et al. (2018) and divides the stratification process into four major sections. The first section is preliminary treatment in which the Eslami–Fellenius soil type (Eslami and Fellenius, 1997) is identified based on the CPT sounding in each depth. In addition, the total number of associated soil types (NAS) shall be known in this section. Then, the UI matrix is formed and calculated through the moving window scheme. After that, stratification process and the detection of primary layers are done in the third step. Finally, all the layers are checked to cover the entire depth. Some modifications should sometimes be applied to the uppermost and lowermost layers in the CPT sounding. These layers may not start from the ground surface or not end at the final depth of sounding. This modification depends on the thickness of the isolated layers, which could be

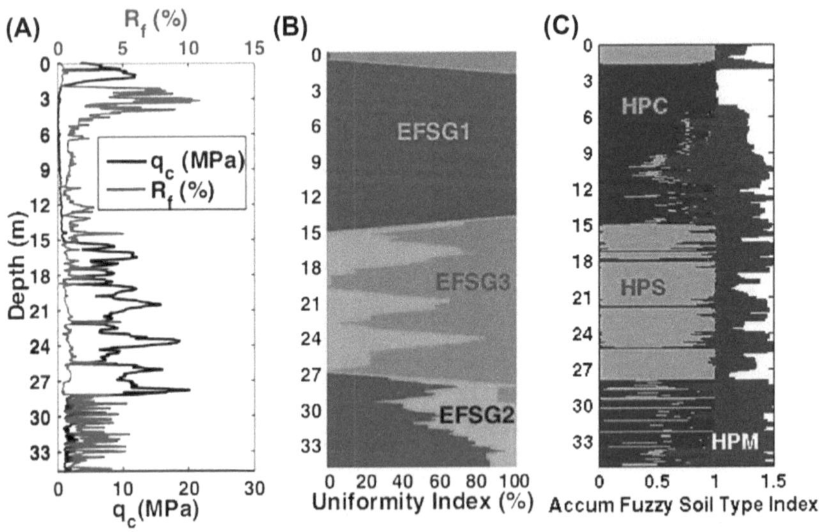

Figure 11.8 Comparison between the proposed and the Fuzzy soil classification methods for CPT data of Fig. 1: (A) CPT profile; (B) proposed soil classification system; and (C) fuzzy soil classification system (Jamshidi Chenari et al., 2018).

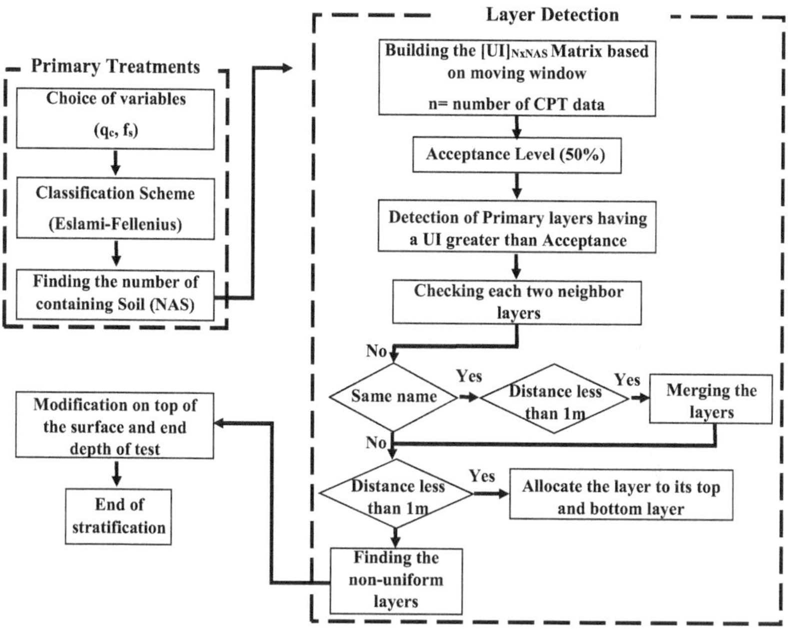

Figure 11.9 Algorithm of stratification based on CPT profiles (Jamshidi Chenari et al., 2018).

considered as new layers, provided they have a thickness greater than 1 m. Otherwise they are merged together.

For the CPT sounding illustrated in Fig. 11.7, a stratifying analysis is performed as shown in Fig. 11.9 to show the efficiency of stratification. Fig. 11.10 elucidates that sharp changes in the CPT cone tip resistance profile imply transformation in soil type or consequently the layer type. Fig. 11.11 illuminates the CPT data, scattered

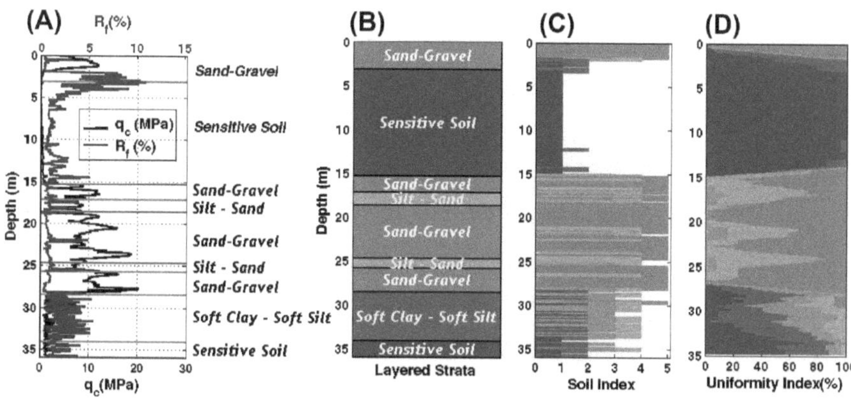

Figure 11.10 Exemplifying the Jamshidi Chenari et al. (2018) method of stratification: (A) CPT profile; (B) layered strata based on proposed method; (C) Eslami–Fellenius soil index; and (D) proposed method of soil classification system (Jamshidi Chenari et al., 2018).

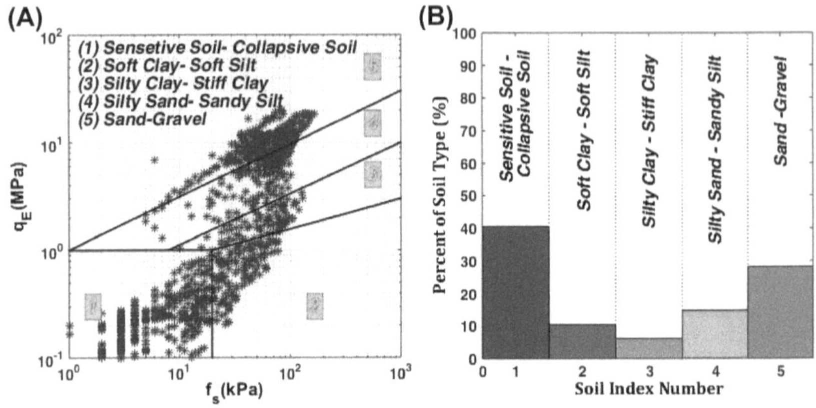

Figure 11.11 Scattering and percentage of contributing soil types for the Case No. 1: (A) Eslami−Fellenius chart and (B) soil-type contribution (Jamshidi Chenari et al., 2018).

in the Eslami−Fellenius classification chart (Eslami and Fellenius, 1997), and the associated percentage of existing soil types through the whole depth. It is shown that the most contributing layers of subsurface strata are sensitive soils−collapsible soil and sand−gravel, respectively. Approximately, 40% of containing soil type belongs to the soil type index 1% and 30% to the soil-type index 5.

One of the controversial issues of the proposed stratification algorithms shown in Fig. 11.9 is the choice of acceptance level for detection of the primary layers. Generally, the stratification process is somehow arbitrarily subject to variation. It largely depends on the engineering judgment of the researcher. In other words, all proposed methods for stratification are not unique and stratification introduced by different methods may render considerable difference. Jamshidi Chenari et al. (2018) elaborated on an acceptance level of 50% in their analyses.

11.4.2.3 Nonstationary realization of CPT data

Disregarding the lithological heterogeneity of CPT profiles may result in inaccurate realizations. Hence, realizing CPT data with a reasonable precision requires consideration of both sources of heterogeneity. Therefore, CPT profiles can hardly be considered as stationary data. From a statistical point of view, the inherent variability has a mean value and a parameter which shows the magnitude of the fluctuations around this mean. The magnitude of the fluctuations can be quantified by the COV and correlation length.

Jamshidi Chenari et al. (2018) proposed an algorithm for nonstationary realization of CPT data (Fig. 11.12). It should be noted that both stationary and nonstationary methods of random field generation share the same structure of simulation, but there are some differences in the details of calculation. Based on this algorithm, it is possible to realize the CPT data using nonstationary random field. The result of stationary and nonstationary CPT production algorithm is illustrated in Fig. 11.13 for two sample cases.

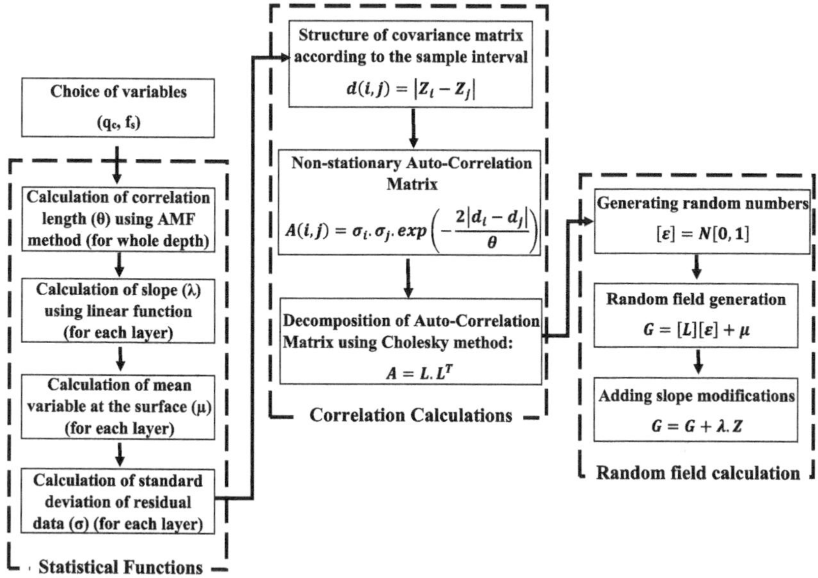

Figure 11.12 Applied nonstationary random field generation algorithm (Jamshidi Chenari et al., 2018).

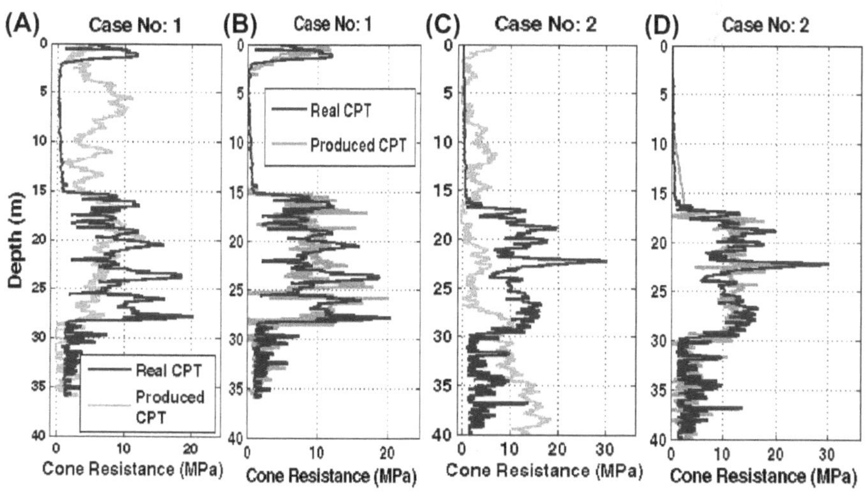

Figure 11.13 The realization of CPT cases: (A) case no. 1—stationary; (B) case no. 1—nonstationary; (C) case no. 2—stationary; and (D) case no.2—nonstationary (Jamshidi Chenari et al., 2018).

11.4.3 Case study No. 3: statistical and probabilistic assessment of model uncertainty for prediction of pile foundation–bearing capacity

Over the years, several methods and approaches have been developed to estimate the axial pile capacity based on various assumptions and approaches. Suffering from over-simplifications, static analysis is one of the first approaches in determining the axial pile–bearing capacity. On the other hand, in situ test–based methods can be realized as complementary tools to static analysis. Moreover, according to the scale of project, it is determined whether performing pile load test is justified or not. Employing different design methods, resulting in a wide range of values, complicates the selection of an appropriate design scheme.

Several studies by Briaud and Tucker (1988), Schneider et al. (2008), Dithinde et al. (2011), Kenarsari, et al. (2011), and Moshfeghi and Eslami (2016) confirm the existence of considerable uncertainty associated with pile-bearing capacity prediction. It is important to accurately quantify the ability of a method to predict axial capacity for (A) assessing the effect of uncertainty on a specific method and (B) allowing comparison of the uncertainty regarding several different methods (Tang and Phoon, 2016).

11.4.3.1 Model parameter

Predictive methods include various assumptions and bases, affecting their performance according to their model uncertainty. Besides, inherent soil variability influences methods similarly due to the direct application of CPT records in prediction process and so does measurement error. Hence, the uncertainties associated with the used predictive models play a prominent role in determining the accuracy and efficiency of a method (Phoon and Kulhawy, 2005). To quantify these uncertainties, a parameter termed the resistance bias factor (i.e., model parameter) is defined. This factor is the ratio of the measured to predicted pile axial capacity ($\lambda_R = Q_m/Q_p$).

11.4.3.2 Database and pile-bearing capacity prediction

The data bank used in this study includes CPT soundings recorded in proximity of pile locations and full-scale pile loading tests for 60 piles driven into different soil types, i.e., 47% of site soils were clayey soil, 30% sandy soil, and 27% mixed soils. Pile lengths vary between 6 and 79 m and pile diameters range from 114 to 1500 mm. In addition, shape and material of the piles differ and contain steel and concrete piles as well as round, square, and H piles. Moreover, all the open-ended (OE) pipe piles have been considered as close-ended (CE) due to potential soil plugging. These 60 case records have been obtained from three sources namely Eslami (1996) (13 cases), Niazi (2014) (30 cases), and AUT-CPT and pile database of Moshfeghi and Eslami (2016) (17 cases). The compiled database pertains to 13 countries from 4 different continents. A summary of case records is also presented in Heidarie Golafzani et al. (2019a).

With regard to the compiled data bank, the performance of static analyses and in situ tests such as SPT- and CPT-based methods is assessed in predicting the axial pile–bearing capacity. Thus, five CPT-based methods, five SPT-based methods, and two static analyses (i.e., one total stress approach (API, 2006) and one effective stress approach (CFEM, 2006)) were considered (Heidarie Golafzani et al., 2019a).

Because the SPT results for the collected data bank were not available, Equations 1 and 2 have been employed to estimate SPT N value. These equations form a simple method to estimate the SPT N value from CPT results (Lunne et al., 1997).

$$\frac{\frac{q_c}{P_a}}{N_{60}} = 8.5\left(1 - \frac{I_C}{4.6}\right) \tag{11.9}$$

where Ic is the Robertson soil type index and Pa is the reference stress = 100 kPa. Jefferies and Davies (1993) suggested that the mentioned method provide a better estimate for the SPT N values than the actual SPT due to the poor repeatability of this test.

11.4.3.3 Statistical and probabilistic assessment of axial pile–bearing capacity

This classification includes assessment of the mean and standard deviation of the model parameter, comparing the best fit line and the root mean square error (RMSE). Comparison of the mean and standard deviation of the model parameter through different methods reveals the accuracy and precision of each method. The closer mean (μ) to *1* and closer standard deviation (σ) to *0* means the method is near to ideal case ($\mu_{\lambda R} = 1$, $\sigma_{\lambda R} = 0$). In addition, the best fit line checks for accuracy of the methods and the closer Q_m/Q_p to 1 signifying better performance. RMSE monitors model errors associated with each method, and the lesser this value, the better is the method. The results of the statistical criteria are presented in Table 11.5.

Probabilistic approach investigates the model parameter via its distribution function. 20% accuracy level, an efficient indicative of the model performance, checks for precisions around unity for model parameter. The more the area between $\lambda_R = 0.8$ and 1.2, the better the performance of the method, i.e., higher accuracy and precision. On the other hand, cumulative distribution function determines the values of the P_{50} and P_{90}. P_{50} value illustrates the tendency of a method in overprediction or underprediction, and the lesser $P_{90}-P_{50}$ value reveals better precision for each method. Confidence interval (CI) demonstrates that for another unspoiled compiled data bank/sample, the mean of the model parameter stands in that interval. In other words, the closer the method to yield values to the ideal case, the lesser this interval will be. The results of the probabilistic criteria are given in Table 11.6.

For reliability-based assessment of these methods, Heidarie Golafzani et al. (2019a,b) implemented the efficiency ratio, i.e., the ratio of LRFD resistance factor to resistance bias factor or model parameter and actual factor of safety, i.e., the product of resistance bias factor to factor of safety; $FS^{\times}\lambda_R$ introduced by Paikowsky et al.

Table 11.5 Statistical assessment of the model parameter (Heidarie Golafzani et al., 2019a).

Methods	Mean and standard deviation			Best fit line		RMSE
	μ	σ	COV	Q_m/Q_p	R^2	
Schmertmann	1.12	0.39	0.35	0.84	0.98	1499
Dutch	1.14	0.47	0.41	0.8	0.98	1866
LCPC	1.29	0.6	0.47	0.68	0.93	3401
Meyerhof (CPT)	0.99	0.44	0.44	0.96	0.92	1714
UniCone	0.99	0.33	0.33	1.13	0.92	1864
Meyrhof (SPT)	1.13	0.59	0.52	0.63	0.92	4196
Shioi and fukui	0.98	0.42	0.43	0.38	0.96	10,572
Bazaara and Kurkur	1.57	0.78	0.5	0.8	0.95	2087
Briaud and Tucker	0.85	0.34	0.4	0.95	0.98	876
Decourt	1.12	0.61	0.54	0.46	0.96	7722
API	1.14	0.5	0.44	0.84	0.98	1486
CFEM	0.87	0.76	0.86	0.52	0.92	6250

Table 11.6 Probabilistic assessment of the model parameter (Heidarie Golafzani et al., 2019a).

Methods	20% accuracy level	Cumulative distribution function			Confidence interval
		P50	P90	P90–P50	
Schmertmann	0.4	1.13	1.59	0.46	0.1
Dutch	0.39	1.13	1.66	0.53	0.12
LCPC	0.33	1.12	2.38	1.26	0.16
Meyerhof (CPT)	0.36	0.92	1.49	0.57	0.11
UniCone	0.43	0.96	1.46	0.5	0.08
Meyrhof (SPT)	0.31	1.09	1.85	0.76	0.15
Shioi and fukui	0.34	0.93	1.51	0.58	0.11
Bazaara and Kurkur	0.25	1.4	2.99	1.59	0.2
Briaud and Tucker	0.35	0.73	1.36	0.63	0.09
Decourt	0.31	0.97	2.28	1.31	0.16
API	0.37	1.09	1.79	0.7	0.13
CFEM	0.19	0.57	1.86	1.29	0.2

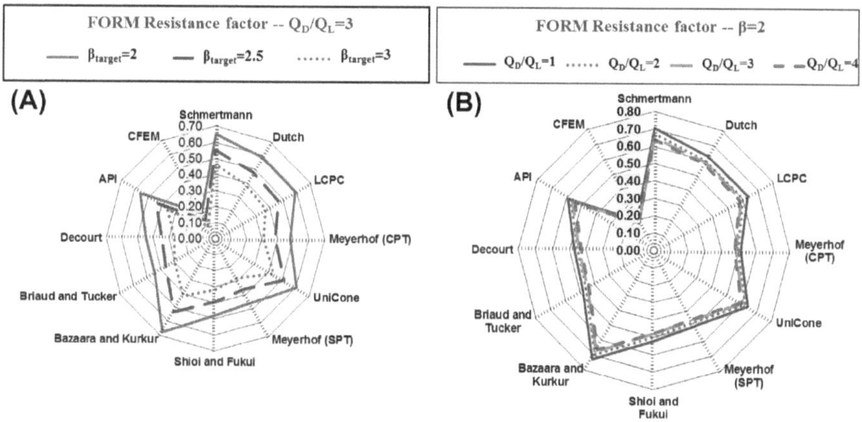

Figure 11.14 FORM resistance factors for different methods: (A) dead to live load ratio effect and (B) target reliability effect (Heidarie Golafzani et al., 2019a).

(2004). Fig. 11.14 depicts the variation of FORM resistance factor with varying dead to live load ratios and target reliability index. Methods such as Bazaraa and Kurkur (1986), Schmertmann (1978), UniCone of Eslami and Fellenius (1997), and LCPC of Bustamante and Gianeselli (1982) render greater resistance factors than other methods.

Fig. 11.15 provides the comparison of efficiency ratios and actual factors of safety for the investigated methods. As depicted by this figure, the more accurate and precise a method (i.e., lower COV value and bias closer to unity) is, the higher its efficiency ratio and the lower its actual factor of safety will be.

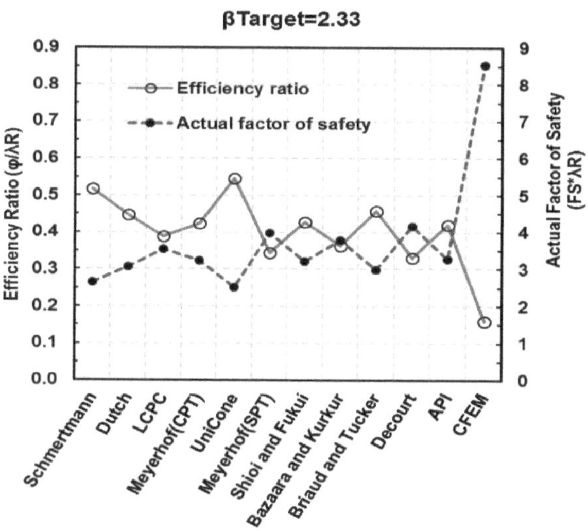

Figure 11.15 Efficiency ratio and actual factor of safety for different methods (Heidarie Golafzani et al., 2019b).

Comparisons show that conservative methods, due to their built-in safety, attain higher resistance factors. However, efficiency ratio, a distinct measure of reliability, considers both resistance factor and resistance bias factor to evaluate performance of different methods. Therefore, the Bazaara and Kurkur (1986) method does not predict the axial pile—bearing capacity efficiently for the investigated database, although it has had the greatest resistance factor. Moreover, the UniCone method (1997) and the Schmertmann (1978) method were the most consistent prediction approach with the lowest COV.

Fig. 11.16 compares the performance of different methods under statistical-, probabilistic-, and reliability-based criteria in radar charts. In this assessment, the scores on each criterion vary from 1 to 12 for the method with the best and the poorest performance, respectively. The method occupying smaller area is realized to perform better than the others.

The better performance of UniCone is seemingly due to several factors such as the excess pore water pressure incorporation to yield long-term behavior of piles, employing a CPT-based soil classification system for better estimation of shaft resistance, correcting the cone stress for the pore water pressure acting on the cone shoulder and the failure zone for estimation of the unit toe resistance.

Another comparison, also, is done among static analyses, SPT- and CPT-based methods, illustrated in Fig. 11.17, confirming more reliability for the direct CPT—based methods than the SPT-based methods and static analyses or indirect methods in other words. The CPT-based methods employ more reliable and repeatable input data (i.e., CPT soundings), recorded continuously under field stresses and boundary conditions. Additionally, these methods directly relate pile unit toe and shaft resistance to qc and fs readings through coefficients, depending mainly on soil and pile type. Consequently, direct CPT—based methods seem to pose less uncertainty in pile capacity prediction.

Although SPT is a low cost, simple and commonly used in situ test, NSPT values are recorded in discontinuous intervals, influenced by many factors and require correction. Consequently, SPT-based methods lack reliability in comparison with the direct CPT—based methods. However, SPT is an in situ test and is prioritized over the static analyses as depicted in Fig. 11.17.

The static analyses employ different empirical correlations to estimate geo-parameters, to use them in simplified geotechnical models. Furthermore, these geo-parameters are considered as deterministic in these methods while SPT and CPT records demonstrate variation of the soil properties in both horizontal and vertical directions. This in turn leads to a higher-level error being incorporated in the estimation of pile axial—bearing capacity.

11.4.4 Case study No. 4: investigation of the effect of various uncertainty sources on a CPT-based axial pile—bearing capacity predictive method

Niazi and Mayne (2016) presented an algorithm to modify the UniCone method (Eslami and Fellenius, 1997) based on a newer data bank. Robertson soil classification

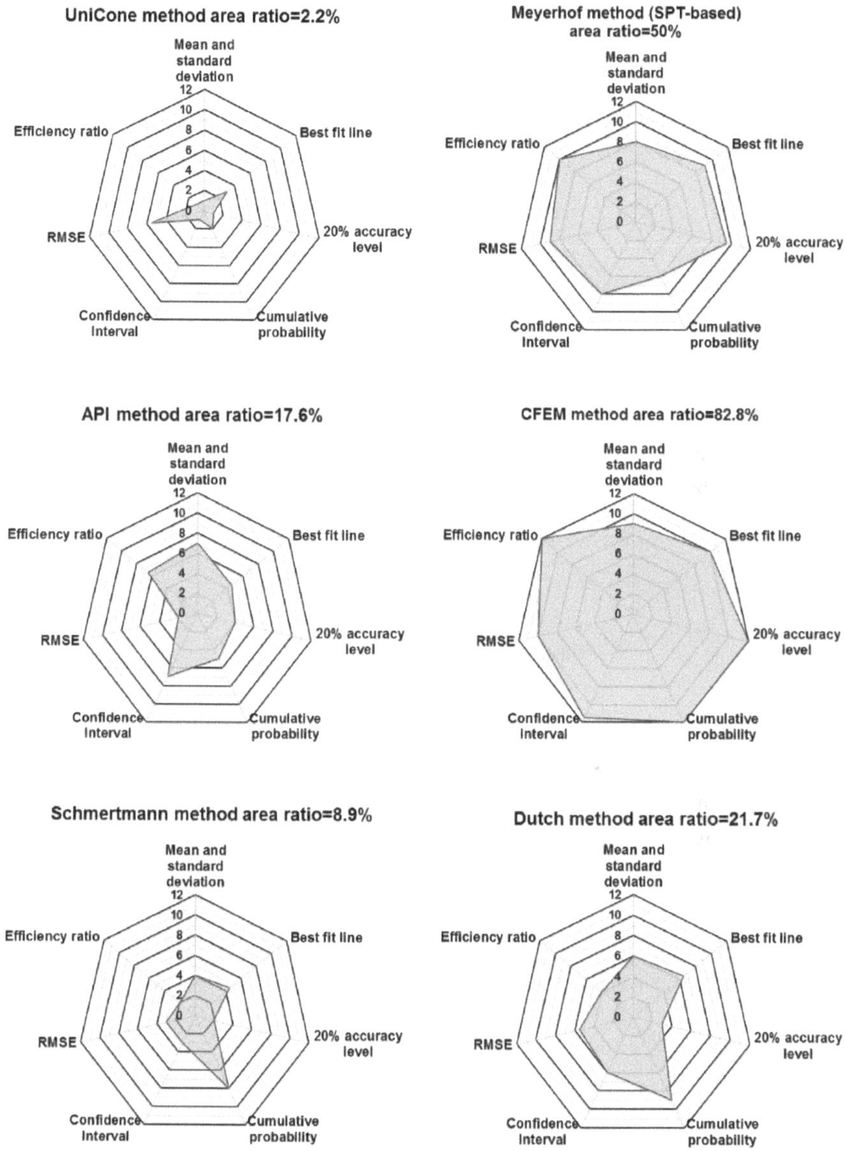

Figure 11.16 Performance of different methods based on various criteria (Heidarie Golafzani et al., 2019b).

system (Robertson, 2010) is applied for soil layering detection and based on these layers the required coefficient for estimation of unit shaft resistance will be determined. Comparison of predicted bearing capacity between UniCone (Eslami and Fellenius, 1997) and Modified UniCone, i.e., approach 1 (Niazi and Mayne, 2016) revealed

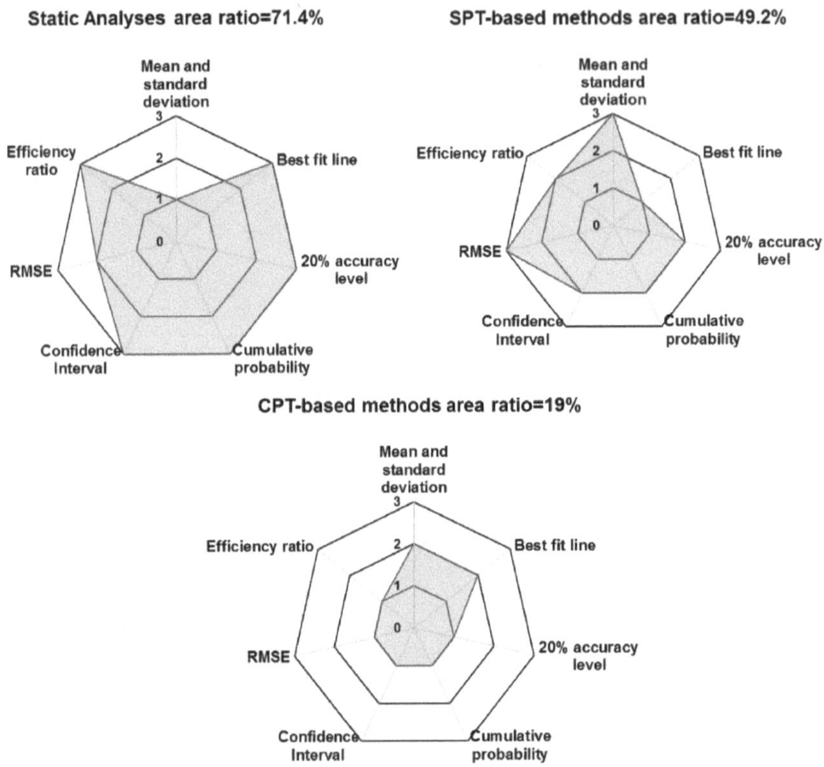

Figure 11.17 Assessment of the three common universal axial pile–bearing capacity prediction approaches (Heidarie Golafzani et al., 2019b).

that this fresh method does not perform as well as its former version for the compiled data bank. To reduce latent uncertainties, a little bit of changes was implied, the soil layering was ignored, and C_{se} was implied for each depth (i.e., approach 2). As demonstrated in Table 11.7 the results grew worse.

Table 11.7 Comparison among conventional and enhanced UniCone methods.

	μ	COV	Median	IQR[a]	Range	RMSE
UniCone (Eslami and Fellenius, 1997)	0.98	0.33	0.95	0.33	1.40	1836
Niazi and Mayne (2016) (A1)	0.96	0.75	0.87	0.65	3.80	20,328
Modified Unicone (A2)	0.83	0.6	0.74	0.55	2.95	22,951
Implemented modified UniCone (A3)	0.97	0.33	0.94	0.47	1.65	2570

[a]IQR stands for Inter Quartile Range. The less this value, the better the performance.

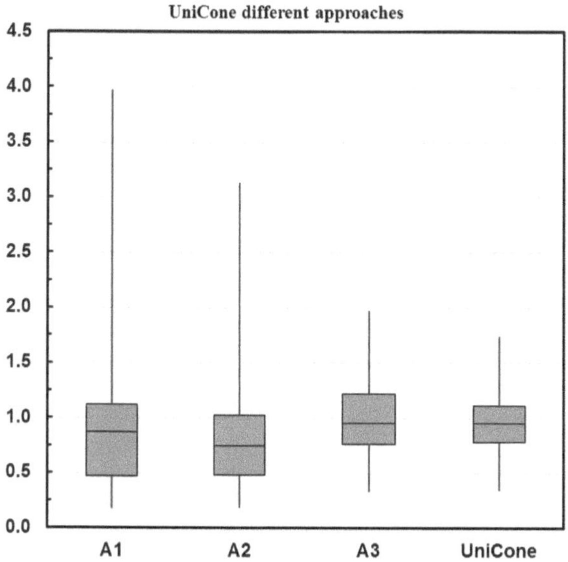

Figure 11.18 Comparison among conventional and enhanced UniCone methods (Heidarie Golafzani, 2018).

Niazi (2014), also, modified Eslami–Fellenius soil classification chart (Eslami and Fellenius, 1997) and suggested the values of C_{se} for each soil type. In this regard, it was decided to implement the modified Eslami–Fellenius classification chart (Niazi, 2014), for each depth and its related C_{se}. Table 11.7 and Fig. 11.18 illuminate the difference between these four approaches for the compiled data bank. The implemented Modified UniCone performs better than the one suggested by Niazi and Mayne (2016) and the other approach. Apparently, implementing Robertson soil classification system and application of experimental correlations for estimating C_{se} (transformation uncertainty effect) and soil layering (disregarding inherent soil variability) are distinct differences among these approaches.

11.5 LRFD application for pile design (new approach)

Although CPT-based methods provide more reliable result for prediction of axial pile–bearing capacity, this question arises which method provides more reliable results for a specific project? Or what factors should be considered for selection of a suitable CPT-based method?

The detailed investigations on different methods reveal that the variety of criteria and assumptions, implemented for each method, lead to a wide range of predictions. These criteria can be categorized as follows:

1. How many CPT readings are considered as input variables in a method? Some methods rely solely on parameter(s) qc or qc and fs, while only limited methods consider u2 in addition to qc and fs.

2. Are qc values modified for the effect of pore water pressure on cone shoulder? Or how is the efficiency affected by implementing qt values instead of qc?
3. What is the model assumption for that method? Some of the methods are based on total stress approach or empirical correlations, while others apply effective stress approach directly or indirectly.
4. Which failure criterion is implemented for interpretation of load tests as the reference load? Does the failure criterion predict short-term bearing capacity or long-term bearing capacity?
5. How long after the pile installation will the pile load test be performed? Do methods consider this time frame? Or, how is the efficiency of a method affected when considering time frame?
6. Do methods distinguishing pile installation methods bring about more reliability than others?
7. Does that specific method consider soil classification along the pile embedment length for calculating shaft capacity? In other words, does classifying soil as sand and clay impact on the uncertainties associated with a specific predictive method? Or implementing more complicated classification schemes based on direct or indirect usage of CPT records?
8. Does friction fatigue affect the reliability of predictive methods? Or how does imposing or ignoring an upper limit for side and tip resistance affect the performance of a method?
9. Do predictive methods take into account the influence of loading direction?
10. Which failure mechanism is considered to govern the influence zone around pile base? And what is the averaging technique figure?
11. Does record interval exert an influence on uncertainty of predictive methods?

The summary of these criteria is presented in Fig. 11.19. All these criteria can be assessed in terms of reliability-based approaches. Considering each criterion, there are several subsections that each method belongs to. In this regard the efficiency ratio for each method, placed in a subsection, is evaluated based on that group efficiency.

Finally, these efficiency ratios are gathered in a radar chart for all these criteria, and the ratio of shaded area to total chart area is introduced as the area ratio. The more the area ratio, the better the performance of that specific method (Fig. 11.20).

In the procedure of pile geotechnical design, either of these two conditions are encountered: (1) sufficient site investigations are performed due to low variability in a specific site, and there is adequate information on site conditions such as CPT records, pile load test results, etc.; and (2) because of economical aspects or project

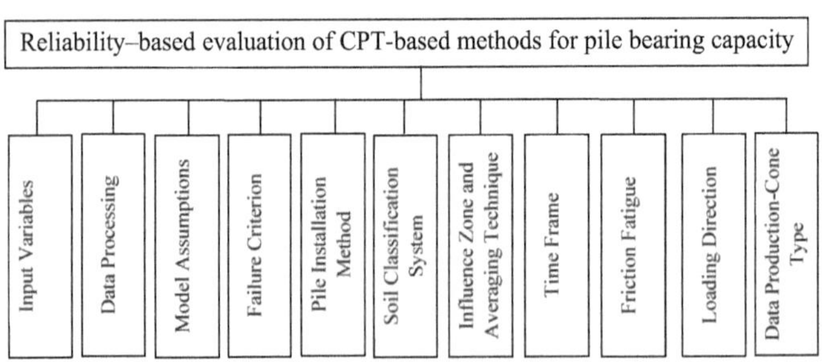

Figure 11.19 Different assessment criteria for reliability-based evaluation of predictive methods (Heidarie Golafzani, 2018).

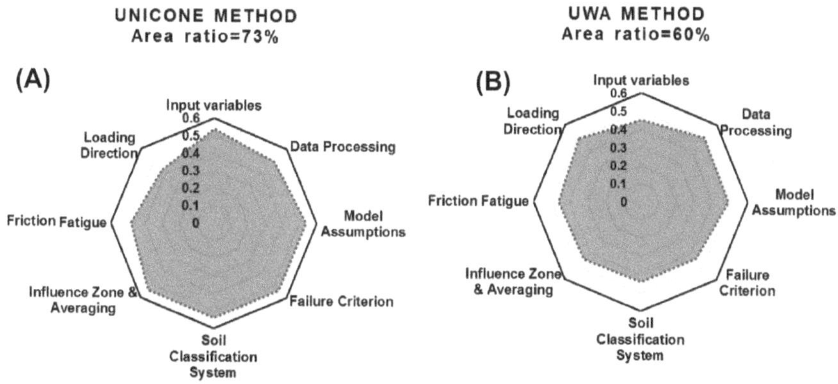

Figure 11.20 Reliability-based assessment of two methods: (A) UniCone method and (B) UWA method (Heidarie Golafzani, 2018).

limitations, it is not possible to perform in situ testing as many that fully represent the site conditions. This especially applies to the site with high inherent variability.

For either case, geotechnical databases can enter the design procedure accompanied by performance-based approaches such as efficiency ratio (as a measure of capacity prediction efficiency), serviceability criteria, value engineering prospects, probability, and reliability assessments. The procedure of employing such geotechnical databases through smart data selection is presented in Fig. 11.21.

The databases can provide the opportunity to reproduce data, such as CPT and pile load–displacement records, similar to the ones encountered in the project site. Considering the availability of these data, the need for performing the time-consuming or expensive loading tests diminishes, and consequently, it can help the efficiency of the project. Moreover, the ability of considering soil inherent variability enhances the geotechnical engineer's understanding of the site conditions.

In the next step, after acquiring necessary data, design approaches must be opted. For commonly used CPT methods of determining pile axial capacity, several evaluation criteria must be employed to assess and compare the performance of each method in a given site condition so as to select the most compatible method with the reasonable performance. In this regard, the more well-known approaches which can used for assessments are as follows:

- Statistical and probabilistic
- Reliability-based
- Risk analysis

Then, by employing the superior method or methods (screening) and LRFD approach, the optimum design can be achieved.

As stated by Fellenius (2015), design of a pile foundation for axial load starts with an analysis of how the load is transferred to the soil, often thought limited to determining only the pile capacity, sometimes separating the capacity on components of shaft and toe resistances. However, the load-transfer is also the basis for a settlement

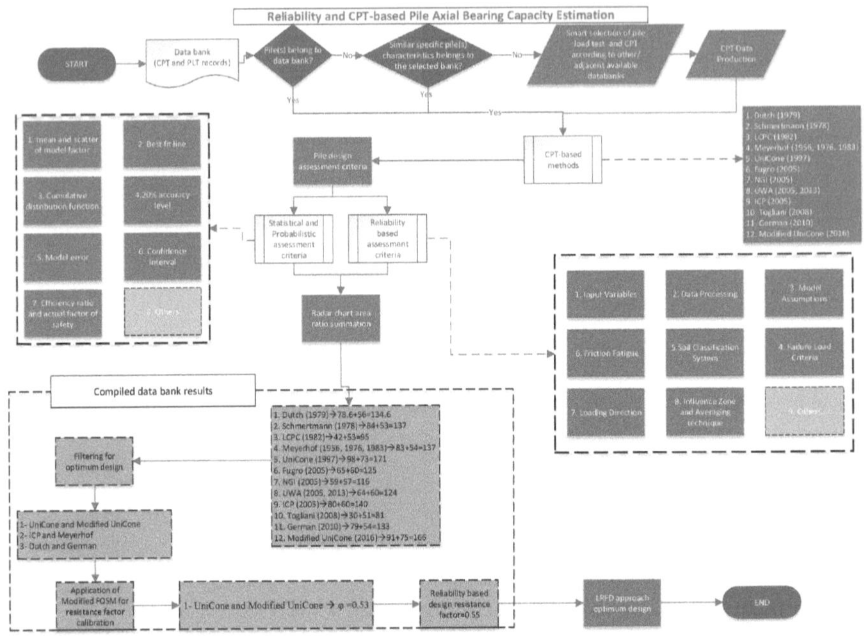

Figure 11.21 The procedure of designing a pile using geotechnical databases (Heidarie Golafzani, 2018; Eslami et al., 2019).

analysis, because in contrast to the design of shallow foundations, settlement analysis of piles cannot be separated from a load-transfer analysis. Therefore, it is evident that evaluating the bearing capacity must be based on load—displacement behavior, which can be considered through the employment of such geotechnical databases.

11.5.1 Case study No. 5: application of the proposed algorithm/ practical application of the proposed algorithm

Shamshirgaran and Ebrahimian (2018) investigated different design methods for an open-ended steel pipe pile (embedment length = 88 m and diameter = 1.5 m). The result of CPTu, done in vicinity of the pile, is shown in Fig. 11.22. The pile was driven into a clayey soil. By searching in the compiled data bank, a pile, TOKYO PORT2, was found in a close condition of either shape, size, embedment length, and soil to this pile.

Based on the analyses, methods such as UniCone (Eslami and Fellenius, 1997), Modified UniCone (Niazi and Mayne, 2016), ICP (Jardine et al., 2005), Meyerhof (1976, 1983), Dutch (De Ruiter and Beringen, 1979), and German (Kempfert and Becker, 2010) perform better than other investigated methods. The result of different predictive methods is presented in Table 11.8. As shown, for long-term capacity methods such as UniCone, Modified UniCone and Dutch result in better estimation of actual axial-bearing capacity.

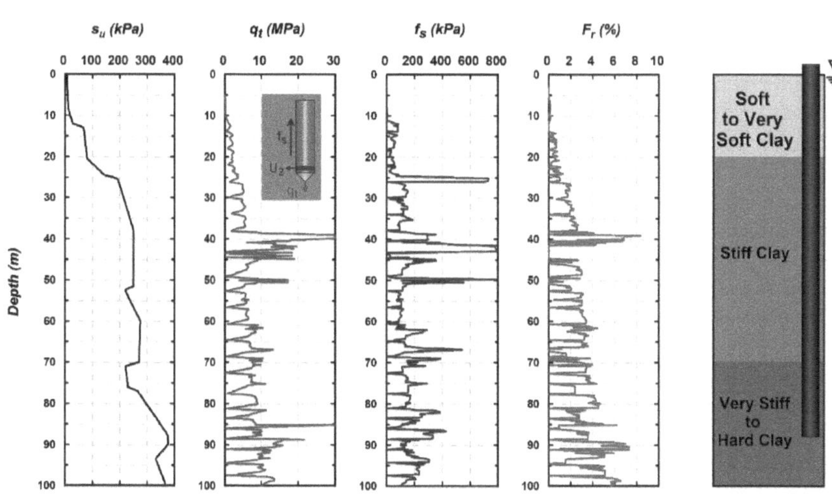

Figure 11.22 CPT records for the investigated pile (Shamshirgaran and Ebrahimian, 2018).

Table 11.8 Model parameter of various methods for the investigated pile (Shamshirgaran and Ebrahimian, 2018).

Method	End-of-Driving Q_p/Q_m (EOD)	Short-term Q_p/Q_m (21-hour BOR)	Medium-term Q_p/Q_m (9-day BOR)	Long-term Q_p/Q_m (263-day BOR)
API	6.29	3.99	2.73	2.03
Fugro	5.379	3.41	2.33	1.73
NGI	5.659	3.59	2.45	1.82
ICP	3.60	2.29	1.56	**1.16**
Modified Unicone	4.61	2.93	2.00	**1.3**
Unicone	3.259	2.07	1.41	**1.05**
Shmertmann	2.52	1.60	1.10	0.81
Dutch	4.08	2.59	1.77	**1.31**
LCPC	1.90	1.20	0.82	0.61

References

American Petroleum Institute (API), 2006. Recommended Practice for Planning, Designing and Constructing Fixed Offshore Platforms-Working Stress Design. API RP 2A-WSD, Washington, DC.

Ang, A.H.S., Tang, W.H., 2007. Probability Concepts in Engineering: Emphasis on Applications in Civil and Environmental Engineering (1&2). Wiley, New York.

Bazaraa, A.R., Kurkur, M.M., 1986. N-values used to predict settlements of piles in Egypt. In: Use of in Situ Tests in Geotechnical Engineering. ASCE, pp. 462–474.

Beacher, G.B., Christian, J.T., 2003. Reliability and Statistics in Geotechnical Engineering. Wiely, ISBN 0-471-49833-5.

Briaud, J.L., Tucker, L.M., 1988. Measured and predicted axial response of 98 piles. Journal of Geotechnical Engineering 114 (9), 984–1001.

Bustamante, M., Gianeselli, L., 1982. Pile bearing capacity predictions by means of static penetrometer CPT. In: Proceedings of 2nd European Symposium on Penetration Testing (ESOPT-II), Amsterdam, pp. 493–500.

Canadian Foundation Engineering Manual (CFEM) (2006) Canadian Geotechnical Society (CGS), Richmond, B.C.

Ching, J.Y., Phoon, K.K., 2017. Characterizing uncertain site-specific trend function by sparse Bayesian learning. Journal of Engineering Mechanics, ASCE 143 (7), 04017028.

De Ruiter, J., Beringen, F.L., 1979. Pile foundations for large North Sea structures. Marine Geotechnology and Georesources 3 (3), 267–314.

Dithinde, M., Phoon, K.K., De Wet, M., Retief, J.V., 2011. Characterization of model uncertainty in the static pile design formula. Journal of Geotechnical and Geoenvironmental Engineering 137 (1), 70–85.

Eslami, A., 1996. Bearing Capacity of Piles from Cone Penetrometer Test Data. Ph.D. Thesis. University of Ottawa, Department of Civil Engineering, 516pp.

Eslami, A., Fellenius, B.H., 1997. Pile capacity by direct CPT and CPTu methods applied to 102 case histories. Canadian Geotechnical Journal 34, 886–904.

Eslami, A., Golafzani, S.H., Chenari, R.J., 2016. Assessment of Babolsar concrete pedestrian bridge failure for 1964 flood event and retrofitting practice. Engineering Failure Analysis 68, 101–112.

Eslami, A., Moshfeghi, S., Heidari, S., Valikhah, F., 2019. AUT: geo-CPT&pile database updates and implementations for pile geotechnical design. Geotechnical Engineering Journal of the SEAGS & AGSSEA 50 (2).

Fellenius, B.H., 2015. Basics of Foundation Design, Electronic Edition. 432 pp.

Haldar, A., Mahadevan, S., 1999. Probability Reliability and Statistical Methods in Engineering Design, first ed. Wiley.

Harr, M.E., 1987. Reliability Based Design in Civil Engineering. McGraw-Hill.

Hasofer, A.M., Lind, N.C., 1974. Exact and invariant second-moment code format. Journal of the Engineering Mechanics Division 100 (1), 111–121.

Heidarie Golafzani, S., 2018. Application of Reliability Method in Processing of CPT Data for Determination of Pile Axial Capacity. Ph.D. Dissertation. University of Guilan, Rasht, Iran.

Heidarie Golafzani, S., Eslami, A., Jamshidi Chenari, R., 2019a. Reliability-based assessment of axial pile bearing capacity: static analysis, SPT and CPT-based methods. In: Under Review for Possible Publication in Georisk:Assessment and Management of Risk for Engineered Systems and Geohazards.

Heidarie Golafzani, S., Eslami, A., Chenari, R.J., 2019b. Probabilistic assessment of model uncertainty for prediction of pile foundation bearing capacity; static analyses, SPT and CPT-based methods. In: Under Review for Possible Publication in Geotechnical and Geological Engineering.

Jamshidi Chenari, R., Kamyab Farahbakhsh, H., 2015. Generating non-stationary random fields of auto-correlated, normally distributed CPT profile by matrix decomposition method. Georisk: Assessment and Management of Risk for Engineered Systems and Geohazards 9 (2), 96–108.

Jamshidi Chenari, R., Kamyab Farahbakhsh, H., Heidarie Golafzani, S., Eslami, A., 2018. Non-stationary realization of CPT data: considering lithological and inherent heterogeneity. Georisk: Assessment and Management of Risk for Engineered Systems and Geohazards 1–14.

Jardine, R.J., Chow, F.C., Overy, R., Standing, J., 2005. ICP Design Methods for Driven Piles in Sands and Clays. Thomas Telford Publishing, London, 105 pp.

Jefferies, M., Davies, M., 1993. Use of CPTu to estimate equivalent SPT N60. ASTM Geotechnical Testing Journal 16 (4), 458–468.

Kempfert, H.G., Becker, P., 2010. Axial pile resistance of different pile types based on empirical values. In: Proceedings of Geo-Shanghai 2010 Deep Foundations and Geotechnical In-Situ Testing (GSP 205). ASCE, Reston, VA, pp. 149–154.

Kenarsari, A.E., Oloomi Dodaran, R., Jamshidi Chenari, R., Eslami, A., 2011. Effect of vertical heterogeneity in soil strength on pile bearing capacity prediction from CPT data. In: Proceedings of the 36th Annual Conference on Deep Foundations DFI 36: Boston, USA (18–21 October 2011).

Kulhawy, F.H., Trautmann, C.H., 1996. Estimation of In-Situ Test Uncertainty. Geotechnical Special Publication.

Look, B.G., 2007. Handbook of Geotechnical Investigation and Design Tables.

Lunne, T., Robertson, P.K., Powell, J.J.M., 1997. Cone Penetration Testing in Geotechnical Practice. Blackie Academic & Professional/Chapman-Hall Publishers, London, 312 pp.

Meyerhof, G.G., 1976. Bearing capacity and settlement of pile foundations. Journal of the Geotechnical Engineering Division 102, 195–228.

Meyerhof, G.G., 1983. Scale effects of pile capacity. Journal of Geotechnical Engineering, ASCE 108 (GT3), 195–228.

Moshfeghi, S., Eslami, A., 2016. Study on pile ultimate capacity criteria and CPT-based direct methods. International Journal of Geotechnical Engineering 1–12.

Niazi, F.S., 2014. Static Axial Pile Foundation Response Using Seismic Piezocone Data. Ph.D. dissertation. Georgia Institute of Technology.

Niazi, F.S., Mayne, P.W., 2016. CPTu-based enhanced UniCone method for pile capacity. Engineering Geology 212, 21–34.

Paikowsky, S.G., Birgisson, B., McVay, M., Nguyen, T., Kuo, C., Baecher, G.B., O'Neill, M., 2004. Load and Resistance Factor Design for Deep Foundations. NCHRP Report 507. Transportation Research Board of the National Academies, Washington DC.

Phoon, K.K., Kulhawy, F.H., 1999. Characterization of geotechnical variability. Canadian Geotechnical Journal 36 (4), 612–624.

Phoon, K.K., Kulhawy, F.H., 2005. Characterization of model uncertainties for laterally loaded rigid drilled shafts. Géotechnique 55 (1), 45–54.

Rackwitz, R., Fiessler, B., 1978. Structural reliability under combined random load sequences. Computers & Structures 9 (5), 489–494.

Robertson, P.K., 2010. Soil behaviour type from the CPT: an update. In 2nd international symposium on cone penetration testing, USA 9–11.

Shamshirgaran, A.H., Ebrahimian, B., 2018. Time effects on the axial compression bearing capacity of piles driven in offshore clays of Persian Gulf – a case study. In: The 3rd Iranian Conference on Geotechnical Engineering, Tehran, Iran.

Schmertmann, J.H., 1978. Guidelines for Cone Penetration Test: Performance and Design (No. FHWA-TS-78-209). Federal Highway Administration, United States.

Schneider, J.A., Xu, X., Lehane, B.M., 2008. Database assessment of CPT-based design methods for axial capacity of driven piles in siliceous sands. Journal of Geotechnical and Geoenvironmental Engineering 134, 1227–1244.

Tang, C., Phoon, K.K., 2016. Model uncertainty of Eurocode 7 approach for the bearing capacity of circular footings on dense sand. International Journal of Geomechanics, ASCE 17 (3), 1–9.

Zhang, Z., Tumay, M.T., 1999. Statistical to fuzzy approach towards soil classification. Journal of Geotechnical and Geenvironmental Engineering, ASCE 125 (3), 179–186.

CPT and CPTu databases in foundation engineering

12

12.1 Introduction

Databases are collections of data which are organized to facilitate access and retrieving information when they are needed. Considering the empirical nature of geotechnical engineering, they are known as helpful tools which allows for approaching more optimum designs in term of both timing and cost. There exist several geotechnical databases as introduced by Long (2001), Kalavar and Ealy (2000), Roling et al. (2011), Anderson and Townsend (2002), etc. A number of existing databases in geotechnical literature are related to these areas:

- In situ testing
- Liquefaction records
- Soil improvement performance
- Retaining structure performance and ground deformations due to excavation
- Slope failure aspects
- Substructure damages
- Soil behavior classification
- Shallow foundations
- Pile axial loading tests
- Pile lateral loading tests

In general, the applications of databases in geotechnical engineering can be divided into two main categories: research and practice. However, these two are not fully independent of each other, and advancing in the former will surely lead to improving the latter.

When it comes to research, geotechnical databases provide the opportunity to evaluate the existing design methods and highlight the pros and cons of them in various conditions. Moreover, they are required during the process of developing a new design method (such as settlement, bearing capacity, stability, etc.) to calibrate and validate new approaches. It is needless to say that it is a must to assess the any new method through a variety of case records to grasp an insight to its performance.

Another implementation of databases is modification of currently existing methods by helping through back analyzing the observed behavior of geotechnical elements such as foundations systems, excavation retaining structures, etc., so that more accurate design procedures can be obtained. In this regard, statistical- and reliability-based approaches are widely used along with geotechnical databases to provide more insight for different occurring scenarios.

In the design procedures, either of these two conditions are encountered: (1) sufficient site investigations are performed due to low variability in a specific

Piezocone and Cone Penetration Test (CPTu and CPT) Applications in Foundation Engineering
https://doi.org/10.1016/B978-0-08-102766-0.00012-2

site, and there is adequate information on site conditions such as CPT records, pile load test results, etc.; and (2) due to economic aspects or project limitations, it is not possible to perform in situ testing as many that fully represent the site conditions. This especially applies to the sites with high inherent variability. For either case, geotechnical databases can enter the design procedure accompanied by performance-based approaches. Additionally, databases can provide the opportunity to reproduce data, such as CPT and pile load–displacement records, similar to the ones encountered in the project site. Considering the availability of these data, the need for performing the time-consuming or expensive loading tests diminishes, and consequently, it can help the efficiency of the project. Moreover, the ability of considering soil inherent variability enhances the geotechnical engineer's understanding of the site conditions. In the next step, after acquiring necessary data, the most compatible method according to assessments must be opted to achieve an optimum design.

With respect to the scope of this book, in this chapter, typical databases in geotechnical engineering including the CPT results are briefly introduced with more concentration on pile and CPT data. Examples of studies done by means of pile and CPT databases are also presented.

12.2 Databases of soil parameters and CPTu

There exist several empirical correlations between CPT data with other geotechnical parameters of geomaterials such as soil relative density, unit weight, soil modulus, over consolidation ratio, etc. An example of a database presented by Mayne and Illingworth (2010) deals with estimating soil shear wave velocity, subsequently, correlating soil unit weight with CPT measurements. The database employed includes the data collected from 44 sites comprising of a wide range of soil types (sand, silt, clay, tilt, etc.). The final proposed correlation employs CPT sleeve friction directly to estimate soil unit weight.

Eslami and Mohammadi (2016) collected a database of CPTu records combined with laboratory test results from various regions in the world. This database was used to verify the accuracy of their approach to estimate the soil shear strength parameters in drained conditions. The records were from eight sites mainly near shoreline (marine and sea).

As described in Chapter 10, the CPT records can be used to monitor and assess the sufficiency of ground improvement in problematic soils. With this aim and by means of a database, Farhadi Nasl (2013) attempted first to examine the ability of existing soil behavior classifications charts to discern the problematic soils and then to illustrate the changes in the soil behavior post ground modification.

Another database pertaining behavior of problematic soils was compiled by Eslami et al. (2017). This database was used to develop a new triangular chart so that it can be a better indicator of behavior of problematic soils, and it included case records of cone penetrometer tests carried out in problematic soils in 54 sites.

Additionally, through the analysis of a database including 11 sites in the Netherlands, Lengkeek et al. (2018) proposed a correlation to estimate soil unit weight. This approach was presented in Chapter 4.

Similarly, other researches such as Eslamizaad and Robertson (1996), Mayne (2006) and Chanmee et al. (2017) correlated CPT data with soil parameters such as shear modulus, shear wave velocity, and over consolidation ratio, respectively. More information on these databases coupled with the outcome of the investigations is presented in Chapter 4.

12.3 Databases for shallow foundations and CPT

Briaud and Gibbens (1999) presented a database of five square spread footings ranging in size from 1 to 3 m being load tested up to the settlement of 150 mm. The footings were located on a medium dense, fairly uniform, silty silica sand with the embedment depth of 0.75m. They have proposed some recommendations regarding how best to measure footing settlement, how to present load test results, new correlations for use in design, creep settlement, effect of cyclic loading and preloading on creep rate, zone of influence under the footing, mode of deformation of the soil mass, and volume change observations.

Eslami and Gholami (2006) compiled a database of loading tests on shallow foundations as well as the CPT profiles performed close to the foundation's locations consisting of 32 case records from seven sites. The cases also included the results of plate load tests. The minimum depth of the CPT profiles was 2m. The soil types were reported as uniform sand, silty sand, clayey sand, and sandy clay with low liquid limit. This database was used to assess the performance of existing methods in determining the bearing capacity of shallow foundations along with proposing a new design method.

Valikhah and Eslami (2017) compiled a database of settlement of shallow foundations constructed on sand and silt soil, accompanied by the records of CPTs conducted adjacent to the foundations location. They used this database to develop a new procedure to estimate the settlement of shallow foundations by means of CPT. This database consists of 46 case studies from eight countries.

More details regarding these two databases were presented in Chapters 6 and 8, respectively.

12.4 Databases for pile and CPT

A number of well-established geotechnical databases including piling and cone penetration testing records in literature are presented and reviewed. Generally, these databases were used to study different aspects of determining pile capacity by means of in situ tests which have illuminated the performance of some of the current methods.

Briaud and Tucker (1988) used 98 case records to evaluate performance of 13 methods of estimating settlement and bearing capacity of piles by SPT, CPT, and

PMT. This database consists of 64 square concrete piles with the widths ranging from 36 to 46 cm, 27 H piles, and 7 drilled shafts with the diameters from 30 to 41 cm. The pile embedment lengths vary between 3 and 25 m with the average of 12.2 m. The pile capacities range from 307 to 2890 kN with the average of 1213 kN.

Alsamman (1995) presented a database of 95 full-scale drilled shaft load tests from 29 sites in 8 countries (mainly in Germany and the United States) to evaluate five methods of bearing capacity estimation and to develop a new procedure of determining capacity by CPT. Records include 48 pile load tests in cohessionless soils, 16 load tests in cohesive soils, and 31 load tests in mixed soils. Diameter of drilled shafts vary between 30 and 213 cm and the shaft lengths between 4.6 and 42 m. CPTs were mainly performed with mechanical cones and contain the cone tip resistance with depth. Ultimate loads were defined at shaft displacement of 5% of the diameter plus the elastic compression of the shaft. For shafts in tension, the axial load at 12.7 mm of displacement was used to define ultimate load.

The Eslami and Fellenius (1997) database consists of 102 case records of 40 sites from 13 countries but mainly in the United States. The soils of the sites consisted of sediments of clay, silt, and sand. Most of the CPT cases included in the data were obtained by electrical cone. Most of the CPT measurements were at a vertical spacing of 300 mm or smaller. Most of the piles have a square or round cross section, and the pile materials are steel and concrete. All but 10 of the piles were installed by driving. The pile embedment lengths range from 5 to 67 m, the pile diameters from 200 to 900 mm, and the pile capacities from 80 to 8000 kN. 14 case records were instrumented, and the toe and shaft resistances were determined separately. This database was used to evaluate and compare six methods of determining pile-bearing capacity by CPT and CPTu.

Van Impe (1999) with aim of clarifying the effect of installation techniques on behavior of pile foundations performed a survey in Waddinxween, The Netherlands. Total number of 55 piles of 6 types were tested including 15 slender precast piles, 10 steel H-beams, 9 vibro-piles, 10 vibrex-piles, 8 CFA piles, and 3 prestressed concrete sheet piles.

Abu-Farsakh and Titi (2004) compiled results of static load tests on 35 square precast prestressed concrete piles to evaluate applicability of 8 CPT-based methods of determining bearing capacity. The piles embedment lengths range from 9 to 38 m, and pile diameters range from 356 to 762 mm. Of the 35 piles, 26 piles were driven in clay soils and other 9 piles were driven in layered soils. The ultimate pile capacity from the load test was determined using Butler–Hoy tangent method (1977).

The UWA Database (Schneider et al., 2008) consists of 77 case records of impact-driven piles and was used to evaluate seven design methods in siliceous sands. All CPT data were digitized to a depth interval of 0.1 m or smaller. The soils include sands with wide range of relative density. Diameters of piles are mainly less than 80 cm and pile lengths vary between 5 and 80 m, most of which are between 10 and 20 m. Pile capacities are mainly less than 5000 kN. In this database, load at displacement of 10% of the pile diameter was defined as pile capacity in compression, if plunging failure was reached, otherwise hyperbolic extrapolation of Chin (1978) was used to estimate pile capacity. Pile capacity in tension was defined as the maximum uplift load less the pile weight.

The Van Dijk and Kolk (2011) database consists of 33 case records of driven piles at 15 different locations. All cases were installed in clay soils. Results of CPTs and equilibrium in situ pore pressure u_0 were determined at 0.1 m depth intervals. Diameters of piles vary between 102 and 812 mm and pile lengths vary between 3.6 and 71.4 m. Among 33 case records, 18 piles were tested in compression and 15 piles in tension. Time interval between pile installation and load tests varies between 0.25 and 134 days. The ultimate capacity of a pile was defined as the maximum resistance measured during the test, corrected for the effective weight of the pile (total weight of the pile minus the weight of the displaced soil).

The Hassani (2010) database includes 70 case records obtained from 16 sources, reporting data from 18 sites in 10 countries. The soils at sites consist of sediments of soft and stiff clay, silt, silty sand, and sand. Most of the CPT cases were obtained by electrical cone. All of the CPT measurements were at a vertical spacing of 25–300 mm. Most of the piles have a round cross section and the pile materials are steel and concrete. All but two of the piles were installed by driving. The pile embedment lengths range from 8.2 to 75 m, the pile diameters range from 270 to 813 mm, and the pile capacities vary from 485 to 6860 kN. This database was used for training Group Method of Data Handling type neural networks to model the effects of effective cone point resistance and cone sleeve friction on pile unit shaft resistance.

Eslami et al. (2011) employed 13 case records of driven pipe piles from 4 sites of Iran marine environments. The soils of the sites were mainly soft sensitive to stiff clay. Results of electrical piezopenetrometer for all cases were available. The pile embedment lengths range from 41 to 85 m, the pile diameters from 685 to 913 mm, and the pile capacities from 4700 to 16,500 kN. This database was used to assess applicability of five CPT- and CPTu-based methods in Iran marine environments, and static capacity derived from both static and dynamic load tests was used as reference test.

Gwizdala and Krasinski (2013) presented a database of bridge structures founded on driven displacement Vibro and Franki piles. They proposed a methodology to derive the load–displacement diagram of piles by means of CPT results. Then, the design assumptions and computational results were verified by static load tests of piles.

ZJU-ICL database was developed by Yang et al. (2015) with cooperation of Zhejiang University and Imperial College London database. It includes 115 records of driven piles in sand and is openly accessible for researchers to use.

Niazi and Mayne (2016) employed a database of piles and piezocone test results to present an enhanced version of the UniCone method (CPTu-based method of determining the axial-bearing capacity of piles). The database consists of 153 case records of piles with various installation types including driven, bored, cast in situ, and jacked. The piles were constructed in both sandy and clayey soils. All the case records contain the measurements of excess pore pressure.

In addition, the methods presented in Chapter 7 were mainly developed, calibrated, and validated by means of CPT and pile databases. For instance, the Schmertmann (1978) and Nottingham (1975) method is based on a summary of the work on model and full-scale piles presented by Nottingham (1975). The De Ruiter and Beringen (1979) method (also called the European method) is based on experience gained from offshore construction in the North Sea. The LCPC method is based on experimental work of Bustamante and Gianeselli (1982) for the French Highway Department. The Tummay and Fakhroo (1982) method is based on an experimental study on a number of different piles in clay soils in Louisiana, USA.

A summary of these CPT and piling databases characteristics is presented in Table 12.1.

Table 12.1 Summary of CPT and pile databases in geotechnical literature.

No.	Database	Number of records	Pile installation technique	Pile surrounding soil
1	Schmertmann (1978)	More than 50	Driven	Sand, clay
2	Bustamante and Gianeselli (1982)	More than 200	Driven, bored	Sand, clay
3	Tummay and Fakhroo (1982)	40	Driven	Clay
4	Briaud and Tucker (1988)	98	Driven, bored	Clay, sand
5	Alsamman (1995)	95	Drilled shaft	Clay, sand
6	Eslami and Fellenius (1997)	102	Driven, bored	Clay, sand
7	Van Impe (1999)	55	Precast piles, steel H-beams, vibro-piles, vibrex-piles, CFA, prestressed concrete sheet piles	Clay, sand
8	Abu-Farsakh and Titi (2004)	35	Prestressed driven	Clay
9	UWA (Lehane et al., 2005)	77	Driven	Sand
10	Hassani (2010)	70	Driven, bored	Clay, sand
11	Van Dijk and Kolk (2011)	33	Circular driven	Clay
12	Eslami et al. (2011)	13	Driven	Clay
13	Gwizdala and Krasinski (2013)	6	Driven displacement vibro and Franki piles	Clay, sand
14	ZJU-ICL (2015)	115	Driven	Sand
15	Niazi and Mayne (2016)	153	Driven, bored cast in situ, jacked	Clay, sand

12.5 AUT: GEO-CPT&Pile database

12.5.1 Establishment

With the primary aim for assessing different approaches of determining pile static capacity based on cone and piezocone penetration tests, AUT: Geo-CPT&Pile database has been compiled mainly from well-published and documented geotechnical engineering sources. It was first introduced in 2015 under the title of "AUT-CPT&Pile database" (Moshfeghi et al., 2015a,b). The database initially consisted of 466 records of pile loading tests as well as the results of cone or piezocone penetration tests carried out in the vicinity of the pile locations.

12.5.2 Updates and accessibility

The AUT: Geo-CPT&Pile database is now further upgraded to the total number of 600 records. The case records have been obtained from 68 sources and are from 24 countries and the majority of cases are located in the United States. The distribution of the case records is depicted in Fig. 12.1.

All the case records include results of axial load tests and CPT records. CPT logs were digitized at depth intervals of 0.05, 0.1, 0.2, 0.3, and rarely 0.5 and 1 m using GetData Graph Digitizer 2.24. About 46% of CPT results include measurements of cone tip resistance (q_c) and sleeve friction (f_s), 29% include only q_c measurement, 21% include q_c, f_s, and pore pressure (u_2 or u_3) and other 4% include measurements of q_c and u_2 or u_3. Site soils include wide ranges of clayey, silty, and sandy deposits.

Load test records include head-down static load tests in compression or tension, static O-cell tests, dynamic load tests (PDA and CAPWAP), and statnamic load tests. Approximately 96% of load test results contain load–displacement diagrams, while for the rest of 4% of cases only the ultimate capacity is reported.

Case records consist of different pile types including driven piles, auger cast piles, drilled displacement piles, continuous flight auger piles (CFA), driven cast in situ piles (DCIS), post grouted piles, jacked piles, vibro-driven piles, and helical piles. The pile materials are steel, concrete, composite (steel and concrete), and CFG (cement fly ash gravel), and the shapes of piles are round, square, pipe, triangular, octagonal, H, X, and

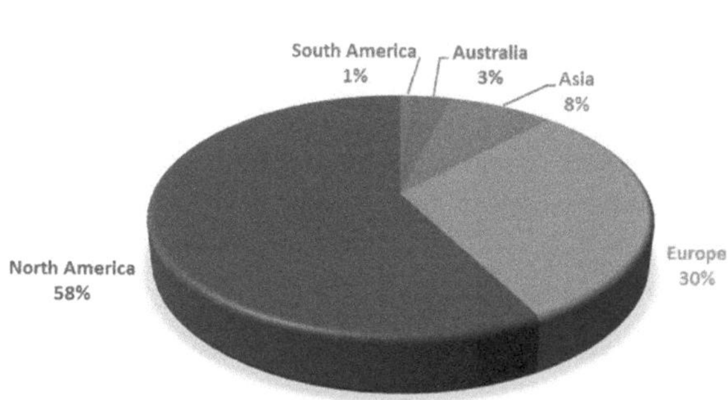

Figure 12.1 Distribution of case records of AUT: Geo-CPT&Pile database.

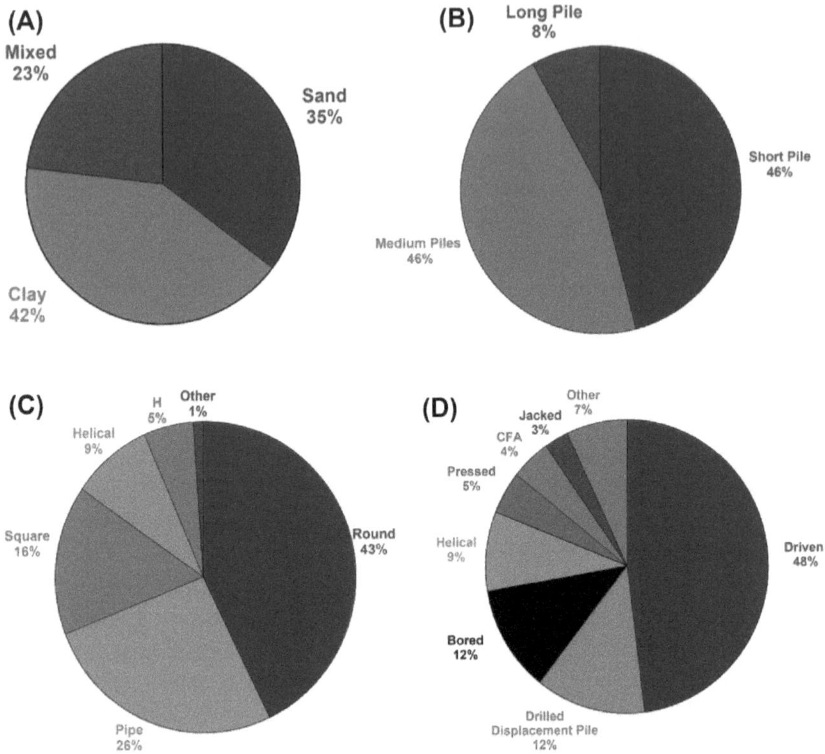

Figure 12.2 Distribution of AUT: Geo-CPT&Pile database case records based on (A) soil types, (B) pile length, (C) shape, and (D) installation procedure.

helical. Embedment lengths of the piles vary between 3.0 and 100 m, but mainly less than 50 m and the pile diameters range from 50 to 2500 mm with an average of 450 mm. The pile capacities range from 50 to 36,000 kN with an average of 1620 kN. The distribution of the case records, in terms of soil and pile characteristics, is presented in Fig. 12.2.

12.5.3 Organization

This database has been organized via Microsoft Access software, consisting of different sections allotted to general information, CPT data, pile characteristics, data sources, as well as a section for searching through the database for data with specified characteristics. Various sections of the database are shown in Figs. 12.3–12.8 below and described in further details:

- Fig. 12.3: ***General records form***: as depicted, the first section of the database includes the list of all the records. Each record is given a unique identification name by which its source, CPT profiles, and pile loading test can be distinguished. In this form, a summary of the records

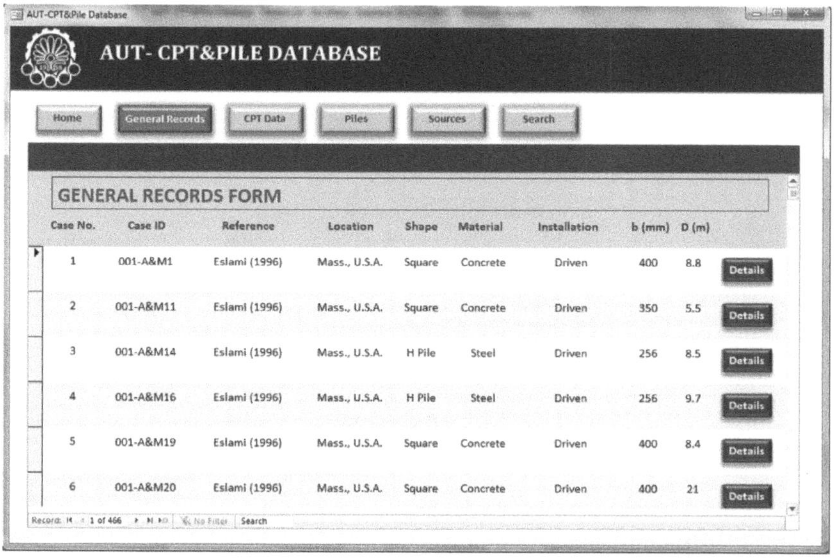

Figure 12.3 AUT: Geo-CPT&Pile database: General records form (Moshfeghi and Eslami, 2016).

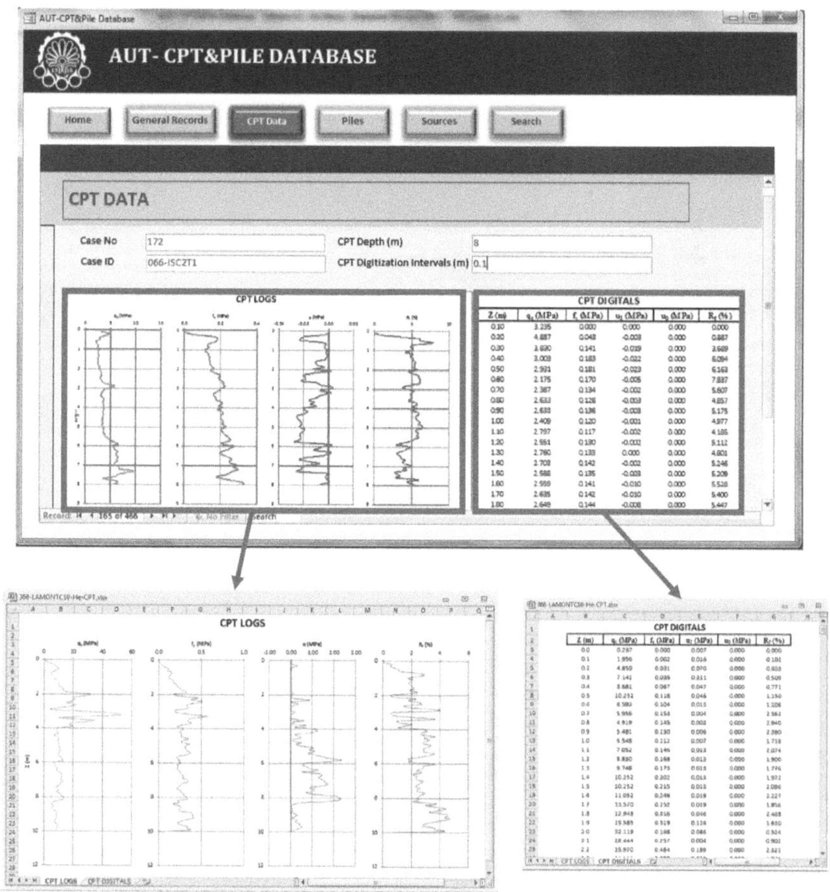

Figure 12.4 AUT: Geo-CPT&Pile database: CPT data.

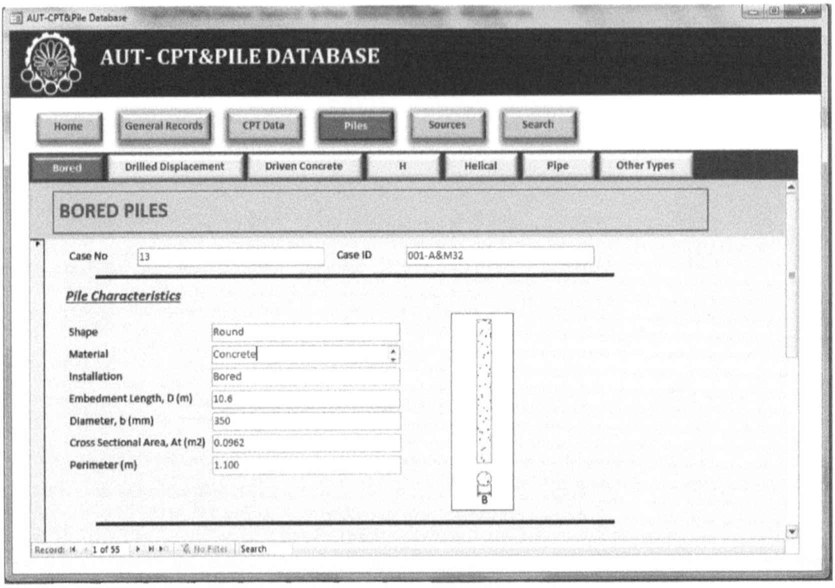

Figure 12.5 AUT: Geo-CPT&Pile database: Pile characteristics (Moshfeghi and Eslami, 2016).

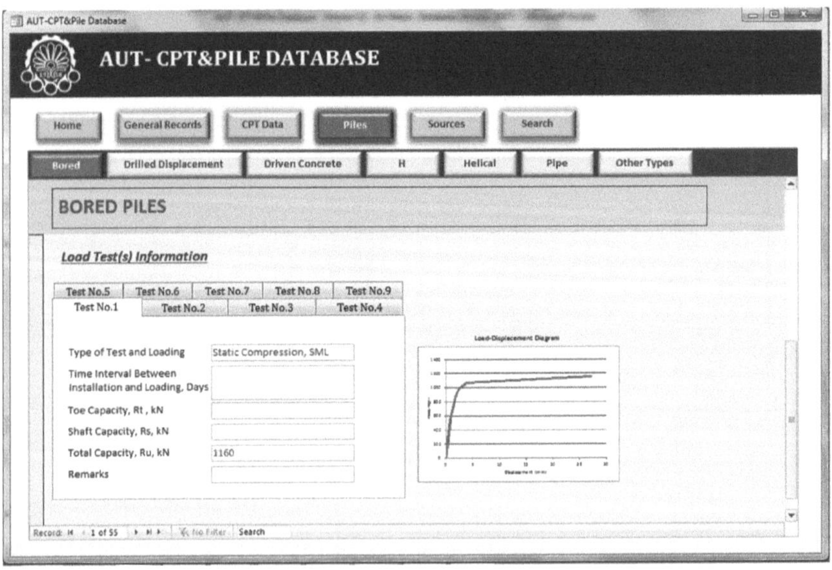

Figure 12.6 AUT: Geo-CPT&Pile database: Pile loading test results (Moshfeghi and Eslami, 2016).

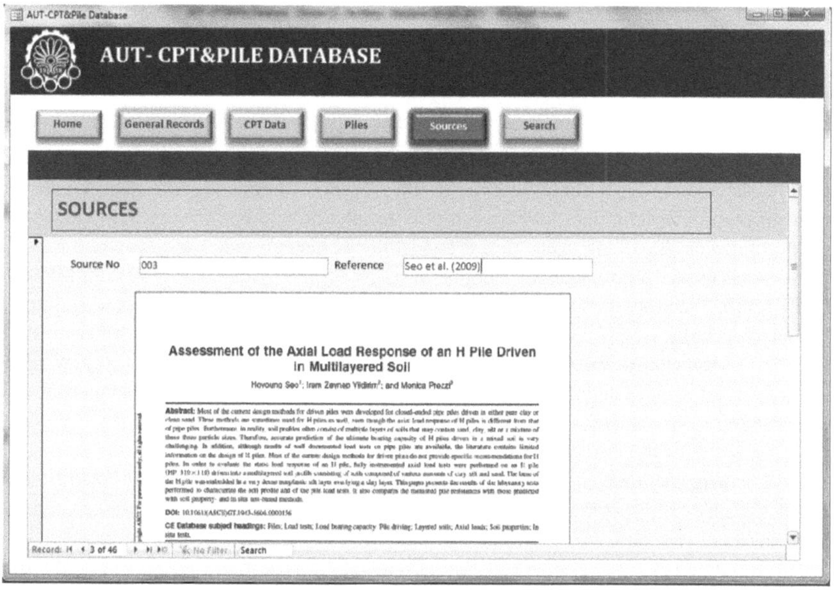

Figure 12.7 AUT: Geo-CPT&Pile database. Data sources (Moshfeghi and Eslami, 2016).

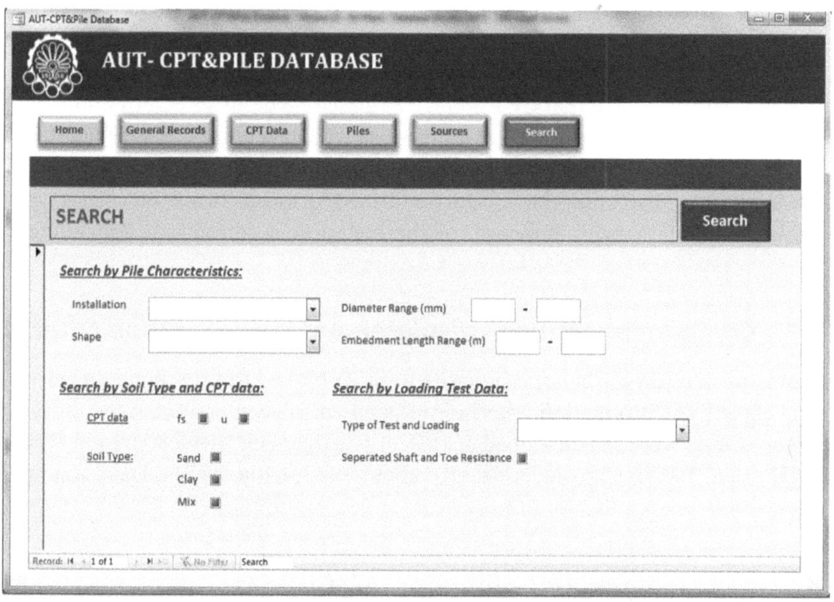

Figure 12.8 AUT: Geo-CPT&Pile database: search section (Moshfeghi and Eslami, 2016).

characteristics is presented as well. More detailed information on each record can be accessed through the "Details" button at the end of each row.

- Fig. 12.4: ***CPT data***: the CPT profiles as well as the digitized results for each case can be found in this section of the database. The data are presented in MS Excel file which can be accessed by clicking on the preview link provided.
- Figs. 12.5 and 12.6: ***Piles characteristics***: this part is allocated to the information on pile geometry, installation type, along with the results of pile loading tests, both diagram and digitized formats. The time interval between pile installation and performing the load tests and the separated shaft and toe capacities, if available, are also provided. As shown in Fig. 12.6, this part is divided into some subsections in terms of pile installation types. Similar to the "CPT results" section, the MS excel files for the pile load test results can be accessed through the preview links.
- Fig. 12.7: ***Information sources***: in this part of the database, the original sources of the data are provided for more detailed reference. It should be mentioned that the name of the reference file for each case is the three digit number at the beginning of the case ID. For instance the source file for the case with the ID "064-SANDPOINT" is "064."
- Fig. 12.8: ***Search form***: in this part, searching through the database is provided based on different factors and specifications including pile type and characteristics, soil type, available CPT and load test data, etc.

12.6 Implementation of the developed database

AUT: Geo-CPT&Pile database, similar to other well-established databases, can be employed as a helpful tool in research and practical applications for optimizing pile analysis and design mainly in soft to medium deposits which dictates deep foundations as the prominent substructure. Considering the extent of the case records in terms of soil diversity, pile installation types, and properties, this database facilitates the evaluation and comparing different CPT and CPTu-based methods of determining pile capacity considering uncertainties, probabilistic- as well as reliability-based approaches, or any other performance-based design approaches.

A number of investigations have been performed by means of sub databases derived from this database. A brief summary of these studies is introduced and referenced below.

12.6.1 Comparison of load–displacement interpretation criteria

43 records of piles driven in sand deposits were employed to investigate the effect of ultimate capacity interpretation criteria from load–displacement diagrams. In this regard, four criteria were selected: load at 10% of pile diameter, Brinch Hansen 80% criterion (Hansen, 1963), Chin-Kondner (Kondner, 1963; Chin, 1971), Davisson (1972) Limit.

12.6.2 Risk analysis and optimum safety factor

The study performed by Moshfeghi and Eslami (2018a) deals with risk, cost optimization approach, and optimum safety factors for the axial bearing capacity of driven piles in sand using CPT-based methods. The database of 76 records is employed to evaluate the performance of 9 commonly used direct CPT-based methods was evaluated against the database.

12.6.3 Nonstationary reproduction of CPT data

Jamshidi Chenari et al. (2018) developed an algorithm for realization of CPT data based on nonstationary random field. The proposed algorithm imposes soil layering alongside soil inherent variability based on Eslami and Fellenius (2004) soil classification chart. After detection of soil layering based on the simplified proposed approach, the statistical characteristics of soil are defined as multicriteria functions, assembled into the nonstationary autocovariance matrix and the routines continue in Monte Carlo scheme.

12.6.4 Statistical and probabilistic assessment

Heidarie Golafzani et al. (2019a,b) investigated the effect of uncertainties (model error and transformation uncertainty) on pile axial–bearing capacity for various methods in terms of efficiency ratio, i.e., the ratio of resistance factor to mean resistance bias factor; φ/λ_R and actual factor of safety, i.e., the product of resistance bias factor to factor of safety; $FS^{x}\lambda_R$ introduced by Paikowsky et al. (2004). To compare the performance of different static analyses, SPT- and CPT-based methods, a database of 60 driven piles was selected including different pile and soil types.

In addition to efficiency ratio and actual factor of safety, several different statistical and probabilistic measures were implemented for assessing the model factor for each method.

12.6.5 Reliability-based assessment of pile capacity

Although CPT-based methods provide more reliable results for prediction of axial pile–bearing capacity, the question arises which method provides more reliable results for a specific project, or what factors should be considered for selection of a suitable CPT-based method. In this regard, a comprehensive investigation has been performed by Heidarie Golafzani et al. (2019a,b) considering a variety of criteria and assumptions, implemented for each method. These criteria include input variables, data processing, model assumptions, failure criterion, time interval between pile installation and pile load test, soil classification system, influence zone, friction fatigue, loading direction, installation method, and CPT record interval. Accordingly, the performance of the 6 methods against a database of 62 driven piles was assessed through each of the abovementioned criteria.

12.6.6 Displacement-based bearing capacity

Valikhah et al. (2018a,b) proposed a new analytical-numerical method to estimate the bearing capacity and axial load–displacement behavior of driven piles in granular soils using CPT records. They used the method of stress characteristics to analyze the stress field below and around the pile and in effect, the failure mechanism. This failure mechanism has been then used by implementation of the kinematical approach of the limit analysis to compute the displacement field. This procedure is employed in a stepwise manner to gradually calculate the stress and displacement field as the pile is assumed to penetrate into the ground.

For this purpose, a database of case histories from the results of 98 full-scale pile load tests was employed with complete information on the soil type and the results of CPT soundings performed close to the pile locations. In the investigated database, all piles are of "driven pile" type. Most of the cases are in sand and some in silt and mixed soils.

12.6.7 Capacity assessment of special piles

12.6.7.1 Helical piles

The feature that CPT prepares the continuous data of soil geotechnical properties per inch in depth was used to modify plate helix locations to achieve higher capacity for helical piles in various soils. In the study carried out by Askari Fateh et al. (2017), axial capacity prediction of 37 cases of helical piles by 10 direct CPT methods was considered, and the results were compared with the measured capacity of the piles from static pile load tests at different sites. The accuracy of initial assumption for the mechanism of failure was determined by the comparison between pile capacity from common static analysis and measured piles' load test under compression or tension. In addition, a new CPT-based method was developed by Askari Fateh et al. (2017) to estimate the bearing capacity of helical piles.

12.6.7.2 Drilled displacement piles

Because the primary approach of determining the bearing capacity of drilled displacement piles are in situ-based methods, a database of 65 records were employed by Moshfeghi and Eslami (2018b) including static load tests on drilled displacement piles in addition to the adjacent CPT profiles.

Statistical reliability—based assessments, focusing on soil—pile specifications, have been conducted for six current CPT-based methods of determining the pile-bearing capacity to investigate the performance as well as the strengths and weaknesses of common methods. They showed that compilation and employment of an extensive database can be beneficial for further assessment of these methods, so as to impose the required modification to reach more optimum and reliable predictions.

References

Abu-Farsakh, M.Y., Titi, H.H., 2004. Assessment of direct cone penetration test methods for predicting the ultimate capacity of friction driven piles. Journal of Geotechnical and Geoenvironmental Engineering 130, 935–944.

Alsamman, O.M., 1995. The Use of CPT for Calculating Axial Capacity of Drilled Shafts (Doctoral Dissertation). University of Illinois at Urbana-Champaign.

Anderson, J.B., Townsend, F., 2002. A laterally loaded pile database. In: Deep Foundations 2002: An International Perspective on Theory, Design, Construction, and Performance, pp. 262–273.

Askari Fateh, A.M., Eslami, A., Fahimifar, A., 2017. Direct CPT and CPTu methods for determining bearing capacity of helical piles. Marine Georesources and Geotechnology 35 (2), 193–207.

Briaud, J.L., Gibbens, R.M., 1999. Behavior of five large spread footing on sand. Journal of Geotechnical and Geoenvironmental Engineering 125 (9), 787–796. ASCE.

Briaud, J.L., Tucker, L.M., 1988. Measured and predicted axial response of 98 piles. Journal of Geotechnical Engineering 114 (9), 984–1001.

Bustamante, M., Gianeselli, L., 1982. Pile bearing capacity predictions by means of static penetrometer CPT. In: Proceedings of 2nd European Symposium on Penetration Testing (ESOPT-II), Amsterdam, pp. 493–500.

Chanmee, N., Chai, J., Hino, T., Wang, J., 2017. Methods for evaluating over consolidation ratio from piezocone sounding results. Underground Space 2 (3), 182–194.

Chin, F.K., 1971. Discussion on pile test-arkansas river project. Journal of Soil Mechanics and Foundation Division 97 (SM6), 930–932. ASCE.

Chin, F.K., 1978. Diagnosis of Pile condition. Geotechnical Engineering 9, 85–104.

Davisson, M.T., 1972. High capacity piles. In: Proceedings, Soil Mechanics Series on Innovation in Foundation Construction, ASCE, Illinois Section, Chicago, pp. 81–112.

De Ruiter, J., Beringen, F.L., 1979. Pile foundations for large North Sea structures. Marine Geotechnology and Georesources 3 (3), 267–314.

Eslami, A., Fellenius, B.H., 1997. Pile capacity by direct CPT and CPTu methods applied to 102 case histories. Canadian Geotechnical Journal 34 (6), 886–904.

Eslami, A., Aflaki, E., Hosseini, B., 2011. Evaluating CPT and CPTu based pile bearing capacity estimation methods using urmiyeh lake causeway piling records. Scientia Iranica 18, 1009–1019.

Eslami, A., Gholami, M., 2006. Analytical model for the ultimate bearing capacity of foundations from cone resistance. International Journal of Science and Technology, Scientica Iranica 13 (3), 223–233.

Eslami, A., Fellenius, B.H., 2004. CPT and CPTu data for soil profile interpretation: review of methods and proposed new approach. Iranian Journal of Science and Technology (IJST) 28, 69–86.

Eslami, A., Alimirzaei, M., Aflaki, E., Molaabasi, H., 2017. Deltaic soil behavior classification using CPTu records—proposed approach and applied to fifty-four case histories. Marine Georesources and Geotechnology 35 (1), 62–79.

Eslami, A., Mohammadi, A., 2016. Drained soil shear strength parameters from CPTu data for marine deposits by analytical model. Ships and Offshore Structures 11 (8), 913–925.

Eslamizaad, S., Robertson, P.K., 1996. Cone penetration test to evaluate bearing capacity of foundations in sands. In: Proceeding of 49th Canadian Geotechnical Society, pp. 429–438.

Farhadi Nasl, F., 2013. Behavioral Classification of Problematic Soils by CPT and CPTu for Improvement Application (MSc. thesis). Amirkabir University of Technology.

Gwizdala, K., Krasinski, A., 2013. Bearing capacity of displacement piles in layered soils with highly diverse strength parameters. In: Proceedings of the 18th International Conference on Soil Mechanics and Geotechnical Engineering. Paper No. 2519.

Hansen, J.B., 1963. Discussion of hyperbolic stress-strain response: cohesive soils. Journal of Soil and Foundations Engineering 89, 241–242. ASCE.

Hassani, A., 2010. Assessment of the Pile Bearing Capacity by CPTu and Artificial Intelligence (M.Sc. thesis). Amirkabir University of Technology.

Heidarie Golafzani, S., Eslami, A., Chenari, R.J., 2019a. Probabilistic assessment of model uncertainty for prediction of pile foundation bearing capacity; static analyses, SPT and CPT-based methods. In: Geotechnical and Geological Engineering. Under review for possible publication in pp. 1-15.

Heidarie Golafzani, S., Jamshidi Chenari, R., Eslami, A., 2019b. Reliability-based assessment of axial pile bearing capacity: static analysis, SPT and CPT-based methods. In: Georisk: Assessment and Management of Risk for Engineered Systems and Geohazards Jun 13, pp. 1−5.

Jamshidi Chenari, R., Kamyab Farahbakhsh, H., Heidarie Golafzani, S., Eslami, A., 2018. Non-stationary realisation of CPT data: considering lithological and inherent heterogeneity. Georisk: Assessment and Management of Risk for Engineered Systems and Geohazards 1−14.

Kalavar, S., Ealy, C., 2000. FHWA Deep Foundation Load Test Database. Geotechnical special publication, pp. 192−206.

Kondner, R.L., 1963. Hyperbolic stress−strain response: cohesive soil. Journal of the Soil Mechanics and Foundations Division 89 (SM1), 115−143. ASCE.

Lehane, B.M., Schneider, J.A., Xu, X., 2005. The UWA-05 method for prediction of axial capacity of driven piles in sand. In: Proceedings of International Symposium on Frontiers in Offshore Geomechanics (ISFOG 2005), Perth, Taylor and Francis, London, pp. 683−689.

Lengkeek, H.J., de Greef, J., Joosten, S., Hicks, M.A., Pisano, F., Peuchen, J., 2018. CPT based unit weight estimation extended to soft organic soils and peat. In: Cone Penetration Testing 2018: Proceedings of the 4th International Symposium on Cone Penetration Testing (CPT'18), 21-22 June, 2018, Delft, The Netherlands. CRC Press, p. 389.

Long, M., 2001. Database for retaining wall and ground movements due to deep excavations. Journal of Geotechnical and Geoenvironmental Engineering 127 (3), 203−224.

Mayne, P.W., 2006. Undisturbed sand strength from seismic cone tests. Geomechanics and Geoengineering 1 (4), 239−257. Taylor & Francis Group, London.

Mayne, P.W., Illingworth, F., 2010. Direct CPT method for footing response in sands using a database approach. In: Proceedings, 2nd International Symposium on Cone Penetration Testing (CPT'10), 3, Huntington Beach, California, pp. 315−322.

Moshfeghi, S., Eslami, A., Mir Mohammad Hosseini, S.M., 2015a. AUT- CPT&Pile database-CPT data and pile loading test records correlation. In: Proceedings of the 4th International Conference on Bridges, Tehran, Iran, 24−26 Jan.

Moshfeghi, S., Eslami, A., Mir Mohammad Hosseini, S.M., 2015b. AUT-CPT&pile database for piling performance using CPT and CPTu records. In: Proceedings of the 40th Annual Conference on Deep Foundations, Oakland, California, USA, 12−15 Oct.

Moshfeghi, S., Eslami, A., 2016. Study on pile ultimate capacity criteria and CPT-based direct methods. International Journal of Geotechnical Engineering 12 (1), 28−39.

Moshfeghi, S., Eslami, A., 2018a. Failure analysis of CPT-based direct methods for axial capacity of driven piles in sand. Georisk: Assessment and Management of Risk for Engineered Systems and Geohazards 1−19.

Moshfeghi, S., Eslami, A., 2018b. Reliability-based assessment of drilled displacement piles bearing capacity using CPT records. Marine Georesources and Geotechnology 1−14.

Niazi, F.S., Mayne, P.W., 2016. CPTu-based enhanced UniCone method for pile capacity. Engineering Geology 212, 21−34.

Nottingham, L.C., 1975. Use of quasi-static friction cone penetrometer data to predict load capacity of displacement piles. Ph. D. thesis, Dept of Civil Engineering, Univ. of Florida, p. 553.

Paikowsky, S.G., Birgisson, B., McVay, M., Nguyen, T., Kuo, C., Baecher, G.B., O'Neill, M., 2004. Load and Resistance Factor Design for Deep Foundations. NCHRP Report 507. Transportation Research Board of the National Academies, Washington DC.

Roling, M.J., Sritharan, S., Suleiman, M.T., 2011. Introduction to PILOT database and establishment of LRFD resistance factors for the construction control of driven steel H-piles. Journal of Bridge Engineering 16 (6), 728–738.

Schneider, J.A., Xu, X., Lehane, B.M., 2008. Database assessment of CPT-based design methods for axial capacity of driven piles in siliceous sands. Journal of Geotechnical and Geoenvironmental Engineering 134 (9), 1227–1244.

Schmertmann, J.H., 1978. Guidelines for Cone Test, Performance and Design. Report FHWA-TS-78209. Federal Highway Administration, Washington, 145pp.

Tummay, M.T., Fakhroo, M., 1982. Friction pile capacity in cohesive soils using electric quasi-static penetration tests. In: Louisiana Department of Transportation and Development Research and Development Section, Baton Rouge, La. Interim Research Rep. No. 1.

Valikhah, F., Eslami, A., 2017. Foundations settlement estimation by non-linear stress-strain approach using CPT records. In: DFI 42nd Annual Conference on Deep Foundations, October 24–26, 2017, New Orleans, LA.

Valikhah, F., Eslami, A., Veiskarami, M., 2018a. Load-displacement behavior of driven piles in sand using CPT-based stress and strain fields. International Journal of Civil Engineering. https://doi.org/10.1007/s40999-018-0388-7 (Published online: 21 Dec. 2018).

Valikhah, F., Eslami, A., Veiskarami, M., 2018b. CPT-based approach to study the load-displacement behavior of driven piles by the new method of stress characteristics. In: Proceedings of China-Europe Conference on Geotechnical Engineering, pp. 1036–1040. August 13-16, 2018, Vienna, Austria.

Van Dijk, B.F.J., Kolk, H.J., 2011. CPT-based design method for axial capacity of offshore piles in clays. In: Gourvenec, S., White, D. (Eds.), Proceedings of the International Symposium on Frontiers in Offshore Geotechnics II. Taylor & Francis Group, London, pp. 555–560.

Van Impe, W.F., 1999. International exercise on piled rafts-Sweden examples. International Society of Soil Mechanics and Geotechnical Engineering. ITC-18 Committee, WVI/HDC/WIV.3.5, Gent University.

Yang, Z.X., Jardine, R.J., Guo, W.B., Chow, F., 2015. A new and openly accessible database of tests on piles driven in sands. Geotechnique Letters 5 (1), 12–20.

Index

'*Note:* Page numbers followed by "f" indicate figures and "t" indicates tables.'

A

Acoustic cone, 66
Algorithm/practical application, 348
Amherst data, 126
Axial-bearing capacity, 189
Axial pile–bearing capacity, 339–345

B

Begemann-type presentations, 112–113, 118, 120

C

Compaction modification, 272–276
Cone penetration test (CPT), 49
- acoustic cone, 66
- calculated parameters, 62–63
- codes, 75–77
- cone pressuremeter, 70
- CPT indirect methods, 161–164, 163f
- data averaging, 150–152, 153f
- data presentation, 61–63
- data processing, 150–152, 153f
- diameter, 152–154, 154f
- direct CPT methods via databases, 174–178
- dual range penetrometer, Brecone, 66
- electrical cones, 58, 59f
- embedment depth, 146–147, 146f
- empirical direct CPT methods, 164–166
- equipment, 60–61, 61f
- factors influencing, 63–65
- graphical presentation, 63, 64f
- influence zone, 147–148, 148f
- lateral stress cone, 66
- measured parameters, 61
- mechanism, 154
- mini-cone, 71–74
- nonhomogeneous condition, 149–150, 149f
- penetration rate, 154
- penetrometers, 57
- physical modeling, 71–74
- piezocone penetration test, 58, 59f
- piezocone with additional sensors, 67–69
- piezovibrocone, 70–71
- procedure, 56f
- resistivity cone, 66–67
- seismic cone, 69
- shaft capacity estimation, 157–159, 160f
- in situ tests, 55
- software, 75–77
- special cones, 65–71
- standard dimensions, 56f
- standards, 75–77
- ultraviolet-induced fluorescence CPT, 71
- vibrocone, 70–71

Constrained modulus, 94–95
Construction phase, 3
CPT and CPTu application, 268–269
CPT correlations, SPT, 97–98
Crude oil contaminated clayey sand behavior, 291–294
Cyclic resistance ratio (CRR), 101
Cyclic stress ratio (CSR), 100–101

D

Data averaging, 150–152, 153f
Databases
- AUT: Geo-CPT&Pile database, 359–364
  - accessibility, 359–360
  - capacity assessment, 366
  - displacement-based bearing capacity, 365–366
  - establishment, 359
  - implementation, 364–366
  - load–displacement interpretation criteria, 364

Databases (*Continued*)
nonstationary reproduction, 365
optimum safety factor, 364–365
organization, 360–364
reliability-based assessment, 365
risk analysis, 364–365
statistical and probabilistic assessment, 365
updates, 359–360
drilled displacement piles, 366
helical piles, 366
pile, 355–358
shallow foundations, 355
soil parameters, 354–355
Data processing, 150–152, 153f
Deep foundation–bearing capacity
axial-bearing capacity, 189
direct CPT–based methods, 194–211, 195f
enhanced UniCone, 210–211
Eslami and Fellenius method, 200–202
Fugro-05 method, 208–209
German method, 210
ICP approach, 207–208
indirect CPT–based approaches, 189–194
Laboratoire Central des Ponts et Chausees method (LCPC), 197
NGI method, 203–204
penpile method, 200
piles, 183
design, 185–186
installation CPT records, 186
unit shaft resistance, 190–194
unit toe resistance, 190–194
Price and Wardle method, 200
Schmertmann and Nottingham method, 195–197
in situ test methods, 188
statics analysis, 188
Tummay and Fakhroo method, 199–200
UniCone, 202
UWA, 205–206
Depth of investigations, 6–10
Diameter, 152–154, 154f
Direct CPT methods via databases, 174–178
Drilling operations, 12

E
Element laboratory testing, 12–13
Embedment depth, 146–147, 146f
Empirical direct CPT methods, 164–166
Empirical methods, 81
Evanston data, 126
Explosive compaction project, 290–291

F
Fine grained soil, 103–105
Fines content correction, 101
First-order reliability method (FORM), 327
First-order second moment method (FOSM), 325–326
Foundation engineering, 82
chemical modification, 36–37
constructability performance requirements, 29
densification, 35–36, 36f
design considerations, 26–29
earthworks, 34–35, 35f
foundation classification, 29–31
ground modification approaches, 33–34
hydraulic modification, 37–38
intermediate trends, 31–39
physical modification, 36–37
reinforcement, 38–39, 39f
semideep foundations, 31–33, 32f
serviceability criterion, 27–28
serviceability limits, 28–29
soil mechanics for, 40–48
stability control, 28
Friction ratio, 111
Frustum confining vessel (FCV) apparatus, 15f
Full-scale testing, 14

G
Geotechnical engineering, 320–323, 321f, 322t–323t
aims, 3–4
applications, 3–4
assessments, 3–4
construction phase, 3
depth of investigations, 6–10
drilling operations, 12
element laboratory testing, 12–13
Frustum confining vessel (FCV) apparatus, 15f

full-scale testing, 14
geophysical test, 11–12
geotechnical data, 10–16
geotechnical exploration, 1
geotechnical reports, 20
instrumentation and monitoring, 15–16
maps and technical literature review, 11
medium-scale testing, 13–14
number and spacing of borings, 6
physical modeling, 13–14
site visit, 11
in situ penetration tests, 16–17
in situ testing, 11
Geotechnical exploration, 1
Geotechnical parameters, CPT codes
constrained modulus, 94–95
CPT correlations, SPT, 97–98
cyclic resistance ratio (CRR), 101
cyclic stress ratio (CSR), 100–101
empirical methods, 81
fine grained soil, 103–105
fines content correction, 101
foundation engineering, 82
geotechnical soil parameters, 105t
internal friction angle, 84–89
liquefaction, 99–105
numerical analyses, 81
overburden stress tip resistance correction, 101
over consolidation ratio, 91
permeability, 98
relative density, 84–89
sandy soil, 101–103
sensitivity, 95–97
shear modulus, 95
shear wave velocity, 95
simplified analytical methods, 81
soft computing, data handling, 81
soil parameters, 82
stiffness, 91–93
undrained shear strength, 90, 90t
unit weight, 82, 83t
Geotechnical soil parameters, 105t
Ground improvement methods
application, 263–268
classification, 264t–267t
compaction modification, 272–276
CPT and CPTu application, 268–269
crude oil contaminated clayey sand behavior, 291–294
explosive compaction project, 290–291
function, 263–268
pre- and postmodification, 278–285
quantitative analysis, 270–276
soil behavior classification approach, 285–294
soil geotechnical parameters, 270–276, 271t–272t
standard penetration test (SPT), 270
vibro compaction, 280–285
vibro replacement, 276

I

Immediate and consolidation settlements, 234
Influence zone, 147–148, 148f
Internal friction angle, 84–89

J

Janbu approach, 236–239

L

Liquefaction, 99–105
Liquefaction potential assessment, 310–316
evolutionary-based approach, 312–314
polynomial model, 311–312
SVM, 315–316
triangular chart mapping, 315–316
Long driven concrete pile, 255
LRFD application, 345–348

M

Medium driven concrete pile, 255
Medium-scale testing, 13–14
Metastable sands, 113
Model parameter, 338
Modified stress exponent, 239–240
Modulus number, 239–240
Mohr–Coulomb failure criterion, 47
Monte Carlo simulation, 327–328

N

Nonhomogeneous condition, 149–150, 149f
Nonlinear stress–strain behavior, 233
Nonstationary realization, 336
North Central Florida, 112

O
Overburden stress tip resistance correction, 101
Over consolidation ratio, 91

P
Penetration rate, 154
Permeability, 98
Pile-bearing capacity, 302–308
  cohesion, 310
  friction angle, 310
  GMDH method, 302–303, 304f
  shaft pile–bearing capacity, 306–308
  shallow foundation settlement, 308–309
  shear stress, 310
  support vector regression method, 303–304
Pile-bearing capacity prediction, 338–339
Pile group settlement, 249
Pile scale effect, 146–157

Q
Quantitative analysis, 270–276

R
Random analysis techniques, 324–328
Relative density, 84–89
Reliability, 21, 49–51, 268, 353, 364–365. *See also* Uncertainty
Reliability index, 324–325

S
Sandy soil, 101–103
Scale effect, 145–146
Scale effect realization, 241–246
Sensitivity, 95–97
Settlement and load–displacement
  case study records, 241
  CPT-based load–displacement behavior of foundations, 249–257
  CPT-based nonlinear stress–strain approach, 241–249
  immediate and consolidation settlements, 234
  Janbu approach, 236–239
  long driven concrete pile, 255
  medium driven concrete pile, 255
  Modified stress exponent, 239–240
  modulus number, 239–240
  nonlinear stress–strain behavior, 233
  pile group settlement, 249
  scale effect realization, 241–246
  shallow foundation, 257
  short driven steel pile, 253–255
  soil stiffness modulus, 235
  stress exponent related to CPT, 246–249
  time-dependent stress–strain behavior, 233
Shaft capacity estimation, 157–159, 160f
Shallow foundation, 257
Shallow foundation bearing capacity, 166–174
Shallow foundation failure, 328–332
Shear modulus, 95
Shear wave velocity, 95
Short driven steel pile, 253–255
Simplified analytical methods, 81
Soft computing, data handling, 81
Soft computing implementations
  GMDH-type neural network method, 299–300
  liquefaction potential assessment, 310–316
    evolutionary-based approach, 312–314
    polynomial model, 311–312
    SVM, 315–316
    triangular chart mapping, 315–316
  pile-bearing capacity, 302–308
    cohesion, 310
    friction angle, 310
    GMDH method, 302–303, 304f
    shaft pile–bearing capacity, 306–308
    shallow foundation settlement, 308–309
    shear stress, 310
    support vector regression method, 303–304
  support vector machines (SVMs), 299, 301–302
  techniques, 299
Soil behavior classification (SBC), 285–294
  analysis, 131–136
  Begemann-type presentations, 112–113, 118, 120
  Eslami–Fellenius, 121–126
  Eslami–Fellenius CPTu profiling, 119–121
  I_{SBC}, 120t
  problematic soils
    case records, 129–131

evaluation, 129
profiling chart, 118f–120f
soil profiling methods, 111–119, 112f, 120t
soil type classification method, 119–121
triangular chart, 136–142
UniCone, 128–129
Soil geotechnical parameters, 270–276, 271t–272t
Soil inherent variability, CPT production, 333–334
Soil parameters, 82
Soil profiling methods, 111–119, 112f, 120t
Soils
classification, 41–42
identification, 41–42
origin, 41
shear strength, 47–48
stresses, 43–45
uncertainty, 48–49
water, 42–43
Soil stiffness modulus, 235
Soil stratification, 334–336
Soil type classification method, 119–121
Standard penetration test (SPT), 270
Statistical and probabilistic assessment, 338–342
Stiffness, 91–93
Stresses
compressibility, 45–47
horizontal stress, 44
induced stress, 44–45
settlement, 45–47
Stress exponent, 246–249
Support vector machines (SVMs), 301–302

T

Time-dependent stress–strain behavior, 233
Triangular chart, 136–142

U

Ultimate capacity condition, 155–157
Uncertainty
algorithm/practical application, 348
axial pile–bearing capacity, 339–345
Case studies, 328–345
database, 338–339
first-order reliability method (FORM), 327
first-order second moment method (FOSM), 325–326
geotechnical engineering, 320–323, 321f, 322t–323t
LRFD application, 345–348
methods, 324–325
model parameter, 338
Monte Carlo simulation, 327–328
nonstationary realization, 336
pile-bearing capacity prediction, 338–339
random analysis techniques, 324–328
reliability index, 324–325
shallow foundation failure, 328–332
soil inherent variability, CPT production, 333–334
soil stratification, 334–336
statistical and probabilistic assessment, 338–342
Undrained shear strength, 90, 90t
UniCone, 128–129
Unit weight, 82, 83t

V

Vancouver data, 126
Vibro compaction, 280–285
Vibro replacement, 276

FSC
www.fsc.org
MIX
Papier | Fördert gute Waldnutzung
FSC® C083411

Kolibri 360 GmbH
Ferdinand-Jühlke-Straße 7
99095 Erfurt, Deutschland
produktsicherheit@kolibri360.de